Producing Fuels and Fine Chemicals from Biomass Using Nanomaterials

Producing Fuels and Fine Chemicals from Biomass Using Nanomaterials

Editor

Manasi Karakre

Producing Fuels and Fine Chemicals from Biomass Using Nanomaterials

Edited by **Manasi Karakre**

Printed in 2017

ISBN: 978-1-68117-036-7

Library of Congress Control Number: 2015931837

© 2016 by
SCITUS Academics LLC,
616, Corporate Way, Suite 2, 4766,
Valley Cottage, NY 10989

www.scitusacademics.com

Contents

Anne De Poulpiquet, Alexandre Ciaccafava,
Saïda Benomar, Marie-Thérèse Giudici-Orticoni, and
Elisabeth Lojou

Preface

Biomass is fuel that is developed from organic materials, a renewable and sustainable source of energy used to create electricity or other forms of power. Biomass power is carbon neutral electricity generated from renewable organic waste that would otherwise be dumped in landfills, openly burned, or left as fodder for forest fires.

When burned, the energy in biomass is released as heat. In biomass power plants, wood waste or other waste is burned to produce steam that runs a turbine to make electricity, or that provides heat to industries and homes.

There are a range of different biomass fuels that can potentially be used for energy applications, derived from a range of different sources.

This book explores the available technologies for the preparation of fuels and chemicals from biomass using nanomaterials. This focus bridges the gap between nanomaterials, energy, and the environment. The book also deals with other important topics related to nanomaterials including toxicity and sustainability and environmental aspects.

Editor

Conversion of Lignocellulosic Biomass to Nanocellulose: Structure and Chemical Process

H. V. Lee, S. B. A. Hamid, and S. K. Zain

Nanotechnology & Catalysis Research Centre (NANOCAT), 3rd Floor, Block A, Institute of Postgraduate Studies (IPS), University of Malaya, 50603 Kuala Lumpur, Malaysia

ABSTRACT

Lignocellulosic biomass is a complex biopolymer that is primary composed of cellulose, hemicellulose, and lignin. The presence of cellulose in biomass is able to depolymerise into nanodimension biomaterial, with exceptional mechanical properties for biocomposites, pharmaceutical carriers, and electronic substrate's application.

However, the entangled biomass ultrastructure consists of inherent properties, such as strong lignin layers, low cellulose accessibility to chemicals, and high cellulose crystallinity, which inhibit the digestibility of the biomass for cellulose extraction. This situation offers both challenges and promises for the biomass biorefinery development to utilize the cellulose from lignocellulosic biomass. Thus, multistep biorefinery processes are necessary to ensure the deconstruction of noncellulosic content in lignocellulosic biomass, while maintaining cellulose product for further hydrolysis into nanocellulose material. In this review, we discuss the molecular structure basis for biomass recalcitrance, reengineering process of lignocellulosic biomass into nanocellulose via chemical, and novel catalytic approaches. Furthermore, review on catalyst design to overcome key barriers regarding the natural resistance of biomass will be presented herein.

INTRODUCTION

Owing to the overconsuming of petroleum resources and increasing demand of fossil-based fuels and chemical, it is necessary to develop renewable resources to produce biofuels and biochemical for economical and sustainable development. Lignocellulosic biomass industry has become green, possible alternative of fossil resources in order to compensate the increasing trend of world's demand for petroleum usage [1]. This type of biomass is the most abundantly available biopolymer in nature. It is estimated that the worldwide production of lignocellulosic biomass is about 1.3×10^{10} metric tons per annum [2, 3]. The lignocellulosic resources included (i) agricultural residues (palm trunk and empty fruit bunch, corncobs, wheat straw, sugarcane bagasse, corn stover, coconut husks, wheat rice, and empty fruit bunches); (ii) forest residues (hardwood and softwood); (iii) energy crops (switch grass); (iv) food wastes; and (v) municipal and industrial wastes (waste paper and demolition wood) [4,5]. The high availability of biomass has appeared to be one of the most potential resources of transportation fuels and chemicals platform. Transformation of cheaper biomass into value-added product by the mean of converting "carbon source" into "carbon sink" indicates that carbon can be fully utilized before it would be released into the atmosphere [6–8]. Reconstruction of low cost lignocellulosic materials to products with superior functions presents a feasible option for improvement of energy security and greenhouse emissions reduction. With the availability of biomass, it is believed that this technology is capable of turning negative cost of biomass (plant waste) into positive-earning materials.

Lignocellulose is a complex carbohydrate polymer, containing polysaccharides built from sugar monomers (xylose and glucose) and lignin, a highly aromatic material. Lignocellulosic biomass fractionation into reactive intermediates, such as glucose, cellulose, hemicellulose, and lignin, is a critical process prior to further development into liquid fuels, chemicals, and other end products. The lignocellulosic biomass consists of defensive inner structure which has contributed to the hydrolytic stability and structural robustness of the plant cell walls and its resistance to microbial degradation. On the other hand, the presence of cross-link between cellulose and hemicellulose with lignin via ester and ether linkages [9–11] leads to the biomass recalcitrance. Thus, it is important to understand the chemistry of biomass in order to deconstruct the material into component that can be chemically or catalytically converted into biomass-derived fuels, chemicals, or reactive intermediate.

This paper provides an overview of lignocellulosic biomass reengineering into nanocellulose reactive intermediate by discussing (i) biomass recalcitrance, (ii) chemical approaches for lignocellulosic biomass fractionation, (iii) nanocellulose synthesis via chemical route, and (iv) new prospects of solid catalyst for nanocellulose synthesis. Finally, conclusions on the catalyst development for targeted cellulosic nanomaterial products from lignocellulosic biomass deconstruction will be drawn based on some literature study.

OVERVIEW OF REFINERY OF LIGNOCELLULOSIC BIOMASS INTO NANOCELLULOSE

Global focus is currently directed towards lignocellulosic biomass valorization which not only is limited to liquid biofuel and chemicals production [12–14] but also involves synthesis of reactive intermediate (nanocellulose) for further end-product processing [15]. Nanocellulose, which is obtained from cellulose, is creating a revolution in biobased materials for diverse applications. The nanocellulose (cellulose nanofibers) is equipped with various superior characteristics which are nanoscale dimension, high surface area, unique optical properties, high crystallinity, and stiffness (comparable to Kevlar and steel) together with the biodegradability and renewability of cellulose. This has made this intermediate a precious green alternative to materials, construction, packaging, automobile, transportation, and biomedical fields [15–18].

Lignocellulosic biomass consists mainly of three biopolymers: (i) cellulose (~30–50% by weight), (ii) hemicellulose (~19–45% by weight), and (iii) lignin (~15–35% by weight) [2, 4, 19]. These polysaccharides are associated with each other in a heteromatrix to different degrees and varying composition depending on the type of biomass, species of plant, and even source of the biomass [9]. The chemical composition of biomass for different types of agriculture, industrial, and forestry wastes is shown in Table 1 [20, 21]. The relative abundance of cellulose, hemicellulose, and lignin is the key factor in determining the feedstock suitability for nanocellulose production. Generally, a biomass pretreatment step is necessary to ensure the separation of cellulose component from tight bond of polymeric constituents (cellulose, hemicellulose, and lignin) in lignocellulosic biomass. The main intention of this fractionation treatment is to increase the accessibility of cellulose fiber to chemical attack prior to mild hydrolysis of isolated cellulose, by cleaving the ether bond between glucose chain in order to produce nanosize cellulose intermediate [22, 23]. However, biomass fractionation is a very complex process as high recovery of polysaccharides (cellulose, hemicellulose, and lignin) is required so that all three components can be fully converted into useful end products. Sometime, the biomass pretreatment shall lead to over depolymerisation of polysaccharide chains and subsequent sugar ring opening, which produce undesirable product such as glucose, acid, alcohol, and aldehyde [24]. But such separation is mandatory step to unlock the stored fiber for effective utilization of nanocellulose in the current nanotechnology field. Chemical treatments are the most popular pretreatment technologies in the isolation of cellulose fibers to nanocellulose. An overall roadmap for lignocellulosic biomass reengineering to nanocellulose intermediate and chemical platform is shown in Figure 1. The conversion included the pretreatment of biomass to separate cellulose from noncellulosic contents and further refinery of cellulose, hemicellulose, and lignin fraction for valuable products.

Table 1: Chemical composition of common agricultural residues and wastes

Types of biomass	Lignocellulosic substrate	Cellulose (%)	Hemicellulose (%)	Lignin (%)
Agriculture waste	Corncobs	45	35	15
	Wheat straw	30	50	15
	Barley straw	33–40	20–35	8–17
	Corn stover	39–42	22–28	18–22
	Nut shells	25–30	25–30	30–40
Energy crops	Empty fruit bunch	41	24	21.2
	Switch grass	45	31.4	12
Forestry waste	Hardwood stems	40–55	24–40	18–25
	Softwood stems	45–50	25–30	25–35
	Leaves	15–20	80–85	0
Industrial waste	Waste papers from chemical pulps	60–70	10–20	5–10
	Organic compound from wastewater solid	8–15	0	0

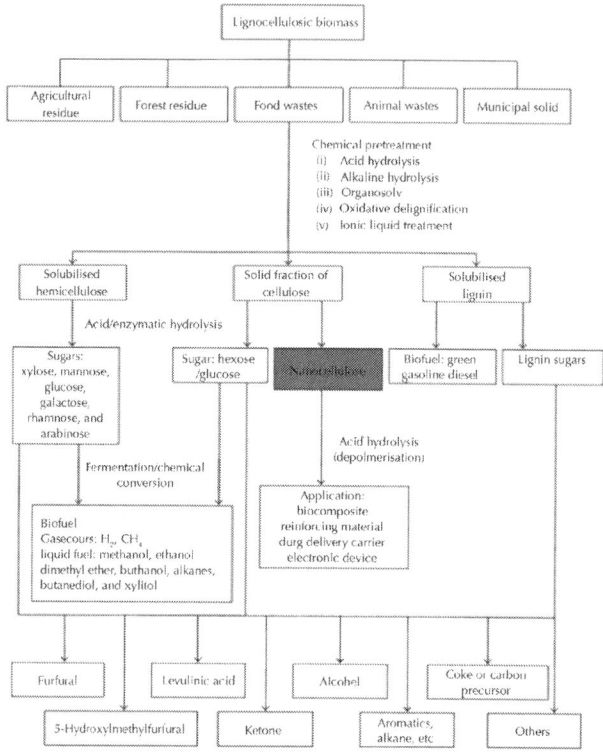

Figure 1: Roadmap of lignocellulosic biomass biorefinery to nanocellulose intermediate and chemicals.

LIGNOCELLULOSIC BIOMASS RECALCITRANCE

Lignocellulose is the primary building block of plant cell walls. The complex hierarchy structure of lignocellulosic biomass is the main obstacle for key components fractionation, where cellulose, hemicellulose, and lignin are hindered by many physicochemical, structural, and compositional factors. Figure 2 showed the plant cell wall structure and different biocompositions. Generally, cellulose fibrils are coated with hemicellulose to form an open network, whose empty spaces are gradually filled up with lignin [66–68]. Several

Factors contributed to biomass recalcitrance are (i) high lignin content; (ii) protection of cellulose by lignin; (iii) cellulose sheathing by hemicellulose; (iv) high crystallinity and degree of polymerization of cellulose; (v) low accessible surface area of cellulose; and (vi) strong fiber strength. Thus, the degree of cellulose isolation with high recovery of hemicellulose and lignin compounds is key for a given biomass fractionation [69, 70].

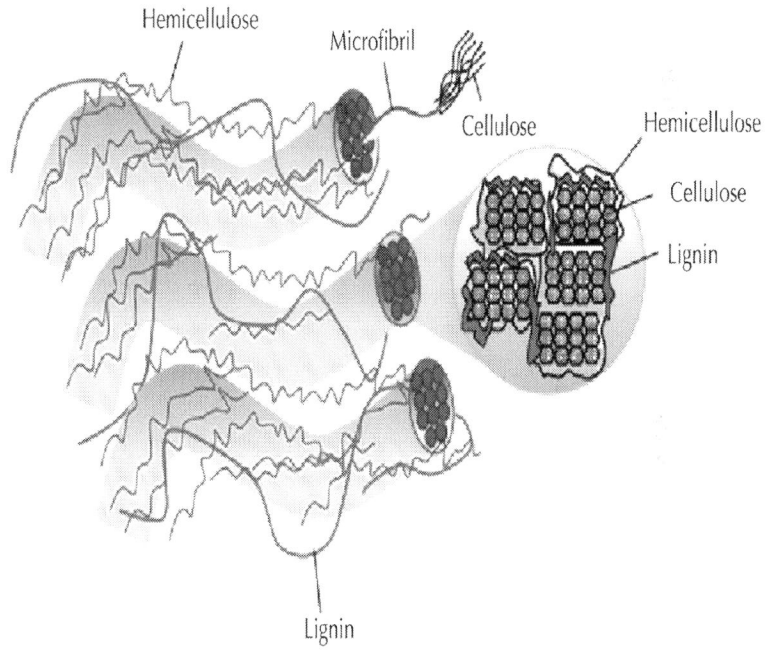

Figure 2: Plant cell wall structure and microfibril cross-section (strands of cellulose molecules embedded in a matrix of hemicellulose and lignin).

Lignin is a complex molecular structure containing cross-linked polymers of phenolic monomers especially p-coumaryl alcohol, coniferyl alcohol, and sinapyl alcohol (Figure 3). The presence of lignin in lignocellulosic biomass is the main obstacle of biomass recalcitrance during separation process. Lignin act as a protective barrier for plant cell permeability and resistance against microbial attacks and thus prevents plant cell destruction. Basically, softwood consists of higher amount of lignin compared to other types of biomass; it makes softwood more recalcitrant and resistant than the other feedstock in cellulose separation step. As a result, removal of lignin is necessary to enhance biomass digestibility up to the point where both hemicellulose and cellulose are exposed for solubilisation process [71, 72].

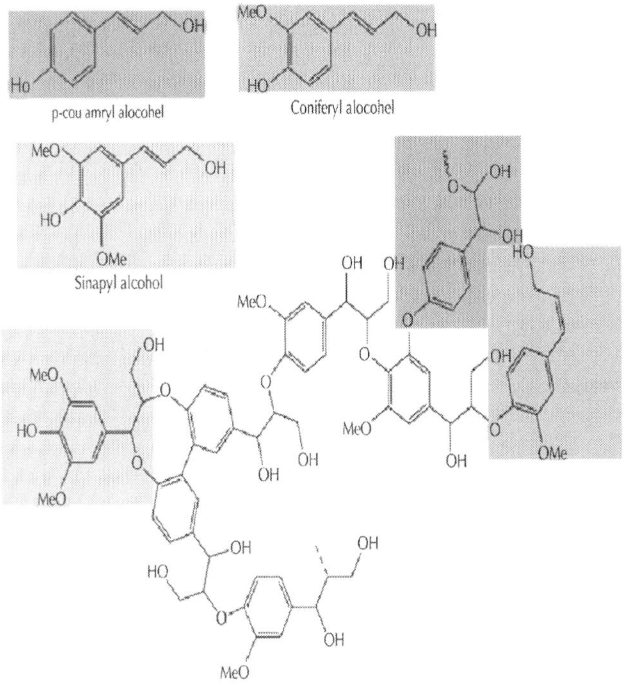

Figure 3: Chemical structures of lignin (p-coumaryl alcohol, coniferyl alcohol, and sinapyl alcohol).

Other than lignin, the accessibility of cellulose is affected by obstructions caused by hemicellulose. Generally, the cellulose fibrils are "coated" with hemicellulose branches with short lateral chains consisting of different sugars (pentoses, hexoses, and acetylated sugars) (Figure 4). It has been suggested that a minimum of 50% of hemicellulose should be removed to extensively increase cellulose digestibility [70]. Comparing to cellulose, hemicellulose can be easily hydrolysed by diluted acid, alkali, or enzymes under mild conditions. Due to its high thermochemical sensitivity, hemicellulose degradation can easily occur to form unwanted coproducts (furfurals and hydroxymethyl furfurals), which inhibits the fermentation process for bioethanol production. However, it does not affect nanocellulose synthesis [70, 73]. For this reason, pretreatment severity is usually a compromise to maximize lignin and hemicellulose recovery during separation process while maintaining cellulose structure for further nanocellulose synthesis.

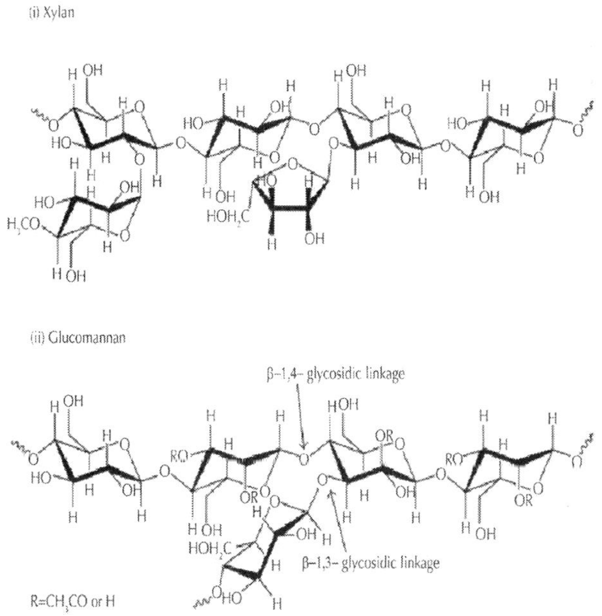

Figure 4: Chemical structure of hemicellulose compounds (xylan and glucomannan are the most existing biopolymer).

Different lignocellulosic biomass pretreatment strategies are currently available with variation in terms of pH, temperature, types of catalyst, and treatment time. These variations affect the severity of the pretreatment and the biomass composition during biomass degradation [21]. Several types of pretreatment that are used to open biomatrix structures are categorized into the following: (i) physical (milling and grinding); (ii) chemical (alkaline, dilute acid, oxidizing agents, and organic solvent); (iii) biological; and (iv) multiple or combinatorial pretreatment of physical and chemical techniques (steam pretreatment/ auto hydrolysis, hydrothermolysis, and wet oxidation) [5, 20, 24, 69, 70, 74–77]. Among the biomass pretreatment, chemical pretreatment proved to be the most efficient method and cost effective for biomass deconstruction with low pretreatment severity. Physical pretreatment includes chipping, grinding, and milling and thermal methods are less efficient and consume more energy than chemical methods, while enzyme for biological pretreatment is expensive and it takes longer pretreatment duration [21]. The detailed chemical pretreatment for lignocellulosic biomass will be further discussed in Section 4.

Cellulose in biomass present in both crystalline and amorphous forms is found in an organized fibrous structure. The long chain cellulose polymers consist of D-glucose subunits which are linked together by -1,4 glycosidic bonds. These linear polymers are linked together by different inter- and intramolecular bonds, which allow them to be packed side by side in planar sheet and bundled into microfibrils (Figure 5). Hence, the cellulose is insoluble in water as the hydroxyl groups in sugar chains are bonded to each other, making a hydrophobic scenario. For this reason, the crystalline domain of microfibrils cellulose, with the presence of extensive intermolecular hydrogen bond and Van Der Waals force makes it another challenge for hydrolysis accessibility to nanocellulose synthesis [78]. In this chemical system, only the cellulosic chains exposed on the surface of the microfibril are easily accessible to solvents, reactants, and chemicals. Thus, the reactivity of cellulose toward hydrolysis is very low.

To improve the performance of cellulose depolymerisation for nanocellulose synthesis, the supramolecular structure of cellulose should be disrupted; that is, some crystalline domains should be converted into amorphous phases. For this purpose, several types of hydrolysis process by chemical or catalytic route have been extensively explored (acid hydrolysis, alkaline hydrolysis, delignification via oxidation, organosolv pretreatment, and ionic liquids pretreatment). Chemistry of depolymerisation of cellulose into nanodimension by acid hydrolysis is another recent interest of study which will be further discussed in Sections 5 and 6.

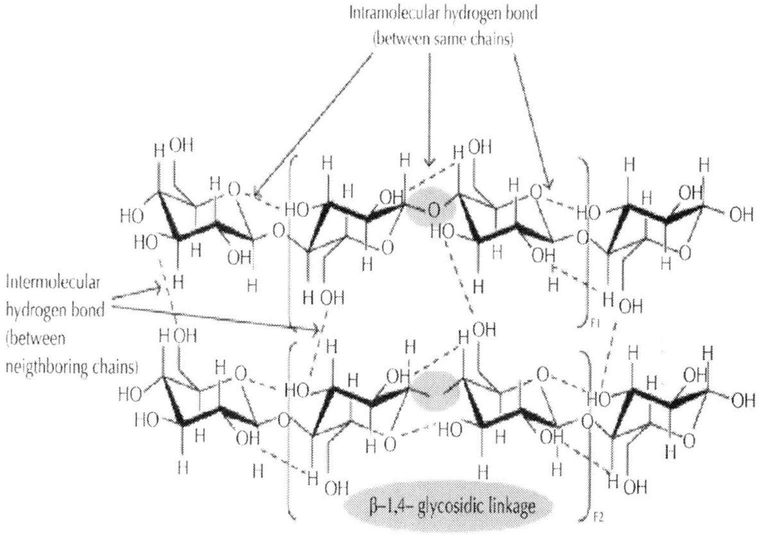

Figure 5: Chemical structures of cellulose chains.

FRACTIONATION OF LIGNOCELLULOSIC BIOMASS VIA CHEMICAL PRETREATMENT

The recalcitrance (resistance of plant cell walls to deconstruction) of lignocellulosic biomass is a major obstacle in the separation of cellulose, hemicellulose, and lignin for different application. The recalcitrance is due to the highly crystalline structure of cellulose which is embedded in a matrix of polymers-lignin and hemicellulose. The main goal of pretreatment is to overcome the recalcitrance which is targeted to alter the size and structure of biomass thru separation of cellulose from the matrix polymers and create access for hydrolysis to turn cellulose into nanocellulose with controllable reaction. Generally, an ideal fractionation of cellulose from lignocellulosic biomass should meet the following requirements: (i) avoid the structure disruption or loss of cellulose, hemicellulose, and lignin content; (ii) be cost effective and reduce energy input; and (iii) minimize production of toxic and hazardous wastes.

The biomass pretreatment is aimed to break the lignocellulosic complex, solubilise the noncellulosic contents (lignin and hemicellulose) but preserve the materials for further valorization, reduce cellulose crystallinity, and increase the porosity of the materials for subsequent depolymerisation process [79, 80] (Figure 6). The present chemical treatments for lignocellulosic biomass degradation are (i) acid hydrolysis, (ii) alkaline hydrolysis, (iii) oxidation agent, (iv) organosolv, and (v) ionic liquids. Table 2 showed operation profiles and degree of fractionation for chemical pretreatments on lignocellulosic biomass degradation. Different types of chemical pretreatment render selective functionality in biomass degradation. Some of the chemical selectively solubilise hemicellulose whilst some chemicals solubilise lignin components. However, all these chemical treatments effectively remove and recover most of the hemicellulose portions as soluble sugars in aqueous solution. The functionalities, advantages, and limitation of different pretreatment are summarized in Table 3. By pointing out that the main barrier in biomass fractionation is the complexity of plant cell structure, it is suggested that the list of chemical pretreatment can be applied individually or in a combination of techniques using several treatment steps to achieve high yield of (nano)cellulose with high recovery of hemicellulose and lignin under low energy input and low severity of process.

Table 2: Degree of biomass degradation by using various types of chemical treatments

Chemical pretreatment	Type of biomass	Concentration	Ratio of BM : chemical	Temperature (°C)	Time (h)	Cᵃ	Hᵇ	Lᶜ	Monomeric sugar
Acid hydrolysis									
(i) Dilute acid									
Phosphoric acid [25]	Rapeseed	0.32% (w/w)	12 : 100	202	0.08	++	++	+	Recovered from hydrolysis of hemicellulose and cellulose
Sulphuric acid [26, 27]	Switch grass	1.2% (w/w)	3 : 100	160	4.3	++	++	+	Recovered from hydrolysis of hemicellulose and cellulose
(ii) Concentrated acid									
Sulphuric acid [28]	Bamboo	75% (w/w)	2.4 : 4.4	59	0.5	++	++	+	Saccharification of BM
Sulphuric acid [29]	Corn stover	65-80% (w/w)	1 : 20	121	1	++	++	+	Saccharification of BM
Alkaline hydrolysis									
Sodium hydroxide [30]	Wheat plant	8% (w/v)	5 : 95	75	1	+	+	++	Recovered from hydrolysis of hemicellulose and cellulose
Calcium hydroxide [31]	Switch grass	0.1 g	1 : 04	121	0.5	+	+	++	Recovered from hydrolysis of hemicellulose and cellulose

	Biomass		Ratio	Temperature	Time	a	b	c	Notes
Organosolv									
Ethyl acetate, ethanol-water [32]	Prairie cordgrass, corn stover, and switch grass	ND	1:10	140	0.33	-	-	++	Limited degradation of sugar
Ionic liquid									
1-Ethyl-3-methylimidazolium acetate ([C2mim][OAc]) [26]	Switch grass	9.7 g	0.3:0.97	160	3	+	+	++	Recovered from hydrolysis of cellulose
1-Butyl-3-methylimidazolium chloride (BMIMCl) [33]	Sugarcane bagasse	ND	1:10	130	0.5	+	+	++	Recovered from hydrolysis of cellulose
Oxidative delignification									
Peracetic acid [34]	Cotton	40 ml/L	1:30	30–70	0.25–4	++	+	++	Recovered from hydrolysis of hemicellulose and cellulose
Peracetic Acid [35]	Aspen Wood	115 mM	1.3:40	60	6	++	+	++	Recovered from hydrolysis of hemicellulose and cellulose

++: hydrolysis towards cellulose, hydrolysis towards hemicellulose, and efficient removal of lignin;

+: less effect towards cellulose hydrolysis, less effect and removal of hemicellulose, and solubilisation of lignin;

–: minor effect toward cellulose, minor effect towards hemicellulose, and less efficient in removal of lignin;

BM: biomass.

aCellulose. bHemicellulose.

cLignin.

Table 3: Functionality, advantages, and limitations for each chemical treatments

Chemical	Mode of action	Advantages	Disadvantages/limitation	Remarks
Dilute acid				
Sulfuric acid, phosphoric acid [36]	(1) Removal of hemicellulose	(1) Higher reaction rates (2) Increase the accessibility of cellulose	(1) Form by-product (fermentation inhibitors) (2) High cost and expensive construction material due to acidic environment (3) Corrosive to reactor	Minimal degradation of lignin and cellulose
Concentrated acid				
Sulfuric acid, phosphoric acid [37]	(1) Solubilisation of hemicellulose and direct hydrolysis of cellulose to glucose	(1) Suitable to all types of biomass	(1) Uncontrolled hydrolysis process (2) Corrosive to reactor	Suitable for the glucose synthesis (saccharification of biomass)
Alkaline hydrolysis				
Sodium hydroxide, calcium hydroxide [38,39]	(1) Removal of lignin (major) (2) Removal of hemicellulose (3) Cellulose swelling	(1) High solubilisation of lignin (2) Low formation fermentation inhibitors	(1) High cost of chemical (2) Alteration of lignin structure	Suitable to use prior to direct fermentation of carbohydrates
Organosolv				

Mixture of organic solvent and water [40]	(1) Extraction of lignin (2) Complete solubilisation of hemicellulose	(1) High recovery of lignin (2) Organic solvent used can be recycled and reused (3) No grinding/ milling of biomass feedstock (4) Selective pretreatment method for lignin extraction	(1) High cost of solvent (2) High energy consumption during solvent recovering process	Suitable for lignin fractionation process where high content of lignin can be recover for specialty chemical synthesis
Ionic liquid				
Imidazolium salts [41]	(1) Extraction of lignin (2) Decrease the cellulose crystallinity index (3) Carbohydrate dissolution	(1) IL is high thermal stability and low volatility	(1) High cost of chemicals.	The effects towards hemicellulose and lignin are depending on the nature of ionic liquid used
Oxidative delignification				
Hydrogen peroxide [42]	(1) Solubilisation of lignin and hemicellulose. (2) Bleaching effect to the pulp	(1) Efficient in removal of lignin (2) Increase biomass digestibility	(1) High costs of chemicals	Suitable for cellulose bleaching where lignin and hemicellulose will degrade in the presence of alkali

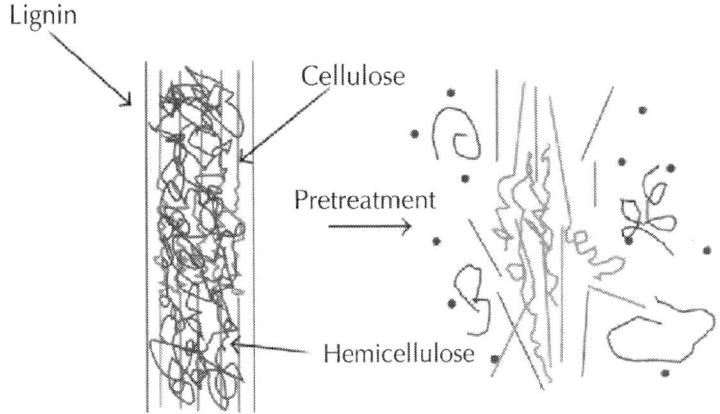

Figure 6: Deconstruction of lignocelluloses into cellulose, hemicellulose, and lignin.

(i) Acid Hydrolysis: Acid pretreatment is a process to break the rigid structure of lignocellulosic material in which hydronium ions breakdown and attack intermolecular and intramolecular bonds among cellulose, hemicellulose, and lignin in biomass hierarchy structure. The acid hydrolysis includes concentrated and dilute acid solutions where different levels of acid severity contribute to various biomass fractionated products [69]. Concentrated acids such as H_2SO_4, HCl, H_3PO_4, and HNO_3 are being used to hydrolyse biomass [28, 33]. It is an effective agent for biomass deconstruction to maximize the yield of monomeric sugars for biofuel production. In nanocellulose synthesis, long chain structure of cellulose feedstock is preferable for further reaction to produce nanocellulose. The concentrated acid used in acid hydrolysis is toxic, hazardous, and corrosive; thus highly corrosion resistant reactor and extreme care in handling are required in the process. This makes acid pretreatment an expensive option. In addition, the concentrated acid must be recovered after pretreatment to make the process economically and environmentally feasible. As a result, the effects for different type of acid, pH, reaction temperature, and reaction time in biomass fractionation process are crucial to influence the maximum yield of targeted cellulose product. Extreme reaction condition will cause uncontrollable degradation of biomass to sugars and subsequently catalyzed the degradation of monosaccharide to undesirable coproducts such as acid and alcohol [21, 79].

For the reasons stated above, dilute-acid hydrolysis has become cost-effective alternative to enhance biomass separation to isolate cellulose, hemicellulose, and lignin which can further be used for nanocellulose synthesis and chemical production. Generally, dilute acid (e.g., concentration of $H_2SO_4 < 4$ wt.%) [26, 27] can dissolve and recover most of the hemicellulose as dissolved sugars up to 100% conversion under low process severity (low temperature and low acid concentration). The remaining fraction of lignin in the residue cellulose from hemicellulose solubilisation process will then be removed via cellulose purification process. Cellulose purification is a process that utilizes alkaline pretreatment, where delignification occurs to separate lignin from cellulose in order to increase accessibility of cellulose for hydrolysis reaction.

(ii) Alkaline Hydrolysis: The major strategy of alkaline pretreatment is to disrupt the lignin structure in biomass, thus improving the susceptibility of the remaining polysaccharides (cellulose and hemicellulose) for other treatment [74]. Alkaline hydrolysis occurs at milder conditions (below 140°C) and lower severity as compared to other pretreatment technologies. The mechanism involves saponification of intermolecular ester bond, which crosslinks xylan (hemicellulose) and lignin [74]. The porosity of the pretreated material will increase with the removal of the crosslinker. Cleavage of this ester linkages are substituted by nucleophilic acyl in the presence of alkaline salt (e.g., NaOH or $Ca(OH)_2$) to form a carboxylic salt and an alcohol [21]. Furthermore, the alkaline degradation of lignin also involved cleavage of two types of aryl ether bonds: $C_{aliphatic} - O - C_{aromatic}$ and $C_{aromatic} - O - C_{aliphatic}$, which produces acids (ferulic acid and p-coumaric acid). Generally, the pretreatment agents that are used for alkaline hydrolysis are NaOH, KOH, $Ca(OH)_2$, hydrazine, and ammonium hydroxide [30, 31]. In the presence of alkali, it also serves the following functions: (i) swelling agent to cellulose; (ii) leading to an increase of internal surface area; (iii) leading to a decrease in the degree of polymerization and crystallinity of cellulose; (iv) leading to partial solvation of hemicellulose; (v) destroying the structural linkages between lignin and carbohydrate by saponification of intermolecular ester bonds; and (vi) disrupting the lignin structure by breaking its glycosidic ether bond. Lignin will fail to act as protective shield to the cellulose after lignin solubilisation step, thus making extracted cellulose more susceptible to nanosynthesis.

These steps are needed to avoid over degradation of cellulose to sugar; it is suggested that alkali pretreatment is implemented after dilute acid treatment to remove hemicellulose component from lignocellulosic material. During hemicellulose solubilisation, there is a strong layer of lignin keeping the cellulose unexposed/inaccessible, which led to less degradation of cellulose to glucose. This process shall maximize the cellulose yield for nanocellulose synthesis. The challenge to complete the biomass delignification is difficult as this component is normally located within the deep cell wall, hydrophobic, physically stiff, strong poly-ring bonds of C–O–C, C–C, and tendency to recondensation [81]. Thus, the remaining cellulose and part of lignin will be subjected for further separation process to solubilise lignin matrix while maintaining a high recovery rate of low crystalline cellulose for nanocellulose production.

(iii) Oxidation Agent: Oxidation agent such as organic peroxide (H_2O_2, $C_2H_4O_3$), ozone, oxygen, or air is another technique used to catalyze delignification process by attacking and cleaving of lignin's ring structure [34, 35]. Normally, oxidation agents are used to enhance the effects of alkaline pretreatments. Under basic condition of pH > 12, the oxygen will reduced to superoxide radical), where the ring will open by this nucleophilic attack [81]. It is an efficient treatment to oxidize aromatic ring of lignin and part of hemicellulose polymer to carboxylic acids compounds (e.g., formic acid, oxalic acid, and acetic acid). This treatment is suitable for extraction of cellulose as the oxidation agent is more aggressive on lignin and partially on hemicellulose, while cellulose is hardly decomposed under this mild condition [82].

(iv) Organosolv: Organosolv pretreatment is the simultaneous process of lignin and hemicellulose degradation, solvation, and solubilisation of lignin fragments from lignocellulosic feedstock with the presence of organic solvents or their aqueous solutions. Organic solvent acts as a dissolving agent by solubilising lignin and some of the hemicellulose under heating condition and leaving a relatively pure solid cellulose residue. The common solvents used for organosolv pretreatment are low boiling point alcohol such as methanol, ethanol, acetone, ethylene glycol, and ethyl acetate [83]. The OH$^-$ ion from alcohol solvent will attack the acid-ester bonds of lignin-hemicellulose compounds. The cleavages of ether linkages from lignin and minor hydrolysis of glycosidic bond in hemicellulose

are important for the breakdown of aromatics and polysaccharides of lignocellulose. An advantage of employing high volatility alcohol is due to the ease of recovery by simple distillation which requires very low energy consumption. In the other hand, these alcohols are lower in cost and soluble in water. Addition of catalyst in organosolv pretreatment such as inorganic (HCl or H_2SO_4) or organic acid (oxalic, acetylsalicylic and salicylic acid) helps to break the internal bonding of lignin-hemicellulose linkage and improve the organosolvation process under lower temperature. Consequently, this pretreatment involve simultaneous prehydrolysis and delignification of lignocellulosic biomass to solubilise noncellulosic components and obtain cellulose fraction. Furthermore, presence of organic solvent was also found to swell the cellulose and reduce crystallinity of cellulose for further application [83, 84].

At present day, the organosolv pretreatment is not economically feasible to be utilized. Extensive washing is needed to wash the pretreated materials with organic solvent prior to water washing in order to avoid the precipitation of dissolved lignin, which leads to cumbersome washing arrangements. Furthermore, recovery of organic solvents causes increase of energy consumption for the whole process [85].

(v) Ionic Liquids: Ionic liquids (ILs) pretreatment is another recent development in chemical-based dissolution pretreatment technology. The tunability of ILs chemistry makes it highly capable of dissolving wide variety of biomass type. Unlike the heterogeneous reaction environment (cellulose in water), ILs makes the catalytic sites highly access the -glycosidic bonds, which facilitates the reaction of biomass fractionation and hydrolysis of cellulose [86, 87]. ILs can dissolve cellulose, hemicellulose, and lignin under considerably mild conditions without degrading the chain's structure. It is reusable liquid salts at room temperature, typically composed of inorganic anion and organic cation, which can be tuned to generate different dissolving capacity for targeted components. The common examples of these ILs include salts of organic cations for cellulose dissolution and biomass pretreatment, such as 1-alkyl-3-methylimidazolium [mim]$^+$; 1-alkyl-2,3-dimethylimidazolium [mmim]$^+$; 1-allyl-3-methylimidazolium [Amim]$^+$; 1-allyl-2,3-dimethylimidazolium[Ammim]$^+$; 1-butyl-3-methylpyridinium [C_4mP_y]$^+$; and tetrabutylphosphonium [Bu4P]$^+$ with

= number of carbons in the alkyl chain [88, 89].

An ideal ILs for lignocellulosic biomass pretreatment process should possess the following: (i) high dissolution capacity for different component by varying organic cation; (ii) low melting point; (iii) low viscosity; (iv) low/no toxicity; and (v) high stability. Considering the fact that fractionation of cellulose from noncellulosic matrix is a complicated process, chemistry of ILs is (anion and cation compositions) needed to be adjusted in order to solubilise hemicellulose and lignin, to ease cellulose fractionation from the biomatrix [90, 91]. For example, 1-ethyl-3-methylimidazolium acetate [Eminm] Ac has high solubility for lignin and low solubility for cellulose. It is able to selectively extract the rigid lignin from the lignocellulose while yielding a highly amorphous cellulose fraction [78, 84].

Other than biomass degradation to isolate cellulose product, ILs pretreatment could reduce crystallinity of cellulose to amorphous nature. Cellulose in the form of fibrils poorly hydrolyse or depolymerise under mild condition due to its intermolecular hydrogen linkages between polysaccharide chains. ILs is capable of disrupting the hydrogen bonds by forming another hydrogen bond between anion of IL with cellulose (sugar hydroxyl protons) in a 1 : 1 ratio. This will break up the cellulose hydrogen bonded structure, thus decreasing the compactness of cellulose and making it more amorphous and susceptible to depolymerisation process [92].

In summary, the chemical pretreatment for fractionation of lignin and carbohydrates (cellulose and hemicellulose) can be achieved through the use of acids, alkalies, solvents, and ionic liquids, which promote selective solubilisation for each biocomponent (Table 3). The presence of acid is responsible for solubilising hemicellulose and cellulose via hydrolysis reaction; thus controllable acid concentration is crucial to promise high recovery of cellulose fibers for further action. Lignin is not hydrolysed by acid, but it can be soluble in alkali treatment and oxidized in the presence of oxidation agent, while organosolv treatment mainly focuses on solubilisation of carbohydrates. Ionic liquid can dissolve both carbohydrates and lignin, which disrupt the intricate network of noncovalent interactions between these polymers. This treatment can reduce lignin content and cellulose crystallinity.

Although chemical process may not be as selective as biological treatment (enzym e), it consisted of many advantages related to operation period, scalability, and process control. Figure 7 showed the chemical pretreatments on selective part of biomass compounds.

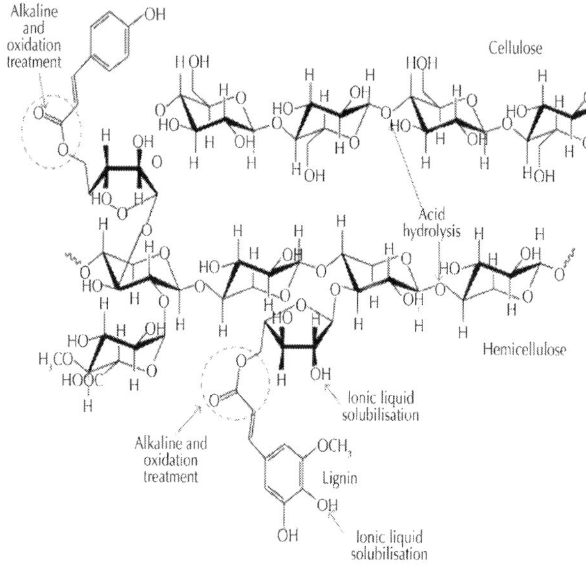

Figure 7: Selectivity of chemical treatments for fractionation of lignocellulosic biomass.

DEPOLYMERISATION OF NATIVE CELLULOSE TO NANOCELLULOSE

Cellulose microfibril contains crystalline and amorphous regions that are randomly distributed along their length. The former, cellulose chain, is packed closely, whereas the latter is in disorder manner, which easily breaks under harsh treatment (Figure 8). Nanocellulose is categorized into nanocrystalline cellulose (NCC) and nanofibrillated cellulose (NFC), with extraordinary properties which was induced by nanoscale effect. Both types of nanocellulose are chemically similar; the dissimilar physical characteristic (different colloidal forms) is cellulosic "rice" and cellulosic "spaghetti" for NCC and NFC, respectively. Nanocellulose has a rigid rod-shaped structure, is 1–100 nm in diameter, and is tens to hundreds of nanometers in length. The relative degree of crystallinity and the geometrical aspect ratio (length to diameter; L/d) are very important parameters controlling the properties of nanocellulose [15, 93].

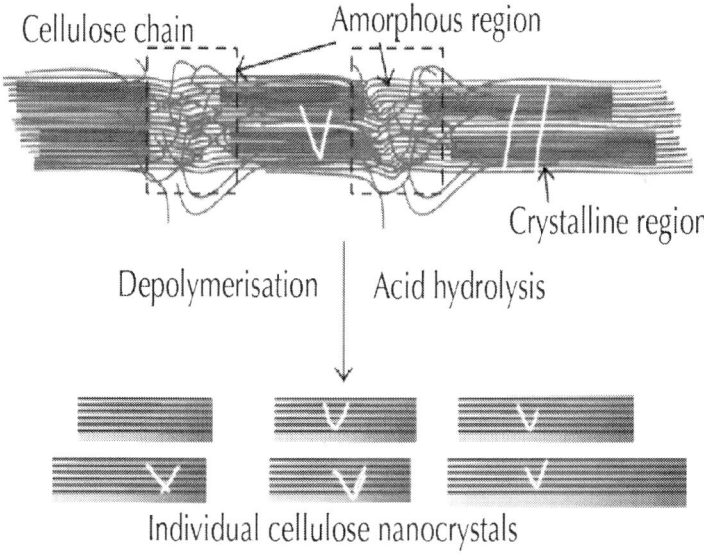

Figure 8: Depolymerisation of cellulose to nanocellulose.

Recently, preparation of nanoscale cellulosic particles under mild conditions is getting attention from the researchers. The top-down destructuring of cellulosic fibers can be conducted by a mechanical reaction (cryocrushing, grinding, and high pressure homogenization) [46, 48, 49, 51, 94, 95], biological reaction (enzymatic treatment) [96, 97], and chemical reaction (oxidation and acid hydrolysis) [43, 45, 46, 93] (Table 4). A major obstacle which needs to be overcome for successful commercialization of nanocellulose is the high energy usage from mechanical disintegration of the fibers into nanofibers, often involving several paths through the disintegration device. Therefore, researchers have combined mechanical pretreatments with chemical techniques to increase efficiency of sizes reduction before homogenization, which helps to lower the energy consumption.

Table 4: Summary of depolymerisation treatments for nanocellulose synthesis

Chemical treatment

Chemical	Cellulose source	Concentration of chemical	Ratio of substrate : acid	Reaction time	Temperature (°C)	Pretreatment	Particle size (nm)*	Yield of NC
Sulfuric acid [43]	Branch barks of mulberry (Morus alba L.)	64% (w/w)	1 : 10	0.5 h	60	Pulping treatment	20-40	N/A
Sulfuric acid [44]	Sugarcane bagasse	60% (w/v)	1 : 20	5 h	50	Pulping treatment	35	N/A
Sulfuric acid [45]	Waste cotton fabrics	63.5% (w/w)	1 : 15	3 h	44	Pulping treatment	20-100	21.50%
TEMPO mediated oxidation [46]	Pure rice straw	0.016 g	1 : 10	N/A	N/A	Oxidation	1.73	20%
Sulfuric acid [47]	Raw cotton linter	60% (w/w)	1 : 20	1 h	45	Grinding	12	N/A

Mechanical treatment

Method	Cellulose source	Pressure/ energy	Temperature (°C)		Pretreatment		Particle size (nm)*	Yield of NC
High pressure homogenizer [48]	Wood pulp	500 bar	55-60		Pulping		ND	N/A
Sonication [49]	Wood powder from poplar trees	400-1200 W	<10		Pulping		5-20	N/A

Method	Cellulose source		Time and temperature	Pretreatment	Particle size (nm)*	Yield of NC
High pressure homogenizer [50]	Sugarcane bagasse	40–140 MPa	130	Bleaching and ionic liquid treatment	20–100	N/A
Sonification [51]	Eucalyptus kraft pulp	50% amplitude, ~80 W	N/A	Pulping treatment	30	N/A
High shear homogenizer [52]	Bleached softwood pulp	22 000 rpm	N/A	Pulping and bleaching	16–28	N/A
High pressure homogenizer [53]	Nonwoody plants (flax, hemp, jute, and sisal)	600 bar	60–70	TEMPO-mediated oxidation	20–50	N/A
Biological treatment						
Enzyme/microorganism	Cellulose source		Time and temperature	Pretreatment	Particle size (nm)*	Yield of NC
Anaerobic microbial [54]	Cotton fibers		7 days and 35°C	Pulping treatment	43 ± 13 & 119 ± 9	12.30%
Hemicell/pectinase and endoglucanase [55]	Curaua and sugarcane bagasse fibers		3 days and 50°C	Pulping treatment	55 ± 21	N/A

*Measured by Transmission Electron Microscope (TEM).

N/A is not available; NC is nanocellulose

Among the cellulose depolymerisation treatments, oxidation pretreatment is one of the common techniques used to disintegrate cellulose into nanocellulose by applying 2,2,6,6-tetramethylpiperidinyl-1-oxyl (TEMPO) radicals. TEMPO-mediated oxidation method generates sinter-fibrillar repulsive forces between fibrils by modifying surface of native cellulose. This led to conversion of primary hydroxyls in cellulose into carboxylate groups, which later become negatively charged and resulted in repulsion of the nanofiber, thus contributing to an easy and fast fibrillation [94, 98].

Acid hydrolysis is a process related to the breakage of -1,4 glycosidic bonds in cellulose. It is well known that acid hydrolysis is the most effective process currently to produce nanocellulose with possibility in lowering energy consumption [20, 22, 95–97, 99]. Commonly, hydrolysis is performed in the presence of mineral acid (50–72% H_2SO_4) for depolymerisation of cellulose. The low density of amorphous regions in native cellulose is more accessible to acid and more susceptible to hydrolytic action than the crystalline domains. These amorphous regions will break up, releasing the individual crystallites cellulose when subjected to acid treatment.

The characteristic of nanocellulose from acid hydrolysis is highly dependent on various factors, such as origin of cellulose sources, types of acid, concentration of acid, reaction time, and temperature for hydrolysis. In the presence of Bronsted acidic media, the acid acts as a catalyst by protonating the oxygen atom of glycosidic bond in cellulose chain [18]. Subsequently, the unstable positive charged group leaves the polymer chain and it is replaced by the hydroxyl group of water (Figure 9). Furthermore, esterification process occurred in between H_2SO_4 and hydroxyl groups to yield "cellulose sulfate" which resulted in negatively charged surface of the cellulose crystallites. The anionic stabilization via the attraction/repulsion forces of electrical double layers is the reason for the stability of colloidal suspensions of crystallites (Figure 10) [97]. However, the produced nanocellulose may be chemically modified by sulfate ester group where further functionalisation of nanocellulose will be limited [100]. Besides, the conventional acid hydrolysis treatment for nanocellulose synthesis is environmentally incompatible and economically unfriendly as extra cost will be generated for effluents treatment [54].

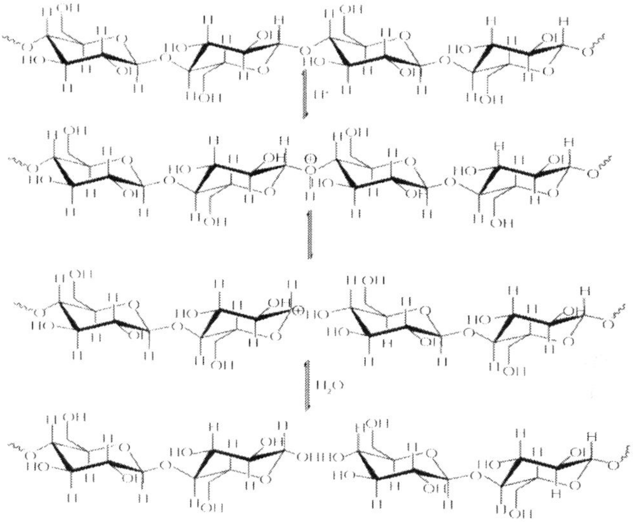

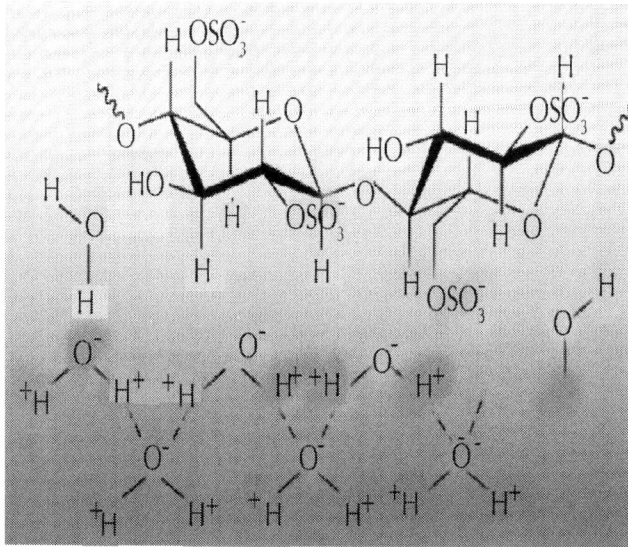

Figure 9: Acid hydrolysis of cellulose in to nanocellulose.

Figure 10: Formation of sulfate group on the nanocellulose surface after hydrolysis reaction.

Thus, researchers have claimed that the use of enzymatic hydrolysis may benefit from the environment point of view on comparing with acid treatment [101]. Biological-based hydrolysing agent cellulase (composed of multicomponent enzyme system) allows restrictive and selective hydrolysis of specified component in the cellulosic fibers. Enzymatic hydrolysis process involved multistep catalyzed reaction in which solid crystal of cellulose is initially disordered at the solid-liquid interface via the synergistic action of endoglucanases and exoglucanases/cellobiohydrolases. Generally, endoglucanases act to cleave the internal bonds (e.g., noncovalent interaction) present in the amorphous structure of cellulose. Besides, exoglucanases/cellobiohydrolases will further attack the terminal glycosidic bond from the end of the exposed cellulose chains generated by endoglucanases. Subsequently, short cellulose chains from initial reaction are accompanied by further hydrolysis, where beta-1,4 glycosidic linkage of cellulose is broken down by cellobiases/beta-glucosidases into nanocellulose or even glucose product.

Although enzymatic route for nanocellulose synthesis offers the potential for higher yields, higher selectivity, lower energy costs, and milder operating conditions than chemical processes, such technology was still hindered by economical (costly cellulase enzyme) and technical (rate limiting step of cellulose degradation with long processing period) obstacles. The slow rate of enzymatic hydrolysis is easily influenced by several reasons: structural features resulting from pretreatment and enzyme mechanism [102–104]. Table 5 showed the limiting factors that affect cellulase hydrolysis performance.

Table 5: Factors that limit hydrolytic efficiency of cellulase on cellulose surface

Factor 1 Biomass structural features	Factor 2 Enzyme mechanism
(I) Physical structure	(I) Enzyme diffusion and accessibility
Accessible surface area, crystallinity, physical distribution of lignin in the biomass matrix, degree of polymerization, pore volume, and biomass particle size	(II) Selectivity of enzyme to component in cellulose fiber
(II) Chemical structure	(III) Enzyme inhibitor
Compositions of cellulose, hemicellulose (xylan), lignin, and acetyl groups bound to hemicellulose	

FUTURE PROSPECT OF ACID CATALYST DEVELOPMENT FOR HYDROLYSIS OF CELLULOSE

Acid hydrolysis can happen in homogeneous or heterogeneous catalyze reaction with first order kinetic rate. Typically, the acid hydrolysis of cellulose that uses homogeneous catalysis is more feasible in terms of mass transfer and reaction efficiency. However, these methods have major drawbacks such as (i) low chemical recovery, (ii) reactor and equipment corrosion, (iii) extra cost for waste effluents treatment, and (iv) uncontrollable selectivity of nanocellulose production. This can conclude that commercial scale of nanocellulose production by using liquid acid (e.g., H_2SO_4 and HCl) is not economically and environmentally sustainable.

To date, a solid acid catalyst is getting attention in selective cellulose hydrolisation. This type of acid catalyst with Bronsted acid active sites is a good alternative to concentrated H_2SO_4 for hydrolysis reaction. It has numerous advantages over liquid H_2SO_4 in terms of activity, selectivity, catalyst lifetime, and reusability. Moreover, the use of solid acid proficiently could reduce acid pollutants and cost of waste water treatment and thus reduce the manufacturing costs [99, 105 112].

The main challenge of solid catalysts is the contact between catalyst active sites with solid macromolecule of cellulose. As both reactant and catalyst are present in solid phase, it is difficult for catalytic acid sites to have a close contact with -1,4 glycosidic linkage in cellulose for hydrolysis reaction. Hence, hydrothermal condition (water as reaction medium) is a good choice to improve the accessibility of catalyst to cellulose. Other than acting as mass transfer medium, water can take part as catalyst for autohydrolysis process. The hydronium ions (H_3O^+) formed on the surface of catalyst lead to promotion of cellulose hydrolysis.

The critical characteristics for ideal acid catalysts for hydrolysis are as follows.

- It must be a water-tolerant acid catalyst.
- It must have strong acidity and high accessibility of acid active sites for reaction.

- It shall ease in separation between solid catalyst and final nanocellulose product in the powder or gel form.

Based on these ideal criteria, several possible green acid catalysts are highly potential to be designed for cellulose hydrolysis (Table 6).

- Nanoparticle hydrolysis catalysts: they disperse in water solution capability, resulting in facile interaction with cellulose and overcome the difficulty of solid-solid reaction. The monodispersed nanoparticles in form of fluid solution are able to promote large number of surface active site for more access to the oxygen atom in the ether linkage of cellulose.

- Magnetic properties acid catalyst: magnetization attraction with external magnetic field will facilitate the separation process and catalyst recovery. Presence of repulsion and attraction effect from magnetic nanoparticle is able to reduce the reaction barrier due to adsorption, agglomeration, dispersion, and viscous effects of the reaction mixture.

- Catalyst with homogeneous, biphasic, and heterogeneous system: catalyst will form liquid-solid phase during hydrolysis reaction through polar or nonpolar solvent and heating. It can be soluble in water and, therefore, more efficient for hydrolysis of solid cellulose. Furthermore, this catalyst can be recovered in solid form after reaction by distilling out the products and solvent.

Table 6: Solid acid catalyzed hydrolysis for nanocellulose production

Feedstock	Catalyst	Acidity	Operation condition	Final product	References
Iron-based catalyst					
Corn stover	$FeCl_3$	pH 1.68	0.1 M $FeCl_3$, 140°C, 20 min	91% removal hemicellulose 91% recovery of cellulose	[56]
Hemicellulose compounds: xylose and xylotriose	$FeCl_3$	pH 1.86	0.8% of $FeCl_3$, 180°C, 20 min	65% degrade of xylose, 78% degrade of xylotriose	[57]

Corn stover	$FeSO_4$	N/A	0.1 mol/L $FeSO_4$, 180°C, 20 min	60.3% degrade of hemicellulose, 89.8% recovery of cellulose	[58]
Cellulose	$FeCl_3$	N/A	10% $FeCl_3$, 110°C, 60 min of reaction time, 180 min of ultrasonic time	22% CNC Diameter = 20–50 nm, Length of 200–300 nm, crystallinity index = 76.2%	[59]
Cellulose	Zn-Ca-Fe based nanocatalyst	N/A	160°C, 12 h	29% of glucose	[60]
MCC	Fe_3O_4@C-SO_3H		140°C, 12 h, 10 mL of water	52.1% of glucose	[61]
Cellobiose	Fe_3O_4-SBA-SO_3H		120°C, 1 h,	98% of glucose	[62]
Heteropoly acid (HPA) catalyst					
Cellulose	HPA	N/A	180°C, 2 h	51% of glucose	[63]
MCC	CS-HPA	N/A	160°C, 6 h	30.1% of total reducing sugar (TRS), 27.2% glucose	[64]
Cation-exchange resin					
MCC	Cation-exchange Resin (NKC-9 cation-exchange resin (NKC-9))	Exchange capacity (mmol/g [H⁺]) 4.7, pearl size of 0.45–1.25 mm, and true wet density of 1.20–1.30 g/mL	Resin : MCC = 10 : 1, 48°C, 189 min under sonication	Yield of NCC is 50.04% Crystallinity index = 84.26%	

Iron-Based Catalyst

Iron metal-based catalyst with its Lewis acid site can perform as catalyst for acid hydrolysis. Several studies reported that transition metal salts could increase hydrolysis rate for biomass fractionation and nanocellulose synthesis. Iron-based inorganic salts ($FeCl_3$ and $Fe_2(SO_4)_3$) had been selected as catalyst to accelerate the degradation of hemicellulose in corn stover. $FeCl_3$ significantly increased the hemicellulose degradation in aqueous solutions heated between 140 and 200°C with high xylose recovery and low cellulose removal, amounting to ~90% and <10%, respectively. In this study, hemicellulose removal increased 11-fold when the corn stover was pretreated with 0.1 M $FeCl_3$ compared to pretreatment with hot water under the same conditions, which was also 6-fold greater than pretreatment with dilute sulfuric acid at the same pH [56]. This finding was supported by Liu's group [57], where $FeCl_3$ significantly increases the hemicellulose (xylose and xylotriose) degradation at 180°C. The result showed that 0.8% of $FeCl_3$ with acid pH of 1.86 rendered 3-fold and 7-fold greater of xylose and xylotriose degradation, respectively, than that for treatment with dilute sulfuric acid with similar pH value. This suggested that different mechanism may exist for the reaction; first, the presence of $FeCl_3$ may induce the formation of H^+ ion from water for hydrolysis; second, $FeCl_3$ may catalyze the dehydration of carbohydrates. $FeSO_4$ catalyst also showed strong effects on the pretreatment of corn stover, where hemicellulose significantly degrades to xylose (60.3%) with high recover of cellulose (89.8%) under temperature of 180°C within 20 min. According to Zhao et al. [58], $FeSO_4$ pretreatment is capable of disrupting the ester linkages between cellulose and hemicellulose. It is a potential pretreatment that is able to enhance further hydrolysis reaction of lignocellulosic biomass by destructing chemical composition and altering structural features. In acid hydrolysis, proton (H^+ from HCl or H_2SO_4) plays an important role to weaken the glycosidic bond energy by attracting electron, making it easier for bone rupture. Metal ions also play same function as proton, which consist of more positive charge to pair with more electrons, but proton can only pair with one electron. This fact was in agreement with the studies above, where $FeCl_3$ (trivalent, Fe^{3+}) rendered better hydrolysis effect than $FeSO_4$ (divalent, Fe^{2+}), while acid solution performed less effect as compared to inorganic salts [113].

Recently, researchers started to use transition metal-based catalyst ($FeCl_3$) for preparing cellulose nanocrystal (CNC) via hydrolysis of cellulose. The results show that 22% of CNC was produced under 10% $FeCl_3$, temperature of 110°C for 60 min, and ultrasonic time of 180 min. CNC produced are shaped rod-like with the diameter of 20–50 nm, the length of 200–300 nm, and crystallinity of 76.2% [59].

Other than iron salts, the mesoporous structure exhibited in iron oxide catalyst allows accessibility of feedstock to active acid sites inside the pores. The characteristics of mesoporous iron oxide for cellulose hydrolysis are [114]

- high surface area;
- tunable pore size to improve accessibility of reactant;
- good thermal stability;
- good acidity;
- nanosize for better dispersion and active sites accessibility.

Paramagnetic-based nanocatalysts have attracted much interest due to their unique properties. The acidity of this nanoiron catalysts suspension (Zn–Ca–Fe based catalyst), capable to tune by adjusting the composition of Fe content, resulted in high accessibility of catalytic active sites to hydrolyse glycosidic bond in cellulose [60]. With the presence of Fe in the catalyst system, the catalyst is capable of physically separating from end product by applying external magnetic field [107]. So far, this catalyst was used as the hydrolysis catalyst for depolymerisation of cellulose into glucose monomer.

Same case goes to Fe_3O_4, although it consists of potential magnetization properties suitable for separation process for most of the catalytic systems; however, it is so far applied in glucose synthesis [115]. Generally, Fe_3O_4 nanoparticles will incorporate with acid carrier such as carbonaceous support ($Fe_3O_4@C-SO_3H$) [61] and mesoporous silica support (Fe_3O_4-SBA-SO_3H) [62] as an acid catalyst while enhancing efficient catalyst recovery system. Zhang et al. [61] reported that the $Fe_3O_4@C-SO_3H$ nanoparticle is composed of magnetic Fe_3O_4 that doped in a sulphonated carbon shell, which render good characteristic in terms of magnetization separation effect, high stability, and good reusability. The acid catalyst showed 48.6% cellulose conversion with 52.1% glucose selectivity under the moderate conditions of 140°C after 12 h reaction. Experiments on the hydrolysis of cellobiose with Fe_3O_4-SBA-SO_3H showed that the magnetic solid acid gave an even better performance than sulfuric acid as well [62]. Thus, supported Fe_3O_4 nanocatalyst provides good access of macrocellulose to SO_3H group with its functional characteristic which allows it to be separated and regenerated after the process.

Heteropoly Acid

Heteropoly acids, HPA (e.g., $H_3PWO_{12}O_{40}$) with its unique physicochemical properties, Bronsted acidity, stability, and high proton mobility, have potential to be used for cellulose hydrolysis. It can be tuned to homogeneous, biphasic, and heterogeneous system at different conditions. HPAs are soluble in water and possess acidic strength similar to sulfuric acid. In theory, HPA acted as H_2SO_4 by donating H^+ ions to oxygen atom in the glycosidic ether bonds of cellulose for depolymerisation process. HPA performed as homogeneous catalyst during hydrothermal hydrolysis, and it can be recovered to solid phase by employing organic solvent such as diethyl ether [63, 107, 116, 117].

The HPA catalyst system in cellulose hydrolysis involved three catalytic modes [118]: (i) surface reaction; (ii) pseudoliquid (bulk-type 1); and (iii) solid bulk-type II. Macrosize cellulose normally hydrolyses on the surface of solid HPA catalyst to reduce the crystallinity of cellulose structure. The smaller molecule of cellulose will later diffuse into the solid catalyst and undergoes reaction with pseudoliquid phase followed by solid bulk-type II diffusion of electrons and protons to assist the redox surface. Jiang's study showed that a remarkably high yield of glucose (50.5%) was achieved via acid hydrolysis of $H_3PW_{12}O_{40}$ under reaction conditions of 180°C and 2 h of reaction time [63]. Furthermore, (Cs-doped HPA) has gained much attention for selective hydrolysis of microcrystalline cellulose (MCC) to sugars in the aqueous phase. Partial substitution of proton by Cs^+ improves the characteristic of HPA, where the surface acidity increases with higher surface area, controlled shape selectivity and good hydrophobility effect (insoluble in water). According to Tian's study, $Cs_1H_2PW_{12}O_{40}$, with the strongest protonic acid site, showed the best catalytic performance in the conversion of MCC for acid hydrolysis, where highest total reducing sugar (TRS) and glucose yields were 30.1 and 27.2%, respectively [64].

Most of the studies used HPA-based catalyst for cellulose hydrolysis to glucose, which was further processed to bioethanol for biofuel application. A control mechanism has to be explored to degrade macromolecule of cellulose to nanomolecular size by using HPA catalyst. It has been suggested that lower reaction temperature and shorter reaction time are the best choice to avoid overdegradation of cellulose to glucose monomer in HPA catalyzed reaction.

Ion-Exchange Resin

Ion-exchange resin commercially acted as solid acid catalysts in reactions such as alkylation with olefins, alkyl halides, alkyl esters, isomerization, transalkylation, nitration, and hydrolysis. Compared with homogeneous liquid, ion-exchange resin is easily separated after reaction, less water washing, low equipment corrosion, and high reusability. Huang Biao's team uses cation-exchange resin as an alternative to liquid acid for the hydrolysis of microcrystalline cellulose (MCC) into nanocrystalline cellulose (NCC) with the aid of ultrasonification treatment. This novel hydrolysis showed that 50.04% of NCC (2–24 nm) was achieved at a ratio of resin: MCC (w/w) of 10, temperature of 48°C, and reaction time of 189 min. The crystallinity of hydrolysed MCC was found to be increased from 75.2% to 84.26%, which indicated the removal of amorphous region and realignment of cellulose molecules [65].

CONCLUSIONS

Lignocellulosic biomass is the most abundant and biorenewable polymer on earth with great potential for sustainable nanocellulose production. The efficient and controlled breakdown of natural cellulose would produce nanocellulose, which is a mother compound for the synthesis of a large number of chemicals for food, energy, advance material, health, and environmental applications. Nanocellulose can be used as an immobilization support for chemical, microbial, and enzyme-catalysts. Synthesis of synthetic rubbers, bioplastic, pharmaceuticals materials, methyltetrahydrofurans, butanediol, and lactones from nanocellulose would be a quantum leap in nanocellulose industry. Nanocellulose and its derivatives could be further used to synthesize conducting polymers which could be used in biosensor appliances as well as molecular sieves. The complex hierarchy structure of lignocellulose is the main obstacle for major components separation. Overcoming the recalcitrance of lignocellulosic biomass is a key step in separating the biopolymer. Existence of lignin and the stability induced by inter- and intramolecular hydrogen bonding of cellulosic materials makes it a challenge for catalyst design. The current methods to convert cellulose into nanocellulose are based on acid, alkali, supercritical water, and thermal hydrolysis which often destroy the hierarchical structure of cellulose microfibrils and subsequently introduce impurities to the final products and produces unwanted by-products. The current methods also consume a lot of energy during and after the process and thus are deemed to be unprofitable and nonenvironmentally friendly. The typically used liquid-based catalysts are mineral acids and alkali. Introduction of novel catalysts such as iron-based catalyst, heteropoly acids, and ion exchange resin offer an integrated approach combining physical-chemical catalysts for the controlled structured degradation of natural cellulose into nanocellulose which promote more reliable methods in degradation of cellulose into nanocellulose. This shall further encourage more researches towards nanocellulosic field as more nanocellulose is readily available for utilization as a result of more efficient cellulose degradation process.

ACKNOWLEDGMENTS

The authors are grateful for cordial support from Ministry of Higher Education Malaysia for Fundamental Research Grant Scheme (FRGS, Project number: FP056-2013B) and High Impact Research (University of Malaya) (HIR F-000032).

REFERENCES

1.   A. A. Refaat, "5.13-biofuels from waste materials," in Comprehensive Renewable Energy, S. Ali, Ed., pp. 217–261, Elsevier, Oxford, UK, 2012.

2.   J. O. Metzger and A. Hüttermann, "Sustainable global energy supply based on lignocellulosic biomass from afforestation of degraded areas," Naturwissenschaften, vol. 96, no. 2, pp. 279–288, 2009.

3.   G. D. Saratale and S. E. Oh, "Lignocellulosics to ethanol: the future of the chemical and energy industry,"African Journal of Biotechnology, vol. 11, no. 5, pp. 1002–1013, 2012.

4.   S. H. Mood, A. H. Golfeshan, M. Tabatabaei et al., "Lignocellulosic biomass to bioethanol, a comprehensive review with a focus on

pretreatment," Renewable and Sustainable Energy Reviews, vol. 27, pp. 77–93, 2013.

5.   A. Limayem and S. C. Ricke, "Lignocellulosic biomass for bioethanol production: current perspectives, potential issues and future prospects," Progress in Energy and Combustion Science, vol. 38, no. 4, pp. 449–467, 2012.

6.   W. P. Q. Ng, H. L. Lam, F. Y. Ng, M. Kamal, and J. H. E. Lim, "Waste-to-wealth: green potential from palm biomass in Malaysia," Journal of Cleaner Production, vol. 34, pp. 57–65, 2012.

7.   J. A. Melero, J. Iglesias, and A. Garcia, "Biomass as renewable feedstock in standard refinery units. Feasibility, opportunities and challenges," Energy and Environmental Science, vol. 5, no. 6, pp. 7393–7420, 2012.

8.   A. Jiménez and O. Chávez, "Economic assessment of biomass feedstocks for the chemical industry," The Chemical Engineering Journal, vol. 37, pp. B1–B15, 1988.

9.   M. E. Himmel, S. Ding, D. K. Johnson et al., "Biomass recalcitrance: engineering plants and enzymes for biofuels production," Science, vol. 315, no. 5813, pp. 804–807, 2007.

10.   J.-P. Lange, "Lignocellulose conversion: an introduction to chemistry," in Catalysis for Renewables, Wiley-VCH, New York, NY, USA, 2007.

11.   J. J. Bozell, C. J. O'Lenick, and S. Warwick, "Biomass fractionation for the biorefinery: heteronuclear multiple quantum coherence-nuclear magnetic resonance investigation of lignin isolated from solvent fractionation of switchgrass," Journal of Agricultural and Food Chemistry, vol. 59, no. 17, pp. 9232–9242, 2011.

12.   S. G. Wettstein, D. M. Alonso, E. I. Gürbüz, and J. A. Dumesic, "A roadmap for conversion of lignocellulosic biomass to chemicals and fuels," Current Opinion in Chemical Engineering, vol. 1, no. 3, pp. 218–224, 2012.

13.   P. Gullón, E. Conde, A. Moure, H. Domínguez, and J. C. Parajó, "Selected process alternatives for biomass refining: a review," The Open Agricultural Journal, vol. 4, pp. 135–144, 2010.

14.   J.-P. Lange, I. Lewandowski, and P. M. Ayoub, "Cellulosic biofuels: a sustainable option for transportation," in Sustainable

Development in the Process Industries, pp. 171–198, John Wiley & Sons, New York, NY, USA, 2010.

15.   I. Siró and D. Plackett, "Microfibrillated cellulose and new nanocomposite materials: a review," Cellulose, vol. 17, no. 3, pp. 459–494, 2010.

16.   F. Ko, V. Kuznetsov, E. Flahaut, et al., "Formation of nanofibers and nanotubes production," inNanoengineered Nanofibrous Materials, S. Guceri, Y. Gogotsi, and V. Kuznetsov, Eds., pp. 1–129, Springer, Dordrecht, The Netherlands, 2004.

17.   W. Bai, J. Holbery, and K. Li, "A technique for production of nanocrystalline cellulose with a narrow size distribution," Cellulose, vol. 16, no. 3, pp. 455–465, 2009.

18.   E. Abraham, B. Deepa, L. A. Pothan et al., "Extraction of nanocellulose fibrils from lignocellulosic fibres: a novel approach," Carbohydrate Polymers, vol. 86, no. 4, pp. 1468–1475, 2011.

19.   R. E. Quiroz-Castañeda and J. L. Folch-Mallol, "Plant Cell wall degrading and remodeling proteins: current perspectives," Biotecnologia Aplicada, vol. 28, no. 4, pp. 205–215, 2011.

20.   P. Kumar, D. M. Barrett, M. J. Delwiche, and P. Stroeve, "Methods for pretreatment of lignocellulosic biomass for efficient hydrolysis and biofuel production," Industrial & Engineering Chemistry Research, vol. 48, no. 8, pp. 3713–3729, 2009.

21.   M. Pedersen and A. S. Meyer, "Lignocellulose pretreatment severity—relating pH to biomatrix opening,"New Biotechnology, vol. 27, no. 6, pp. 739–750, 2010.

22.   A. Cabiac, E. Guillon, F. Chambon, C. Pinel, F. Rataboul, and N. Essayem, "Cellulose reactivity and glycosidic bond cleavage in aqueous phase by catalytic and non catalytic transformations," Applied Catalysis A: General, vol. 402, no. 1-2, pp. 1–10, 2011.

23.   B. L. Peng, N. Dhar, H. L. Liu, and K. C. Tam, "Chemistry and applications of nanocrystalline cellulose and its derivatives: a nanotechnology perspective," Canadian Journal of Chemical Engineering, vol. 89, no. 5, pp. 1191–1206, 2011.

24.   H.-J. Huang and S. Ramaswamy, "Overview of biomass conversion processes and separation and purification technologies in biorefineries," in Separation and Purification Technologies in Biorefineries, pp. 1–36, John Wiley & Sons, 2013.

25. J. C. López-Linares, C. Cara, M. Moya, E. Ruiz, E. Castro, and I. Romero, "Fermentable sugar production from rapeseed straw by dilute phosphoric acid pretreatment," Industrial Crops and Products, vol. 50, pp. 525–531, 2013.

26. C. Li, B. Knierim, C. Manisseri et al., "Comparison of dilute acid and ionic liquid pretreatment of switchgrass: biomass recalcitrance, delignification and enzymatic saccharification," Bioresource Technology, vol. 101, no. 13, pp. 4900–4906, 2010.

27. S. T. Moe, K. K. Janga, T. Hertzberg, M.-B. Hägg, K. Øyaas, and N. Dyrset, "Saccharification of lignocellulosic biomass for biofuel and biorefinery applications-a renaissance for the concentrated acid hydrolysis?" Energy Procedia, vol. 20, pp. 50–58, 2012.

28. Z. Sun, Y. Tang, T. Iwanaga, T. Sho, and K. Kida, "Production of fuel ethanol from bamboo by concentrated sulfuric acid hydrolysis followed by continuous ethanol fermentation," Bioresource Technology, vol. 102, no. 23, pp. 10929–10935, 2011.

29. Z.-S. Liu, X.-L. Wu, K. Kida, and Y.-Q. Tang, "Corn stover saccharification with concentrated sulfuric acid: effects of saccharification conditions on sugar recovery and by-product generation," Bioresource Technology, vol. 119, pp. 224–233, 2012.

30. M. Taherdanak and H. Zilouei, "Improving biogas production from wheat plant using alkaline pretreatment," Fuel, vol. 115, pp. 714–719, 2014.

31. J. Xu, J. J. Cheng, R. R. Sharma-Shivappa, and J. C. Burns, "Lime pretreatment of switchgrass at mild temperatures for ethanol production," Bioresource Technology, vol. 101, no. 8, pp. 2900–2903, 2010.

32. I. Cybulska, G. Brudecki, K. Rosentrater, J. L. Julson, and H. Lei, "Comparative study of organosolv lignin extracted from prairie cordgrass, switchgrass and corn stover," Bioresource Technology, vol. 118, pp. 30–36, 2012.

33. Z. Zhang, I. M. O›Hara, and W. O. S. Doherty, "Pretreatment of sugarcane bagasse by acid-catalysed process in aqueous ionic liquid solutions," Bioresource Technology, vol. 120, pp. 149–156, 2012.

34.   E. S. Abdel-Halim and S. S. Al-Deyab, "Low temperature bleaching of cotton cellulose using peracetic acid,"Carbohydrate Polymers, vol. 86, no. 2, pp. 988–994, 2011.

35.   D. Y. Yin, Q. Jing, W. W. AlDajani et al., "Improved pretreatment of lignocellulosic biomass using enzymatically-generated peracetic acid," Bioresource Technology, vol. 102, no. 8, pp. 5183–5192, 2011.

36.   Y. Zheng, C. Lee, C. Yu et al., "Dilute acid pretreatment and fermentation of sugar beet pulp to ethanol,"Applied Energy, vol. 105, pp. 1–7, 2013.

37.   N. Sathitsuksanoh, Z. Zhu, and Y.-. P. Zhang, "Cellulose solvent-based pretreatment for corn stover and avicel: concentrated phosphoric acid versus ionic liquid [BMIM]Cl," Cellulose, vol. 19, no. 4, pp. 1161–1172, 2012.

38.   S. McIntosh and T. Vancov, "Enhanced enzyme saccharification of Sorghum bicolor straw using dilute alkali pretreatment," Bioresource Technology, vol. 101, no. 17, pp. 6718–6727, 2010.

39.   D. L. Sills and J. M. Gossett, "Assessment of commercial hemicellulases for saccharification of alkaline pretreated perennial biomass," Bioresource Technology, vol. 102, no. 2, pp. 1389–1398, 2011.

40.   H. Wang, C. Zhang, H. He, and L. Wang, "Glucose production from hydrolysis of cellulose over a novel silica catalyst under hydrothermal conditions," Journal of Environmental Sciences, vol. 24, no. 3, pp. 473–478, 2012.

41.   D. Fu and G. Mazza, "Aqueous ionic liquid pretreatment of straw," Bioresource Technology, vol. 102, no. 13, pp. 7008–7011, 2011.

42.   J. A. D. C. Correia, J. E. M. Júnior, L. R. B. Gonçalves, and M. V. P. Rocha, "Alkaline hydrogen peroxide pretreatment of cashew apple bagasse for ethanol production: Study of parameters," Bioresource Technology, vol. 139, pp. 249–256, 2013.

43.   R. Li, J. Fei, Y. Cai, Y. Li, J. Feng, and J. Yao, "Cellulose whiskers extracted from mulberry: a novel biomass production," Carbohydrate Polymers, vol. 76, no. 1, pp. 94–99, 2009.

44.   A. Mandal and D. Chakrabarty, "Isolation of nanocellulose from waste sugarcane bagasse (SCB) and its characterization," Carbohydrate Polymers, vol. 86, no. 3, pp. 1291–1299, 2011.

45. R. Xiong, X. Zhang, D. Tian, Z. Zhou, and C. Lu, "Comparing microcrystalline with spherical nanocrystalline cellulose from waste cotton fabrics," Cellulose, vol. 19, no. 4, pp. 1189–1198, 2012.

46. F. Jiang and Y. Hsieh, "Chemically and mechanically isolated nanocellulose and their self-assembled structures," Carbohydrate Polymers, vol. 95, no. 1, pp. 32–40, 2013.

47. J. P. S. Morais, M. D. F. Rosa, M. D. S. M. de Souza Filho, L. D. Nascimento, D. M. Do Nascimento, and A. R. Cassales, "Extraction and characterization of nanocellulose structures from raw cotton linter,"Carbohydrate Polymers, vol. 91, no. 1, pp. 229–235, 2013.

48. N. Quiévy, N. Jacquet, M. Sclavons, C. Deroanne, M. Paquot, and J. Devaux, "Influence of homogenization and drying on the thermal stability of microfibrillated cellulose," Polymer Degradation and Stability, vol. 95, no. 3, pp. 306–314, 2010.

49. W. Chen, H. Yu, Y. Liu, P. Chen, M. Zhang, and Y. Hai, "Individualization of cellulose nanofibers from wood using high-intensity ultrasonication combined with chemical pretreatments," Carbohydrate Polymers, vol. 83, no. 4, pp. 1804–1811, 2011.

50. J. Li, X. Wei, Q. Wang et al., "Homogeneous isolation of nanocellulose from sugarcane bagasse by high pressure homogenization," Carbohydrate Polymers, vol. 90, no. 4, pp. 1609–1613, 2012.

51. G. H. D. Tonoli, E. M. Teixeira, A. C. Corrêa et al., "Cellulose micro/nanofibres from Eucalyptus kraft pulp: preparation and properties," Carbohydrate Polymers, vol. 89, no. 1, pp. 80–88, 2012.

52. J. Zhao, W. Zhang, X. Zhang, X. Zhang, C. Lu, and Y. Deng, "Extraction of cellulose nanofibrils from dry softwood pulp using high shear homogenization," Carbohydrate Polymers, vol. 97, no. 2, pp. 695–702, 2013.

53. S. Alila, I. Besbes, M. R. Vilar, P. Mutjé, and S. Boufi, "Non-woody plants as raw materials for production of microfibrillated cellulose (MFC): a comparative study," Industrial Crops and Products, vol. 41, no. 1, pp. 250–259, 2013.

54. P. Satyamurthy and N. Vigneshwaran, "A novel process for synthesis of spherical nanocellulose by controlled hydrolysis of microcrystalline cellulose using anaerobic microbial consortium," Enzyme and Microbial Technology, vol. 52, no. 1, pp. 20–25, 2013.

55. A. de Campos, A. C. Correa, D. Cannella et al., "Obtaining nanofibers from curauá and sugarcane bagasse fibers using enzymatic hydrolysis followed by sonication," Cellulose, vol. 20, no. 3, pp. 1491–1500, 2013.

56. L. Liu, J. Sun, C. Cai, S. Wang, H. Pei, and J. Zhang, "Corn stover pretreatment by inorganic salts and its effects on hemicellulose and cellulose degradation," Bioresource Technology, vol. 100, no. 23, pp. 5865–5871, 2009.

57. C. Liu and C. E. Wyman, "The enhancement of xylose monomer and xylotriose degradation by inorganic salts in aqueous solutions at 180 °C," Carbohydrate Research, vol. 341, no. 15, pp. 2550–2556, 2006.

58. J. Zhao, H. Zhang, R. Zheng, Z. Lin, and H. Huang, "The enhancement of pretreatment and enzymatic hydrolysis of corn stover by $FeSO_4$ pretreatment," Biochemical Engineering Journal, vol. 56, no. 3, pp. 158–164, 2011.

59. L. Q. Lin, T. L. Rong, Y. H. Juan et al., "Environmentally-friendly and efficient preparation of cellulose nanocrystals by $FeCl_3$-catalyzed hydrolysis of cellulose," Science & Technology Review, vol. 32, pp. 56–60, 2014.

60. F. Zhang, X. Deng, Z. Fang, H. Zeng, X. Tian, and J. A. Kozinski, "Hydrolysis of microcrystalline cellulose over Zn-Ca-Fe oxide catalyst," Petrochemical Technology, vol. 40, no. 1, pp. 43–48, 2011.

61. C. Zhang, H. Wang, F. Liu, L. Wang, and H. He, "Magnetic core-shell Fe_3O_4@C-SO_3H nanoparticle catalyst for hydrolysis of cellulose," Cellulose, vol. 20, no. 1, pp. 127–134, 2013.

62. D. M. Lai, L. Deng, Q.-X. Guo, and Y. Fu, "Hydrolysis of biomass by magnetic solid acid," Energy & Environmental Science, vol. 4, no. 9, pp. 3552–3557, 2011.

63. J. Tian, J. Wang, S. Zhao, C. Jiang, X. Zhang, and X. Wang, "Hydrolysis of cellulose by the heteropoly acid $H_3PW_{12}O_{40}$," Cellulose, vol. 17, no. 3, pp. 587–594, 2010.

64. J. Tian, C. Fang, M. Cheng, and X. Wang, "Hydrolysis of cellulose over $Cs_xH_{3-x}PW_{12}O_{40}$ (X = 1–3) Heteropoly acid catalysts," Chemical Engineering and Technology, vol. 34, no. 3, pp. 482–486, 2011.

65. H. Biao, T. Li-rong, D. Da-song, O. Wen, L. Tao, and C. Xue-rong, Preparation of Nanocellulose with CationExchange Resin Catalysed Hydrolysis.

66. L. Pereira Ramos, "The chemistry involved in the steam treatment of lignocellulosic materials," Química Nova, vol. 26, no. 6, pp. 863–871, 2003.

67. W. Gindl, H. S. Gupta, T. Schöberl, H. C. Lichtenegger, and P. Fratzl, "Mechanical properties of spruce wood cell walls by nanoindentation," Applied Physics A, vol. 79, no. 8, pp. 2069–2073, 2004.

68. S. P. S. Chundawat, B. S. Donohoe, L. Da Costa Sousa et al., "Multi-scale visualization and characterization of lignocellulosic plant cell wall deconstruction during thermochemical pretreatment," Energy and Environmental Science, vol. 4, no. 3, pp. 973–984, 2011.

69. A. T. W. M. Hendriks and G. Zeeman, "Pretreatments to enhance the digestibility of lignocellulosic biomass," Bioresource Technology, vol. 100, no. 1, pp. 10–18, 2009.

70. V. B. Agbor, N. Cicek, R. Sparling, A. Berlin, and D. B. Levin, "Biomass pretreatment: fundamentals toward application," Biotechnology Advances, vol. 29, no. 6, pp. 675–685, 2011.

71. J. H. Grabber, "How do lignin composition, structure, and cross-linking affect degradability? a review of cell wall model studies," in Proceedings of the CSSA Annual Meeting Lignin and Forage Digestibility Symposium, vol. 45, pp. 820–831, Denver, Colo, USA, 2003.

72. J. Pérez, J. Muñoz-Dorado, T. De La Rubia, and J. Martínez, "Biodegradation and biological treatments of cellulose, hemicellulose and lignin: an overview," International Microbiology, vol. 5, no. 2, pp. 53–63, 2002

73. H. Yang, R. Yan, H. Chen, C. Zheng, D. H. Lee, and D. T. Liang, "In-depth investigation of biomass pyrolysis based on three major components: hemicellulose, cellulose and lignin," Energy & Fuels, vol. 20, no. 1, pp. 388–393, 2006.

74. Y. Sun and J. Cheng, "Hydrolysis of lignocellulosic materials for ethanol production: a review," Bioresource Technology, vol. 83, no. 1, pp. 1–11, 2002.

75. G. Brodeur, E. Yau, K. Badal, J. Collier, K. B. Ramachandran, and S. Ramakrishnan, "Chemical and physicochemical pretreatment of lignocellulosic biomass: a review," Enzyme Research, vol. 2011, Article ID 787532, 17 pages, 2011.

76. C. E. Wyman, B. E. Dale, R. T. Elander, M. Holtzapple, M. R. Ladisch, and Y. Y. Lee, "Coordinated development of leading biomass pretreatment technologies," Bioresource Technology, vol. 96, no. 18, pp. 1959–1966, 2005.

77. N. Mosier, C. Wyman, B. Dale et al., "Features of promising technologies for pretreatment of lignocellulosic biomass," Bioresource Technology, vol. 96, no. 6, pp. 673–686, 2005.

78. H. P. S. A. Khalil, A. H. Bhat, and A. F. I. Yusra, "Green composites from sustainable cellulose nanofibrils: a review," Carbohydrate Polymers, vol. 87, no. 2, pp. 963–979, 2012.

79. P. Harmsen, Literature Review of Physical and Chemical Pretreatment Processes for Lignocellulosic Biomass, Wageningen UR, Food & Biobased Research, 2010.

80. K. Karimi, M. Shafiei, and R. Kumar, "Progress in physical and chemical pretreatment of lignocellulosic biomass," in Biofuel Technologies, V. K. Gupta and M. G. Tuohy, Eds., pp. 53–96, Springer, Berlin, Germany, 2013.

81. O. Sánchez, R. Sierra, and C. J. Alméciga-Díaz, Delignification Process of Agro-Industrial Wastes an Alternative to Obtain Fermentable Carbohydrates for Producing Fuel, 2011.

82. J. Miron and D. Ben-Ghedalia, "Effect of hydrolysing and oxidizing agents on the composition and degradation of wheat straw monosaccharides," European Journal of Applied Microbiology and Biotechnology, vol. 15, no. 2, pp. 83–87, 1982.

83. X. Zhao, K. Cheng, and D. Liu, "Organosolv pretreatment of lignocellulosic biomass for enzymatic hydrolysis," Applied Microbiology and Biotechnology, vol. 82, no. 5, pp. 815–827, 2009.

84. T. J. McDonough, The Chemistry of Organosolv Delignification, 1992.

85. Y. Zheng, Z. Pan, and R. Zhang, "Overview of biomass pretreatment for cellulosic ethanol production,"International Journal of Agricultural and Biological Engineering, vol. 2, no. 3, pp. 51–68, 2009.

86. Y. Xiong, Z. Zhang, X. Wang, B. Liu, and J. Lin, "Hydrolysis of cellulose in ionic liquids catalyzed by a magnetically-recoverable solid acid catalyst," Chemical Engineering Journal, vol. 235, pp. 349–355, 2014.

87. S. Morales-delaRosa, J. M. Campos-Martin, and J. L. G. Fierro, "High glucose yields from the hydrolysis of cellulose dissolved in ionic liquids," Chemical Engineering Journal, vol. 181-182, pp. 538–541, 2012.

88. H. Tadesse and R. Luque, "Advances on biomass pretreatment using ionic liquids: an overview," Energy and Environmental Science, vol. 4, no. 10, pp. 3913–3929, 2011.

89. M. Zavrel, D. Bross, M. Funke, J. Büchs, and A. C. Spiess, "High-throughput screening for ionic liquids dissolving (ligno-) cellulose," Bioresource Technology, vol. 100, no. 9, pp. 2580–2587, 2009.

90. S. H. Lee, T. V. Doherty, R. J. Linhardt, and J. S. Dordick, "Ionic liquid-mediated selective extraction of lignin from wood leading to enhanced enzymatic cellulose hydrolysis," Biotechnology and Bioengineering, vol. 102, no. 5, pp. 1368–1376, 2009.

91. M. Mora-Pale, L. Meli, T. V. Doherty, R. J. Linhardt, and J. S. Dordick, "Room temperature ionic liquids as emerging solvents for the pretreatment of lignocellulosic biomass," Biotechnology and Bioengineering, vol. 108, no. 6, pp. 1229–1245, 2011.

92. H. Liu, K. L. Sale, B. M. Holmes, B. A. Simmons, and S. Singh, "Understanding the interactions of cellulose with ionic liquids: a molecular dynamics study," The Journal of Physical Chemistry B, vol. 114, no. 12, pp. 4293–4301, 2010.

93. N. Lavoine, I. Desloges, A. Dufresne, and J. Bras, "Microfibrillated cellulose—its barrier properties and applications in cellulosic materials: a review," Carbohydrate Polymers, vol. 90, no. 2, pp. 735–764, 2012.

94. T. Saito and A. Isogai, "TEMPO-mediated oxidation of native cellulose. The effect of oxidation conditions on chemical

and crystal structures of the water-insoluble fractions," Biomacromolecules, vol. 5, no. 5, pp. 1983–1989, 2004.

95. C. Zhou, X. Xia, C. Lin, D. Tong, and J. Beltramini, "Catalytic conversion of lignocellulosic biomass to fine chemicals and fuels," Chemical Society Reviews, vol. 40, no. 11, pp. 5588–5617, 2011.

96. Q. Zhang, P. Zhang, Z. J. Pei, and D. Wang, "Relationships between cellulosic biomass particle size and enzymatic hydrolysis sugar yield: analysis of inconsistent reports in the literature," Renewable Energy, vol. 60, pp. 127–136, 2013.

97. L. Brinchi, F. Cotana, E. Fortunati, and J. M. Kenny, "Production of nanocrystalline cellulose from lignocellulosic biomass: technology and applications," Carbohydrate Polymers, vol. 94, no. 1, pp. 154–169, 2013.

98. T. Saito, S. Kimura, Y. Nishiyama, and A. Isogai, "Cellulose nanofibers prepared by TEMPO-mediated oxidation of native cellulose," Biomacromolecules, vol. 8, no. 8, pp. 2485–2491, 2007.

99. J. Lee and T. W. Jeffries, "Efficiencies of acid catalysts in the hydrolysis of lignocellulosic biomass over a range of combined severity factors," Bioresource Technology, vol. 102, no. 10, pp. 5884–5890, 2011.

100. Y. Habibi, "Key advances in the chemical modification of nanocelluloses," Chemical Society Reviews, vol. 43, pp. 1519–1542, 2014.

101. B. Yang, Z. Dai, S. Ding, and C. E. Wyman, "Enzymatic hydrolysis of cellulosic biomass," Biofuels, vol. 2, no. 4, pp. 421–450, 2011.

102. K. Igarashi, T. Uchihashi, A. Koivula et al., "Traffic jams reduce hydrolytic efficiency of cellulase on cellulose surface," Science, vol. 333, no. 6047, pp. 1279–1282, 2011.

103. L. Zhu, J. P. O'Dwyer, V. S. Chang, C. B. Granda, and M. T. Holtzapple, "Structural features affecting biomass enzymatic digestibility," Bioresource Technology, vol. 99, no. 9, pp. 3817–3828, 2008.

104. L. T. Fan, Y.-H. Lee, and D. H. Beardmore, "Mechanism of the enzymatic hydrolysis of cellulose: effects of major structural features of cellulose on enzymatic hydrolysis," Biotechnology and Bioengineering, vol. 22, pp. 177–199, 1980.

105. Y. Lin and G. W. Huber, "The critical role of heterogeneous catalysis in lignocellulosic biomass conversion," Energy and Environmental Science, vol. 2, no. 1, pp. 68–80, 2009.

106. R. Rinaldi and F. Schüth, "Design of solid catalysts for the conversion of biomass," Energy & Environmental Science, vol. 2, no. 6, pp. 610–626, 2009.

107. F. Guo, Z. Fang, C. C. Xu, and R. L. Smith Jr., "Solid acid mediated hydrolysis of biomass for producing biofuels," Progress in Energy and Combustion Science, vol. 38, no. 5, pp. 672–690, 2012.

108. P. Lanzafame, D. M. Temi, S. Perathoner, A. N. Spadaro, and G. Centi, "Direct conversion of cellulose to glucose and valuable intermediates in mild reaction conditions over solid acid catalysts," Catalysis Today, vol. 179, no. 1, pp. 178–184, 2012.

109. S. Dutta, "Catalytic materials that improve selectivity of biomass conversions," RSC Advances, vol. 2, no. 33, pp. 12575–12593, 2012.

110. P. Yang, H. Kobayashi, and A. Fukuoka, "Recent developments in the catalytic conversion of cellulose into valuable chemicals," Chinese Journal of Catalysis, vol. 32, no. 5, pp. 716–722, 2011.

111. G. Feng and Z. Fang, "Solid- and nano-catalysts pretreatment and hydrolysis techniques," in Pretreatment Techniques for Biofuels and Biorefineries, Z. Fang, Ed., pp. 339–366, Springer, Berlin, Germany, 2013.

112. C. Perego and D. Bianchi, "Biomass upgrading through acid-base catalysis," Chemical Engineering Journal, vol. 161, no. 3, pp. 314–322, 2010.

113. Y. Yan, T. Li, Z. Ren, and G. Li, "A study on catalytic hydrolysis of peat," Bioresource Technology, vol. 57, no. 3, pp. 269–273, 1996.

114. G. Feng and Z. Fang, "Solid-and nano-catalysts pretreatment and hydrolysis techniques," in Pretreatment Techniques for Biofuels and Biorefineries, pp. 339–366, Springer, New York, NY, USA, 2013.

115. Y. Xiong, Z. Zhang, X. Wang, B. Liu, and J. Lin, "Hydrolysis of cellulose in ionic liquids catalyzed by a magnetically-recoverable solid acid catalyst," Chemical Engineering Journal, vol. 235, pp. 349–355, 2014.

116. Y. Kamiya, M. Sadakane, and W. Ueda, "7.08-heteropoly compounds," in Comprehensive Inorganic Chemistry II (Second Edition), R. Jan and P. Kenneth, Eds., pp. 185–204, Elsevier, Amsterdam, The Netherlands, 2013.

117. R. Palkovits, K. Tajvidi, A. M. Ruppert, and J. Procelewska, "Heteropoly acids as efficient acid catalysts in the one-step conversion of cellulose to sugar alcohols," Chemical Communications, vol. 47, no. 1, pp. 576–578, 2011.

118. M. Misono, "Unique acid catalysis of heteropoly compounds (heteropolyoxometalates) in the solid state,"Chemical Communications, no. 13, pp. 1141–1152, 2001.

Chapter 2

Advances on Waste Valorization: New Horizons for a more Sustainable Society

Rick Arneil D. Arancon[1,2], Carol Sze Ki Lin[2], King Ming Chan[3], Tsz Him Kwan[3], and Rafael Luque[4,5]

[1]Department of Chemistry, School of Science and Engineering, Ateneo de Manila University, Loyola Heights, Quezon City, Philippines

[2]School of Energy and Environment, City University of Hong Kong, Hong Kong

[3]Environmental Science Program, School of Life Sciences, Chinese University of Hong Kong, Hong Kong

[4]Departamento de Quimica Organica, Universidad de Cordoba, Campus Universitario de Rabanales, Edificio Marie Curie (C3), E-14014, C ordoba, Spain

[5]Department of Chemical and Biomolecular Engineering (CBME), Hong Kong University of Science and Technology, Clear Water Bay, Kowloon, Hong Kong.

ABSTRACT

Increasingly tighter regulations regarding organic waste, and the demand for renewable chemicals and fuels, are pushing the manufacturing industry toward higher sustainability to improve cost-effectiveness and meet customers' demand. Food waste valorization is one of the current research areas that has attracted a great deal of attention over the past few years as a potential alternative to the disposal of a wide range of residues in landfill sites. In particular, the development of environmentally sound and innovative strategies to process such waste is an area of increasing importance in our current society. Landfill, incineration and composting are common, mature technologies for waste disposal. However, they are not satisfactory to treating organic waste due to the generation of toxic methane gas and bad odor, high energy consumption and slow reaction kinetics. In fact, research efforts have also been oriented on novel technologies to decompose organic waste. However, no valuable product is generated from the decomposition process. Instead of disposing and decomposing food waste, recent research has focused on its utilization as energy source (e.g., for bioethanol and biodiesel production). Organic waste is also useful to generate useful organic chemicals via biorefinery or white biotechnology (e.g., succinic acid and/or bio-plastics). This article is aimed to summarize recent development of waste valorization strategies for the sustainable production of chemicals, materials, and fuels through the development of green production strategies. It will also provide key insights into recent legislation on management of waste worldwide as well as two relevant case studies (the transformation of corncob residues into functionalized biomass-derived carbonaceous solid acids and their utilization in the production of biodiesel-like biofuels from waste oils in Philippines, as well as the development of a bakery waste based biorefinery for succinic acid and bioplastic production in Hong Kong) to illustrate the enormous potential of biowaste valorization for a more sustainable society. Future research directions and possible sustainable approaches will also be discussed with their respective proofs of concept.

INTRODUCTION

Climate change, energy crisis, resource scarcity, and pollution are major issues humankind will be facing in future years. Sustainable development has become a priority for the world's policy makers since humanity's impact on the environment has been greatly accelerated in the past century with rapidly increasing population and the concomitant sharp decrease of ultimate natural resources. Finding alternatives and more sustainable ways to live, in general, is our duty to pass on to future generations, and one of these important messages relates to waste. Waste from different types (e.g., agricultural, food, industrial) is generated day by day in extensive quantities, generating a significant problem in its management and disposal. A widespread feeling of "environment in danger" has been present everywhere in our society in recent years, which, however, has not yet crystallized in a general concienciation of cutting waste production in our daily lives. Many methods could achieve sustainable development, methods that could not only improve waste management but could also lead to the production of industrially important chemicals, materials, and fuels, in essence, valuable end products from waste.

Waste valorization is the process of converting waste materials into more useful products including chemicals, materials, and fuels. Such concept has already existed for a long time, mostly related to waste management, but it has been brought back to our society with renewed interest due to the fast depletion of natural and primary resources, the increased waste generation and landfilling worldwide and the need for more sustainable and cost-efficient waste management protocols. Various valorization techniques are currently showing promise in meeting industrial demands. One among such promising waste valorization strategies is the application of flow chemical technology to process waste to valuable products. A recent review of Ruiz et al. [1] highlighted various advantages of continuous flow processes particularly for biomass and/or food waste valorization which included reaction control, ease of scale-up, efficient reaction cycles producing more yield, and no required catalyst separation. Although flow chemistry has been known to be used in industries for other processing methodologies, it still remains to be used in biomass/waste valorization – a limitation caused by the large energy needed to degrade highly stable biopolymers and recalcitrant compounds (e.g., lignin). The deconstruction of such biopolymers, most of the time, requires extreme conditions of pressure and temperature – conditions achieved by microwave heating, which is another green valorization technology. These requirements are not simple to satisfy and various techniques (e.g., microwave irradiation) need to be combined to satisfy the prerequisites for a successful transformation of waste. However, the main challenge for this combination is on the scale-up itself. As conceptualized by Glasnov et al. [2] microwave and flow chemistries maybe coupled by attaching back-pressure regulators to flow devices. This approach can revolutionize industrial valorization since it will synthesize products fast (due to microwave heating) on one continuous run (flow process). Although the approach presented is possible, the main challenge of temperature transfer from microwave to flow remains to be solved. A buildup of temperature gradient inside the instrument could lead to various instrument inefficiencies.

Another valorization strategy is related to the use of pyrolysis in the synthesis of fuels. This involves biomass heating at high temperatures in the absence of oxygen to produce decomposed products [3, 4]. Although pyrolysis is a rather old method for char generation, it has been recently utilized to produce usable smaller molecules from very stable biopolymers. This method has been particularly employed in the production of Bio-Oil (a liquid, of relatively low viscosity that is a complex mixture of short-chain aldehydes, ketones, and carboxylic acids). In a study by Heo et al.[5], several conditions for the fast pyrolysis of waste furniture sawdust were studied, and it was found that bio-oil yields do not necessarily increase with temperature. The optimized pyrolysis temperature was set at 450°C (57% bio-oil yield) using a fluidized bed reactor. The reason for the nonlinear dependence bio-oil yield/temperature is the possible decomposition of small molecules into simpler gases. This theory is supported by the increase in the amount of gaseous products found at increasing temperatures. A separate study by Cho et al. [6] employed fast pyrolysis under a fluidized bed reactor to recover BTEX compounds (benzene, toluene, ethylbenzene, and xylenes) from mixed plastics. The highest BTEX yield was obtained at 719°C. The pyrolysis of cotton stalks was also reported to produce second generation biofuels [7]. The study found that at much higher temperatures of pyrolysis the amounts of H_2 and CO collected increased, while CO_2 levels lowered. The decrease in

CO_2 production could be due to the degradation of the gas at much higher temperatures producing CO and O_2. More recently, synergy between these first proposed technologies (microwave and pyrolysis) has been also reported to constitute a step forward toward more environmentally friendly low temperature pyrolysis protocols for bio-oil and syngas production [8]. Microwave-assisted pyrolysis of a range of waste feedstock can provide a tuneable and highly versatile option to syngas with tuneable H_2/CO ratios or bio-oil-derived biofuels via subsequent upgrading of the pyrolysis oil [8].

Aside from energy applications, pyrolysis can also be used to produce advanced materials including carbon nanotubes and graphene-like materials, which have a wide range of applications. These studies along with many others in literature illustrate the potential of pyrolysis to convert waste materials into valuable chemicals.

A third green method of valorization would be on the use of biological microorganisms to degrade complex wastes and produce fuel. The method is used by taking advantage of cellulose (or any biopolymer) degrading enzymes by microorganisms as demonstrated by Wulff et al. [9]. In their work, cellulase Xf818 was isolated from the plant pathogen *Xylella fastidiosa* (known to cause citrus variegated chlorosis in plants). The gene responsible for the enzyme was also probed and then later on expressed on *Escherichia coli*. Such enzyme was found to be mostly active in the hydrolysis of carboxymethyl celluloses, oat spelt xylans, and wood xylans.

Bioconversion has been under intensive research for the past years, and one of the most significant advances in the field relates to the possibility of a synthetic control of microorganisms' metabolic pathways to produce favorable metabolic processes, which will in turn increase the yield of products. A notable example is the use of a bioengineered *E. coli* to produce higher alcohols including is butanol, 1-butanol, 2-methyl-1-butanol, 3-methyl-1-butanol and 2-phenylethanol from glucose [10]. The protocol was amenable for the conversion of 2-ketoacid intermediates (from amino acid biosynthesis) into alcohols by amplifying expression of 2-ketoacid decarboxylases and alcohol dehydrogenases.

To model the design for isobutanol, the gene *ilvIHCD* was over-expressed with the P_LlacO_1 promoter in a plasmid to amplify 2-ketoisovalerate biosynthesis. Other genes were tested such as alsS gene (from *Bacillus subtilis*) to further improve the alcohol yield while some genes responsible for by-product formation (*adh*E, *ldh*A, *frd*AB, *fnr*, *pta*) and pyruvate competition (*pfl*B) were silenced. Overall, the isobutanol yield reached ~300 mmol/L (22 g/L) under microaerobic conditions.

The three presented strategies (microwave, pyrolysis, and bioengineering) represent some of the most important valorization methodologies. With the rapid advancement of these fields in waste valorization, it is expected that most industrial sustainability practices will have a different focus in various future scenarios.

Waste valorization is currently geared toward three sustainable paths: one would be on the production of fuel and energy to replace common fossil fuel sources and in parallel on the production of high-value platform chemicals as well as useful materials. Fossil-based fuels are clearly diminishing in supply and this has caused a global environmental concern due to rapidly rising emissions of fossil fuel by-products (both for processing and actual use). Because of this, waste valorization for energy and fuels are not only geared toward a sustainable fuel source but also toward a more benign fuel fit for an industrial up-scale. According to the Netherlands Environmental Assessment Agency, global CO_2 emissions reached an all-time high in 2011 at around 34 billion tones of greenhouse gases (GHGs) [3]. Close to 90% of these emissions derive from fossil fuel combustion. Other toxic gases such as volatile organic compounds, nitrogen oxides (precursors of toxic ozone) and particulates come together with GHGs. In a more than likely scenario of a minimum of 2.5% energy demand growth per year, it is necessary to substitute fossil fuels progressively with cleaner fuel sources. Biomass combustion for electricity and heat production was reported to be less costly, providing at the same time a larger CO_2-reduction potential [11]. Many studies also have shown convincing proof that the use of biomass for energy applications could be a highly interesting solution and cleaner technology for the future [12-14].

Another direction of waste valorization aims to produce high-value chemicals from residues including succinic acid (SA) [15], furfural and furans [16], phenolic compounds [17], and bioplastics [18]. These can be produced via chemical, chemo-enzymatic, and biotechnological approaches (e.g., solid state fermentation) but depending on the type of residue some compounds (e.g., essential oils, chemicals, etc.) can even be produced upon extraction and isolation [19]. The production of biomass-derived chemicals is a sustainable approach since it maximizes the use of resources and, at the same time, minimizes waste generation.

The major strength of biotechnology is its multidisciplinary nature and the broad range of scientific approaches that it encompasses. Among the broad range of technologies with the potential to reach the goal of sustainability, biotechnology could take an important place, especially in the fields of food production, renewable raw materials and energy, pollution prevention, and bioremediation. At present, the major application of biotechnology used in the environmental protection is to utilize microorganisms to control environmental contamination. Developing biotechnology could be a solution for these problems – this will also be given emphasis in this review.

Although waste valorization is an attractive approach for sustainability, on a large scale perspective, the purification, processing, and even the degradation of stable natural polymers (e.g., lignin) into simple usable chemicals still remain a significant challenge (Fig. 1).

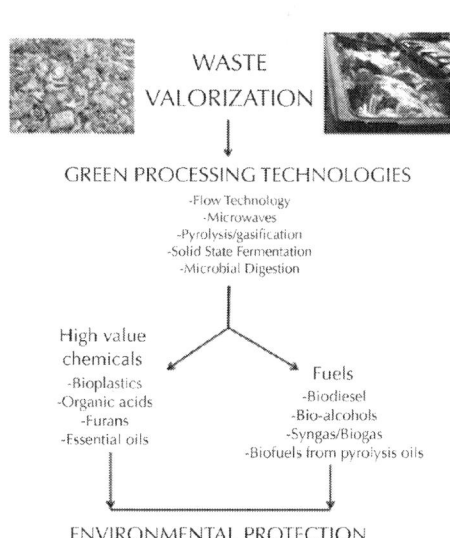

Figure 1: Valorization is essentially a concept of recycling waste into more usable industrial chemicals. Using established Green Processing technolo-

gies, various types of waste can be converted into high-value chemicals and fuels with the purpose of minimizing waste disposal volumes and eventually protecting the environment.

In recent years, there have been increasing concerns in the disposal of food waste. The amount of food waste generated globally accounts for a staggering 1.3 billion tonnes per year. Apart from causing the loss of a potentially valuable food source or the regenerated resource, there are problems associated with the disposal of food waste into landfills. With this imminent waste management issue, food waste should be diverted from landfills to other processing facilities in the foreseeable future. In Hong Kong, there are 3600 tonnes of food waste generated (Table 1), 40% of which is made up of municipal solid waste (MSW). Fifty two percent (52%) of the MSW generated is dumped into landfills [20]. It is estimated that by 2018, all current landfill sites in Hong Kong will be exhausted.

Table 1: Composition of waste in Hong Kong [20]

Waste	Tonnes/day
Municipal solid waste	9000
Domestic waste (including food waste)	6000 (2550)
Commercial and industrial waste (including food waste)	3000 (1050)
Construction waste	3350
Sewage sludge	950
Other waste	200
Total	13,500

Although the problem of food waste is commonly found over the world, the systems of food waste processing can only be formulated at the local community level with the consideration of the area-specific characteristics. These include regional characteristics and composition of waste, land availability, people's attitude and so forth. However, due to the lack of the local study concerning the suitability of food waste processing technologies for Hong Kong, this review is important to provide a few suggestions for the authorities to contemplate the adoption of a strategy on food waste disposal.

This contribution has been conceived to provide an overview on recent development of waste valorization strategies (with a particular emphasis on food waste) for the sustainable production of chemicals, materials, and fuels, highlighting key examples from recent research conducted by our groups. Reports on the development of green production strategies from waste and key insights into the recent legislation on management of wastes worldwide will also be discussed. The incorporation of these processes in future biorefineries for the production of value-added products and fuels will be an important contribution toward the world's highest priority target of sustainable development.

WASTE: PROBLEMS AND OPPORTUNITIES

In recent years, problems associated with the disposal of food waste to landfills lead to increased interest in searching for innovative alternatives due to the high proportion of organic matter in food waste,. First generation food waste processing technologies include waste to energy (e.g., anaerobic digestion), composting, and animal feed. Based on the characteristics of food waste, an integrated approach should be adopted with the focus on food waste reduction and separation, recycling commercial and industrial food waste, volume reduction of domestic food waste and energy recovery from food waste.

Sources, Characterization, and Composition of Waste

The large amounts of waste generated globally present an attractive sustainable source for industrially important chemicals. Food waste including garbage, swill, and kitchen refuse [21], can be generally described as any by-product or waste product from the production, processing, distribution, and consumption of food [22].

The definition of food waste is, however, different in different countries or cities. In the European Union, food waste is defined as "any food substance, raw or cooked, which is discarded, or intended or required to be discarded." The United States Environmental Protection

Agency (EPA), on the other hand, defines food waste as "Uneaten food and food preparation waste from residences and commercial establishments such as grocery stores, restaurants, and produce stands, institutional cafeterias and kitchens, and industrial sources including employee lunchrooms." In the United Nations, "Food waste" and "Food loss" are distinguished. Food losses refer to the decrease in food quantity or quality, which makes it unsuitable for human consumption [23] while food waste refer to food losses at the end of the food chain due to retailers' and consumers' behavior [24]. All in all, food waste includes not just wasted foodstuffs, but also uncooked raw materials or edible materials from groceries and wet market.

Food waste is generally characterized by a high diversity and variability, a high proportion of organic matter, and high moisture content. Table 1summarizes some reported characteristics of food waste, indicating moisture content of 74–90%, volatile solids to total solids ratio (VS/TS) of 80–97%, and carbon to nitrogen ratio (C/N) of 14.7–36.4 [25]. Due to these properties, food waste disposal constitutes a significant problem due to the growth of pathogens and rapid autoxidation [26]. As there are already many different microorganisms in food waste, the high rate of microbial activity and the amount of nutrients in food wastes facilitate the growth of pathogens, which cause the concern for foul odor, sanitation problems, and could even lead to infectious diseases. The high moisture contents [23] also increase the cost of food waste transportation. Food waste with high lipid content is also susceptible to rapid oxidation. The release of foul-smelling fatty acids also adds difficulties to the storage of treatment of food waste (Table 2).

Table 2: Characteristics of reported domestic food waste [18]

Source	Characteristics			Country
	Moisture content (%)	Volatile solid/total solid (%)	Carbon/nitrogen	
A dining hall	80	95	14.7	Korea
University's cafeteria	80	94	NA[a]	Korea
A dining hall	93	94	18.3	Korea
A dining hall	84	96	NA	Korea

Mixed municipal sources	90	80	NA	Germany
Mixed municipal sources	74	90–97	NA	Australia
Emanating from fruit and vegetable, markets, house-hold and juices centers	85	89	36.4	India

[a] NA, not available

According to a study commissioned by the United Nations Food and Agriculture Organization (UNFAO) in 2011, 1.3 billion tonnes of food waste is generated per year and roughly one third of food produced for human consumption is lost or wasted globally. The report also noted that food waste of industrialized countries and developing countries have different characteristics. Firstly, increasingly important quantities of food waste are generated in industrialized countries as compared to volumes observed in developing countries on a per capita basis. Figure 2 shows the per capita food loss in Europe and North America is 280–300 kg/year. In contrast, the food loss per capita in sub-Saharan Africa and South/Southeast Asia accounts for 120–170 kg/year. Also, food waste is mainly generated at retail and consumer levels in industrialized countries. Comparatively, food waste is generated in developing countries mainly at postharvest and processing levels, supported by the per capita that is, food waste generated by consumer levels in Asia is only 6–11 kg/year [27].

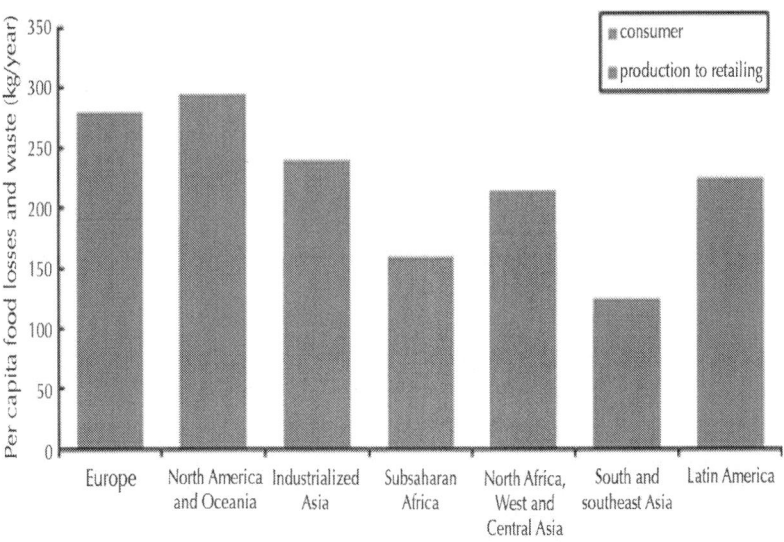

Figure 2: Per capita food losses and waste, at consumption and pre-consumption stages, in different regions [19].

Interestingly, the amount of food waste generated for example in Hong Kong is staggering. Figure 3 shows an increasing trend of the food waste generated daily from 3155 tonnes in 2002 to 3484 tonnes in 2011. Although the disposal of food waste in landfills was found to be the most economical option [28], it causes numerous problems in landfill sites. As landfilling disposal generally buries and compacts waste under the ground, the decomposition of food waste produces methane, a GHG that is twenty-one times powerful than carbon dioxide (CO_2) under anaerobic environment conditions. Such production can in fact remarkably affect the environment in the area as some reports indicated that around 30% of GHG produced in Hong Kong are generated in landfill sites [29]. Methane is also flammable and may lead to fires and explosions upon accumulation at certain concentrations. In addition, the decomposition of food waste develop unpleasant odor as well as leachates and organic salts that could damaging landfill liners, leaching out heavy metals and resulting in contamination of ground waters [30].

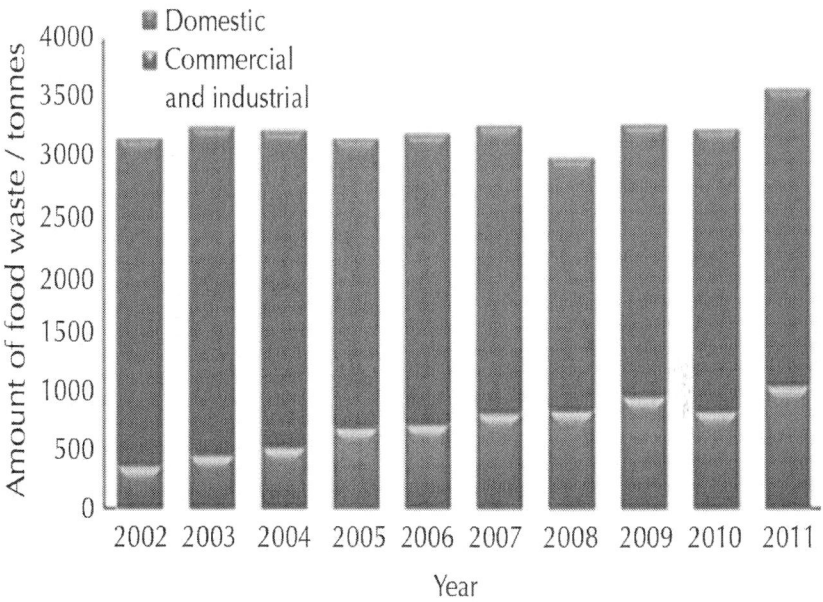

Figure 3: The amount of food waste generated daily in Hong Kong from 2002 to 2011 [25].

Valorization research has evolved through the years, with many techniques and developments achieved in recent decades. Waste feedstock including bread, wheat, orange peel residues, lignocellulosic sources, etc. are currently explored as sources of chemicals and fuels. On a recent review by Pfaltzgraff et al. [31], it was noted that the valorization of food wastes into fine chemicals is a more profitable and less energy consuming as compared to its possibilities for fuels production. Because of this, related waste processing technologies, particularly related to the production of fuels, have also been proposed to address energy efficiency and profitability from a range of different feedstocks. Toledano et al.[32, 33] reported a lignin deconstruction approach using a novel Ni-based heterogeneous catalyst under microwave irradiation. Different hydrogen donating solvents were explored for lignin depolymerization, finding formic acid as most effective hydrogen donating reagent due to the efficient generation of hydrogen for hydrogenolysis reactions (from its decomposition into CO, CO_2, and H_2) and its inherent acidic character that induces acidolytic cleavage of C-C bonds in lignin at the same time. The heterogeneous acidic support also acted as a Lewis acid, coordinating to lignin thereby promoting acidic protonation, and eventually dealkylation and deacylation reactions (Fig. 4). Figure 5 shows the structural complexity of the lignin biopolymer. Lignin deconstruction to simple aromatics including syringaldehyde, mesitol, and related compounds could serve the basis for a new generation of renewable gasolines [34].

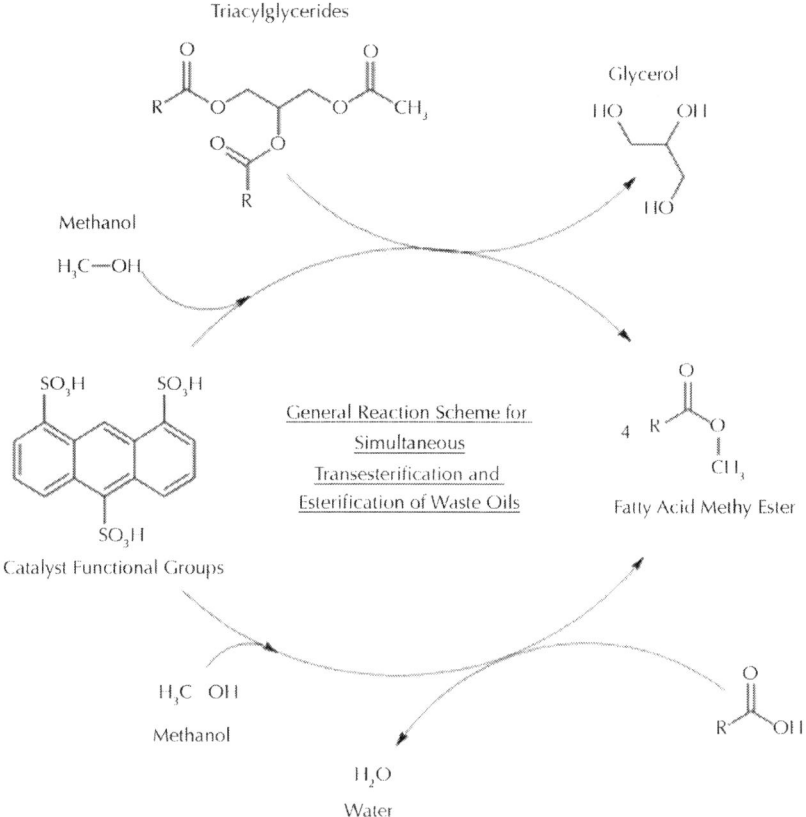

Figure 4: Simultaneous transesterification and esterification of waste oils using solid acid catalysts produced fatty acid methyl esters (a nonpolar component) along with water and glycerol (polar compounds) that separate out spontaneously from the reaction mixture forming two phases.

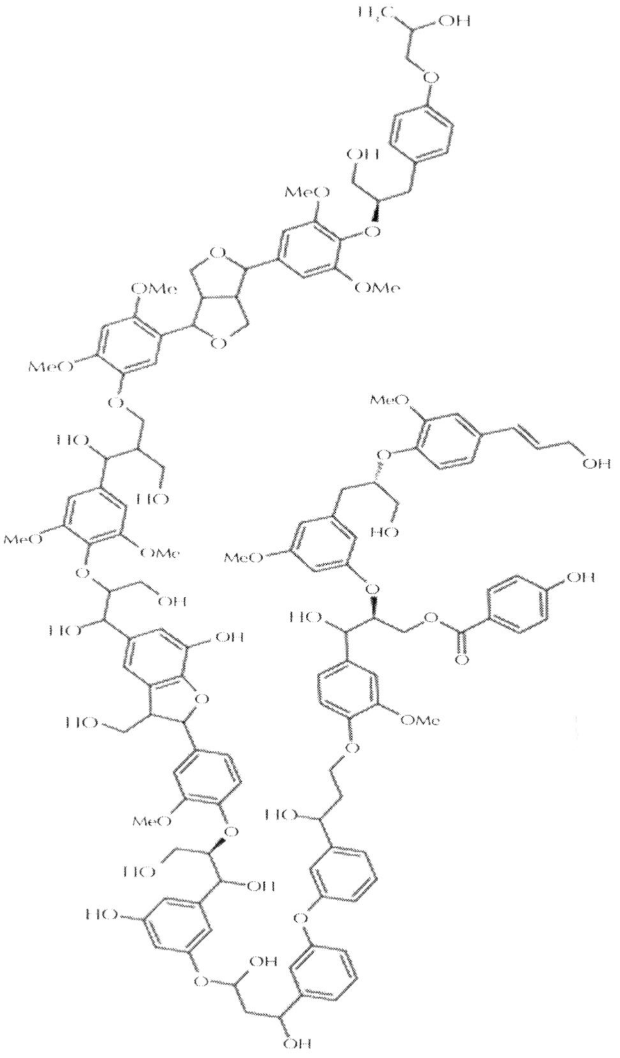

Figure 5: The structural complexicity of lignin being composed of aromatic compounds show its potential in different applications such as for fuel, and in the production of high-value chemicals (Image adapted from Stewart et al., [40]).

Simple phenolic compounds with potential antioxidant properties can also be derived from cauliflower by-products [35]. The proposed valorization strategy comprised a combined solvent-extraction step using an organic solvent together with a polystyrene resin (Amberlite XAD – 2) to recover most phenolics prior to high performance liquid chromatography (HPLC) analysis. Kaempferol-3-O-sophoroside-7-O-glucoside and its sinapoyl derivative kaempferol-3-O-(sinapoylsophoroside)-7-O-glucoside were obtained as main extracted components. A separate study by Sáiz et al.[36] also proved near-infrared (NIR) spectroscopy was a highly useful technique to characterization online of alcohol fermentation from onions. Along with multivariate calibration, this technique can lead to the analysis of samples with complex matrices without a prior sample preparation. One approach to a greener characterization method would be the coupling of a chromatographic technique to a flow instrument. This coupling has been shown to work in studies on metal analysis [37], online derivatization and separation of aspartic acid enantiomers [38], as well as for an enzyme inhibition assay [39], but it has not yet been shown to be successful for waste valorization.

Development of Greener Valorization Strategies

There are numerous options for waste processing and/or recycling in the world. Composting, regenerated animal feed and bedding, incineration, anaerobic digestion, and related first generation strategies have been proposed and investigated for a long time. Some of these techniques have been successful in making their way to commercialization. Considering the storage problem and the large amount of food waste generated every day, food waste processing facilities have to be in a mega-scale size with enough treatment capacity to handle numerous tonnes of food wastes daily. It definitely requires a large initial investment for setting up the industrial scale facilities. Also, in case of off-site processing, the large volume and great weight of food waste adds difficulty since the collection of food waste significantly increases the transportation cost and time. Besides, the variation in composition of food wastes, affects the quality of regenerated products, such as compost and animal feed. Therefore, it decreases the product's competiveness in the market.

As demonstrated in the above-mentioned examples, valorization may be carried out under different conditions depending on the target components needed. Before reaching an industrial upscale, enhancement of valorization product yield may be done by careful variation of the valorization strategies, in particular advanced protocols able to diversify on feedstock and end products obtained from them. Currently, an active area of research relates to catalytic valorization strategies using solid acid catalysts [41, 42]. One example of a green protocol on valorization of waste oils to biodiesel was provided by Fu et al. [43], in which a super acid was prepared by adding a sulfuric acid solution to zirconium hydroxide powder. Under optimum reaction conditions, 9:1 MeOH/oil molar ratio, 3% (w/w) catalyst, and 4 h reaction time at 120°C, biodiesel yield reached 93.2%. Apart from metal supports functionalized with acids, carbon-based catalysts for waste valorization are also attractively developed protocols. Aside from being an easily separable reaction component, functionalized carbonaceous materials can also be recyclable. In a study by Clark et al. [44], carbonaceous materials from porous starches (Starbons®, Department of Chemistry, University of York, York, UK) functionalized with sulfonated groups were found to have a catalytic activity 2–10 times greater to those of common microporous carbonaceous catalysts in a range of chemistries including biodiesel production from waste oils and SA transformations in a fermentation broth. A separate study by Luque et al. [45] employed carbonaceous residues of biomass gasification as catalysts for biodiesel synthesis. The results showed good ester conversion yields from fatty acids to methyl esters. The above mentioned examples demonstrate that designer catalysts can be attractive options in the valorization of a range of waste feed stocks.

A promising sustainable approach would also be the use of ionic liquid-type compounds which can be derived from renewable feedstock such as the so-called deep-eutectic solvents [46-49] and even selected designer ionic liquids. These compounds are salts in their liquid states with very unique properties such as very low vapor pressure, thermal stability, and tunability based on different applications. A study by Ruiz et al. [50] presented a $-SO_3H$ functionalized Bronsted acid ionic liquids catalyzed synthesis of an important chemical precursor such as furfural from C_5 sugars under microwave heating. Furfural yield varied from 40% to 85% depending on the type of Ionic liquid used and the feedstock employed in the process. It was shown that the ionic liquid 1-(4-Sulfonylbutyl)pyridinium tetrafluoroborate produced a yield of 95% for xylose conversion and 85% for furfural. Importantly, the protocol was amenable to the utilization of a biorefinery-derived syrup enriched in C5 oligomers, from which a 40–45% of furfural yield could be derived. A separate study by Zhang et al. [51] showed that the direct conversion of monosaccharides, and polysaccharides to 5-hydroxymethylfurfural (5-HMF) may be accomplished using ionic liquids in the presence of Germanium (IV) chloride. Yields of the reaction could go as high as 92% depending on the reaction conditions used. The mechanism proposed by the researchers indicate the role of the $GeCl_4$ as a Lewis acid catalyst for the ring opening of the sugars, which is immediately followed by several dehydration steps to produce 5-HMF. As alternative to these catalytic strategies, photo catalytic approaches to waste valorization could also serve the basis of innovative and highly attractive future valorization protocols. A recent review by Colmenares et al. [52] addressed the potential and opportunities of photocatalysis to convert lignin biomass into fine chemicals using designer TiO_2 nanocatalysts. These nanomaterials featuring doping agents (to lower the band gap of titania) have been shown to be effective in water splitting experiments (to form H_2 and O_2) to harness the potential of hydrogen as fuel. One of the earliest promising works of light-mediated degradation was shown by Stillings et al. [53] when they were able to degrade cellulose using Ultraviolet radiation. However, this has not been shown to be possible using visible light due to energy considerations. A photo catalytic approach to degradation may also be accomplished using functionalized graphenes (monolayers of sp^2 carbon atoms in a honeycomb lattice known to have ballistic electron transport properties). Functionalized graphenes and composites with other semiconductors have been shown to exhibit degradation properties [54, 55], but this concept has not yet been applied to waste valorization strategies.

RECENT LEGISLATION ON WASTE MANAGEMENT

Philippines

In Metro Manila at Philippines, almost 3.5 kg of solid waste is generated per capita every day. This amount includes food/kitchen waste, papers, polyethylene terephthalate bottles, metals, and cans. Although most Metro Manila residents do not practice the open burning of waste, a necessary waste segregation is performed for ease of collection. Being the country's capital, and one of the world's most densely populated cities, Metro Manila generates over 2400 tons of waste every day, which equates to a government spending of Php 3.4 billion (63 Million Euros) in collection and disposal. Not much legislation is available in the Philippines in terms of waste management. Although Republic Act 9003 (Solid Waste Management Act) has been passed last 2000, a recent 2008 study showed that it has not been properly implemented [56].

Hong Kong

One third of the food waste generated in Hong Kong come from the Commercial and Industry (C&I) sector, with the remaining percentage coming from households. In recent years, the amount of disposal of food waste from C&I sectors remarkably increased by 280% from 373 tonnes in 2002 to 1050 tonnes in 2011. It is anticipated that the food waste generated in Hong Kong will continue to rise, driven by the significant increase of the C&I food waste generation. The disposal of food waste (an organic waste which decomposes easily) to landfills is not sustainable, as it leads to rapid depletion of the limited landfill space. From the 2013 Policy Address by the Office of the Chief Executive in Hong Kong [57], there was a special emphasis on "Reduction of Food Waste" as stated in Section 142 below:

"Food waste imposes a heavy burden on our landfills as it accounts for about 40% of total waste disposed of in landfills. In addition, od our from food waste creates nuisance to nearby residents. The Government has recently launched the "Food Wise Hong Kong Campaign" to mobilise the public as well as the industrial and commercial sectors to reduce food waste. We will build modern facilities in phases for recovery of organic waste so that it can be converted into energy, compost and other products." [57].

The Environmental Protection Department (EPD) has planned to develop Organic Waste Treatment Facilities (OWTF). Such facilities will adopt biological technologies – composting and anaerobic digestion to stabilize the organic waste and turn it into compost and biogas for recovery. The first phase of the OWTF will be constructed at Siu Ho Wan with a daily treatment capacity of 200 tonnes of source separated organic waste (Fig. 6). The second phase will be located at Sha Ling of North District with a daily treatment capacity of 300 tonnes of organic waste.

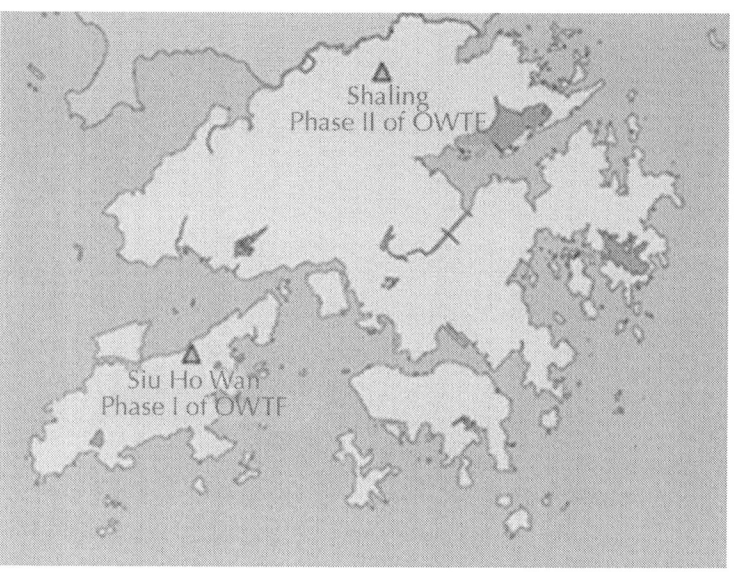

Figure 6: Map of Hong Kong indicates the location of the two organic waste treatment facility (OWTF) in Siu Ho Wan (Phase I) and Shaling (Phase II) [16].

Waste reduction at source should be the top priority so as to reduce the amount of food waste generated. Successful examples for the implementation of MSW charging scheme in Asian cities such as Taipei, Taiwan, and Seoul, South Korea could effectively reduce the total amount of MSW by 50% in 10 years [20]. These governments introduced quantity/volume-based charging scheme to create financial incentives to change public's food waste-generating behavior to achieve waste reduction at source. In addition, they introduced prepaid designated food waste bag charging system so as to achieve source separation. Food waste together with plastic bags can undergo treatment without extra separation step in the treatment facilities.

WASTE VALORIZATION STRATEGIES: CASE STUDIES

Biological treatment technologies including anaerobic digestion and composting have been reported extensively in past years. Under anaerobic digestion, biogas is generated as main product. Takata et al. [58] reported the production of 223 m^3 biogas from 1 tonne of food waste. However, Bernstad et al. [59] reported that the yield of biogas production may vary depending on the composition of waste and the existence of detergent. Numerous studies show that the lack of enough nutrients limits the ability of enzymes to digest waste [60, 21]. This can divert waste from landfill, and thus prevent the emission of GHG to the environment. Also, the solid residues can be used as compost, which can reduce the amount of used chemical fertilizers. Economically, anaerobic digestion can generate electricity on-site and may reduce energy cost. Also, it can be adopted in sewage treatment facilities, thereby eliminating transportation costs. Another way to valorize waste is by incineration for energy recovery. However, burning food waste is an energy intensive process and may remove important functional groups from the treated feedstocks. The following sections report case studies of different feedstock in different countries to illustrate the potential of waste valorization for the production of materials, chemicals, and products.

Utilization of Bakery Waste in the Biotechnological Production of Value-Added Products

Based on the large quantities of food waste generated at Hong Kong on a daily basis, Lin et al. have been recently focused on the valorization of unconsumed bakery products to valuable products via bio-processing in collaboration with retailer "Starbucks Hong Kong". Research was initially set on the production of bio-plastics poly(3-hydroxybutyrate) (PHB) and platform chemicals (e.g., SA) via enzymatic hydrolysis of non pretreated bakery waste, followed by fungal solid state fermentation to break down carbohydrates into simple sugars for subsequent SA or PHB fermentation. In the proposed biotechnological process, bakery waste serve as the nutrient source, including starch, fructose, free amino nitrogen (FAN), and trace amount of subsidiary nutrients. The nutrient content is listed in Table 3 below.

Table 3: Bakery waste composition (per 100 g) [61, 62]

Content	Pastry	Cake	Wheat bran
Moisture	34.5 g	45.0 g	N/A
Starch (dry basis)	44.6 g	12.6 g	N/A
Carbohydrate	33.5 g	62.0 g	15.0 g
Lipids	35.2 g	19.0 g	6 g
Sucrose	4.5 g	22.7 g	N/A
Fructose	2.3 g	11.9 g	N/A
Free sugar			1.5 g
Fiber	N/A	N/A	50 g
Protein (TN × 5.7) (dry basis)	7.1 g	17.0 g	14.0 g
Total phosphorus (dry basis)	1.7 g	1.5 g	N/A
Ash (dry basis)	2.5 g	1.6 g	N/A

N/A, data not available.

In general, pastries have larger starch and lipid content to those of cakes; whereas cakes have higher sugar (fructose and sucrose) and protein content. Nevertheless, both types of bakery waste were proved to serve as excellent nutritional substrates for fermentative production of SA or bioplastics after hydrolysis. Our groups previously demonstrated that SA could be produced from wheat-based renewable feedstock [63-65] and bread waste [66] via fermentation. Similarly, production of biopolymers from various types of food industrial waste and agricultural crops was shown to be techno-economically feasible for replacing petroleum-derived plastics [67].

The key components in the project are illustrated in Figure 7. In the upstream processing, the bakery waste was collected from a Starbucks outlet in the Shatin New Town Plaza. A mixture of fungi comprising *Asperillus awamori* and *Asperillus oryzae* were utilized for the production of amylolytic and proteolytic enzymes, respectively. Macromolecules including starch and proteins contained in bakery waste were hydrolysed, expected to enrich the final solution in glucose and FAN. This hydrolysate was subsequently used as feedstock in a bioreaction by two different types of microorganisms (*Actinobacillus succinogenes* and Halomonas boliviensis) to produce (SA) and PHB, respectively.

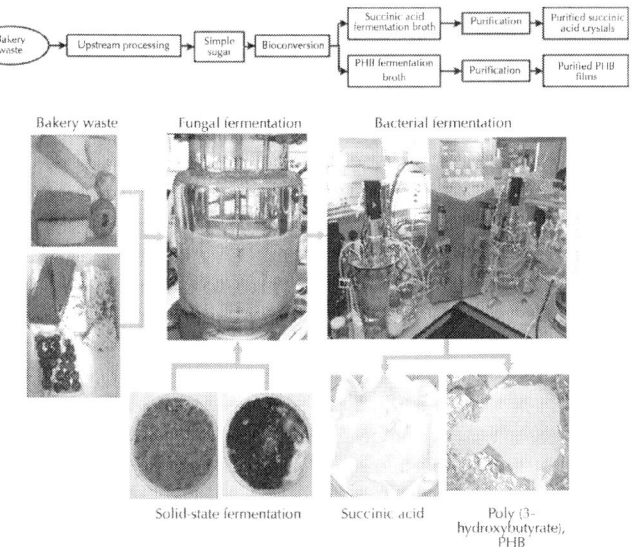

Figure 7: Flow chart of a bakery-based food waste biorefinery development, from bakery waste as raw material to succinic acid and poly(3 hydroxybutyrate), PHB as final products.

Although food waste is a no-cost nutritional source, the application of commercial enzymes in upstream processing might not be cost-efficient. To reduce process costs, the degradation of bread and bakery waste has been previously studied [61, 66]. In these studies, *A*. awamori and A. oryzae were the fungal secretors of glucoamylase protease and phosphatase as well as a range of other hydrolytic enzymes that does not require any external addition of commercial enzymes.

According to Figure 8, glucose (54.2 g/L) and FAN concentrations (758.5 mg/L) were achieved at 30% (w/v) pastry waste after enzymatic hydrolysis. On the other hand, sucrose present in cake was hydrolyzed to form 1 mole of glucose and 1 mole of fructose. The glucose (35.6 g/L), fructose (23.1 g/L), and FAN concentrations (685.5 mg/L) were achieved at 30% (w/v) cake waste. Among all, waste bread hydrolysate contained the highest glucose and FAN concentrations, which were 104.8 g/L and 492.6 mg/L, respectively. These results clearly demonstrate the potential of utilizing bakery hydrolysate as generic feedstock for fermentations.

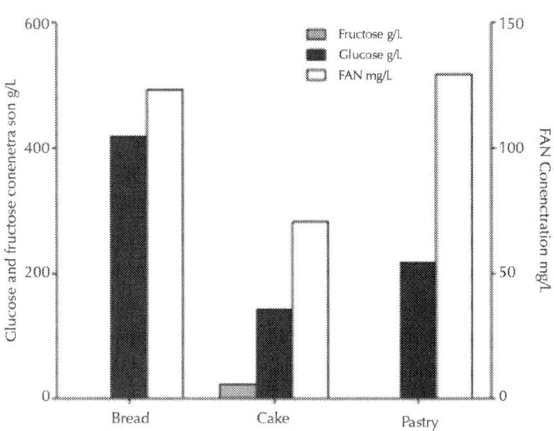

Figure 8: Sugars and FAN concentrations achieved from enzymatic hydrolysis using different bakery waste (30%, w/v) with *Aspergillus awamori* and *Aspergillus oryzae*.

Batch fermentations on enzymatic hydrolysates were subsequently carried out to investigate the cell growth, glucose consumption as well as SA production. Cake hydrolysate consisting an initial sugar content of 23.1 g/L glucose and 18.5 g/L fructose, and pastry hydrolysate with an initial sugar content of 44.0 g/L glucose were both utilized as fermentation feedstock. At the end of fermentation, the remaining glucose was 5.2 g/L whereas fructose was 3.7 g/L. A final SA concentration of 24.8 g/L was obtained at the end point, which corresponded to a yield of 0.8 g SA/g total sugar and a productivity of 0.79 g/L.h (Fig. 9). The overall conversion of waste cake into SA was 0.28 g/g cake.

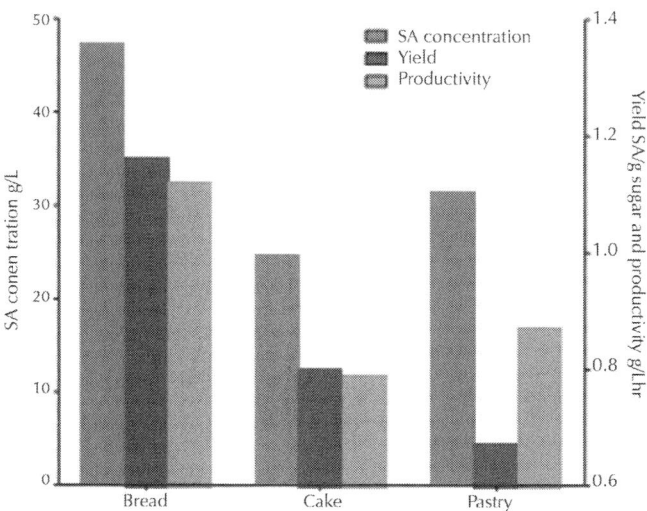

Figure 9: Succinic acid concentration, yield, and productivity in *Actinobacillus succinogenes* fermentations using different bakery hydrolysates.

Compared with cake hydrolysates, pastry hydrolysates possessed larger concentrations of initial glucose (44.0 g/L). SA concentration continuously increased until sugar was depleted after 44 h. At the end of fermentation, the SA concentration reached 31.7 g/L, which corresponded to a yield of 0.67 g SA/g glucose and a productivity of 0.87 g/L.h.

SA production achieved from various food waste residues has been compared in Table 4. It is clear that SA yields obtained when using cake and pastry wastes as feedstock were comparable or higher to those of other food waste-derived media.

Table 4: Comparison of succinic acid yields achieved using different food waste substrates with Actinobacillus succinogenes

Substrate	SA yield (g SA/g TS)	Overall SA yield (g SA/g substrate)	References
Wheat	0.40	0.40	[63]
Wheat flour milling by-product	1.02	0.087	[64]

Potatoes	N/A	N/A	[68]
Corncob	0.58	N/A	[69]
Rapeseed meal[a]	0.115	N/A	[70]
Rapeseed meal[b]	N/A	N/A	[71]
Orange peel	0.58	Negligible	[72]
Bread	1.16	0.55	[66]
Cake	0.80	0.28	[61]
Pastry	0.67	0.35	[61]

N/A, not available.

[a]Rapeseed meal is treated by diluted sulfuric acid hydrolysis and subsequent enzymatic hydrolysis of pectinase, celluclast, and viscozyme.

[b]Rapeseed meal is treated by enzymatic hydrolysis using *A. oryzae*.

Biotechnological PHB Production Using Bakery Waste and Seawater

Halomonas boliviensis has been utilized in fermentations for the bioconversion of bakery hydrolysate into PHB. This microorganism is a moderate halophilic and alkali tolerant bacterium that can produce PHB through fermentative processes under aerobic condition [73]. It was isolated from a Bolivian salt lake, and the rod-shaped *H. boliviensis* is able to survive and synthesize PHB under salty environment.

Table 5 shows a summary of PHB fermentation results using defined medium and bakery hydrolysates, namely cake and pastry hydrolysates. PHB yields suggested that a defined fermentation medium (40 g/L glucose, 5 g/L yeast extract) can provide an optimum PHB yield (72%). The lowest PHB yield (1–2%), as expected, was obtained under bakery hydrolysate fermentation media. This demonstrates that a defined medium with 40 g/L glucose and around 5 g/L yeast extract could provide sufficient nutrients for *H. boliviensis* to produce PHB efficiently.

Table 5: The overall performance of both defined medium and bakery hydrolysate fermentation for PHB production in terms of fermentation conditions and results

Batch no.	Fermentation medium	Fermentation mode	Feeding media	Fermentation time (h)	Glucose consumption (g)	CDW (g)	PHB production (g)	PHB content (%)
1	Defined (40 g/L glucose, 2 g/L yeast extract)	Batch	NIL	64.0	13.0	NIL	NIL	NIL
2	Defined (40 g/L glucose, 5 g/L yeast extract)	Batch	NIL	88.0	24.0	24.9	17.4	17.4
3	Defined (40 g/L glucose, 8 g/L yeast extract)	Batch	NIL	75.0	59.9	9.2	4.3	4.3
4	Pastry hydrolysate	Batch	NIL	23.5	32.8	NIL	NIL	NIL
5	Pastry hydrolysate	Fed-batch	Glucose solution	135.5	112.3	5.7	2.1	2.1
6	Pastry hydrolysate	Fed-batch	Pastry hydrolysate	67.0	208.8	38.2	3.6	3.6
7	Pastry hydrolysate	Fed-batch	Pastry hydrolysate	87.0	359.9	15.6	0.6	0.59
8	Cake hydrolysate	Fed-batch	Cake hydrolysate	63.0	200.5	11.6	2.9	2.9

CDW. cell dry weight.

The overall glucose consumption for defined medium fermentation in the batch mode ranged from 13 to 60 g. High initial nitrogen source could hinder PHB production by 10 times, as indicated from PHB yield obtained by defined medium. A similar effect could possibly lead to the low PHB yield observed in bakery hydrolysate fermentation. With the continuous supply of nitrogen source, *H. boliviensis* consumed glucose in a faster rate for PHB production, maintenance, and synthesis of other metabolites (six times higher overall glucose consumption with the feeding of bakery hydrolysate). *Halomonas boliviensis* synthesizes ectonie and hydroxyectonie as osmolytes as NaCl concentration increases in the cell's environment. Van-Thuoc et al. [74] reported the co-production of ectonie and PHB in a combined two-step fed-batch culture. Similarly, the formation of other primary metabolites such as ectoines in the bakery hydrolysate fermentation was observed in these studies. This consequently led to a lower PHB production when bakery hydrolysate and seawater were used as fermentation feed stocks. The highest overall yield of PHB production for the defined medium (with less glucose consumption and higher PHB production) was about 17% as compared to a rather low 3.5% observed for bakery hydrolysate.

In summary, this project is currently demonstrating the green credentials in the development of advanced food waste valorization practices to valuable products, which also include GHG reductions as well as the production of other air pollutants. Such a synergistic solution may be feasible for adoption by the Hong Kong Government as part of their strategy for tackling the food waste issue as well as for the environmentally friendly production of alternative platform chemicals and biodegradable plastics.

Chemical Valorization of Food Waste for Bioenergy Production

The valorization of waste to important chemicals can be accomplished through different approaches as discussed. Another potentially interesting approach to advanced valorization practices would be the chemical utilization of various waste raw materials for conversion into high-value products.

A case study of such integrated valorization is a recent study on the conversion of corncob residues into functional catalysts for the preparation of fatty acid methyl esters (FAME) from waste oils [75]. The design of the catalyst involved an incomplete carbonization step under air to partially degrade the lignin materials mostly present in corncobs, followed by subsequent functionalization via sulfonation to generate $-SO_3H$ acidic sites. The solid acid catalyst was then subjected to conditioning prior to its utilization in the conversion of waste cooking oil with a high content of free fatty acids (FFA) to biodiesel-like biofuels. The advantage of the designed solid acid catalyst, apart from being derived from food waste, is the possibility to conduct a simultaneous esterification of FFA present in the waste oil as well as transesterification of the remaining triglycerides also present in the oil (Fig. 4).

In this approach, the generation of two valuable products (a cheap solid acid catalyst and biodiesel-like biofuels) can be achieved starting from two food waste feedstock (corncobs and waste cooking oils). The solid acid catalysts were characterized using a range of techniques. Fourier transform infrared spectroscopy (FTIR) showed the presence of different functional groups including C=O, C-O, C-S, and aromatic C=C in the materials (Table 6). The catalytic activity of the solids also showed remarkable activity toward the conversion of waste cooking oils into biodiesel-like biofuels. A maximum of 98% yield to methyl esters could be obtained without prior purification of the oils. Importantly, kinetics of the transesterification reaction was significantly slower to those of the esterification of FFA present in the oil. Despite a low $-SO_3H$ loading (1 wt% S, 0.16 mmol/g $-SO3H$), the catalytic activity was still high, indicating a possibly different surface functionality (Fig. 10).

Table 6: Summary of bands observed in IR analysis for all sulfonated samples

Frequency of band	Corresponding functional group
1700 cm^{-1}	C=O
1597 cm^{-1}	C=C aromatic
1219 cm^{-1}	S=O
1029 cm^{-1}	C-S

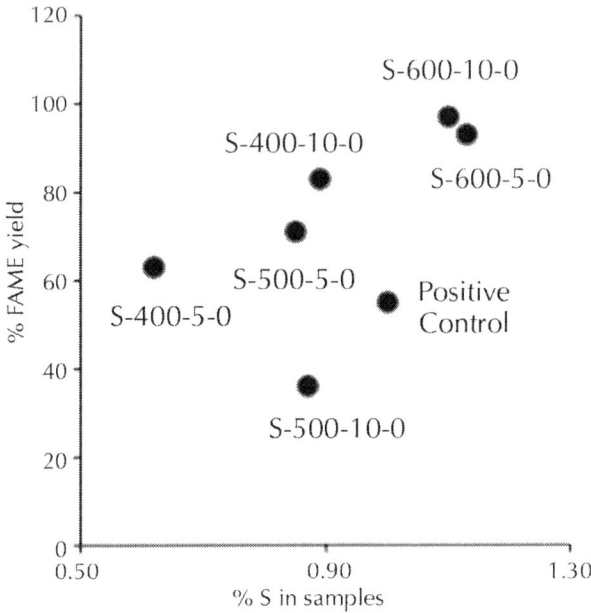

Figure 10: Plot of the FAME yield of the samples versus the %S content (A, material carbonized at 400°C for 5 h; B, carbonized at 400°C for 10 h; C, carbonized at 500°C for 5 h; D, carbonized at 500°C for 10 h; E, carbonized at 600°C for 5 h; F, carbonized at 600°C for 10 h). A higher degree of functionalization (higher %S) generally leads to improved FAME yields. Results also highlight the superior catalytic activities of sulfonated carbonaceous materials compared to blank (no conversion, data not shown) and the positive control referring to the homogeneously H_2SO_4 catalyzed reaction.

The recyclability of the solid acid catalysts still, however, needs to be further optimized. Catalysts were found to deactivate quickly (after two uses) due to the aqueous promoted decomposition and hydrolysis of the sulfonated groups in the material [75]. Materials should be tested under different conditions of temperature, carbonizing atmosphere, and even pressure to improve the stability and robustness of the catalyst for the selected process. Nevertheless, this study provides a promising proof of concept of the potential of an integrated valorization of various waste raw material into valuable end products and biofuels as it avoids the pretreatment of the waste oils (generally required to reduce the high FFA content to allow the conventional base-transesterification process to take place avoiding the formation of undesirable soaps and emulsions) and generates a relatively pure biofuel from a residue using an environmentally friendly and cheap solid acid catalyst.

Tailored-Made Healing Biopolymers from the Meat Industry

The meat industry generates enormous quantities of solid waste [76]. Managing such residues entails a significant problem for the sector as many of the generated by-products and residues are prone to degradation and microbial contamination. However, an important part of these residues are rich in various added value products, which upon extraction could constitute a source of interesting revenues for these industries. Among the most promising compounds from meat industry-derived by products, we can include oily fats and collagen [77]. Valorization of the aforementioned waste fats from slaughterhouses and meat processing industries to biodiesel-like biofuels has been studied via esterification/transesterification using different types of catalysts and protocols, which entailed in some cases a pretreatment and refining of the fat [78]. In principle, these will be, however, conducted in a similar way to that reported in the previously showcased study of waste cooking oils valorization.

Comparatively, collagen-containing residues (e.g., bovine hides) are increasingly important residues from the meat industry that are often derived to leather processing companies. Interestingly, the significant amounts of collagen present in such samples are not that well known [77].

Collagen is a ubiquitous and most relevant biopolymer in vertebrates [77-79], which possesses a highly interesting versatility to be employed in a wide range of applications in different areas from regenerative medicine to cosmetics and veterinary. Extraction and stabilization of collagenic biopolymers from waste, particularly related to their physical properties, (e.g., via cross-linking) constitutes an innovative pathway toward the production of novel potentially industrial products (e.g., tissue engineering, wound healing, antimicrobial aposits, etc.). Cross-linking methodologies can in principle generate additional bond formation to stabilize polymers with additional benefits on physical properties including swelling and flexibility. In the light of these premises, a recent example on the extraction, cross-linking and purification of collagenic biopolymers from splits (pickled hides) and the so-called wet-white hides (from tannery-derived hides treated with glutaraldehyde or phenolic compounds for chromium-free leather production) demonstrated the possibility to obtain valuable end products based on tailored-made biocollagen with improved mechanical properties, stabilized structures, and desired molecular weight ranges, which could be employed in wound healing acceleration in rats [80, 81]. Interestingly, these biopolymers could be easily shaped into various forms including fibers, sponges, and/or films, paving the way to the development of potentially novel biomaterials for different biomedical applications (Fig. 11). A simple hydrolytic process was able to extract the collagen, followed by subsequent cross-linking to stable biopolymers or direct application upon purification by ultrafiltration as unguent for induced wounds in rats [80, 81].

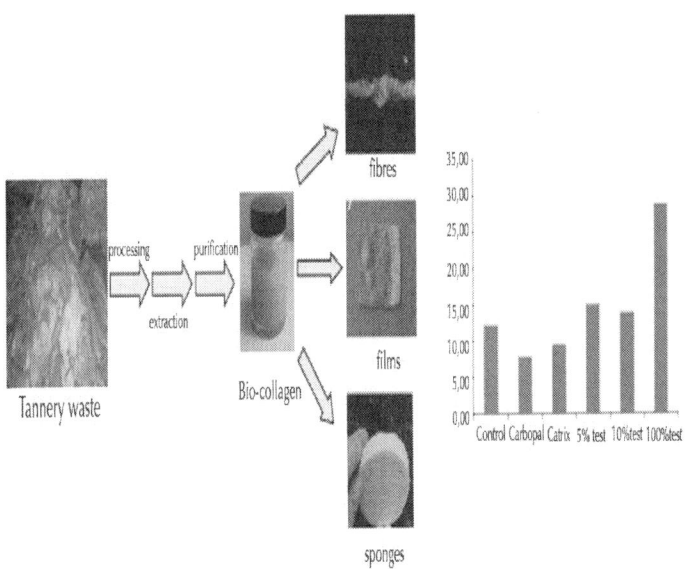

Figure 11: Meat and tannery-derived residues can be valorized to valuable collagenic biopolymers that can be formed into fibers, films, and sponges for various applications. Right plot depicts a comparison of activity in induced wounds in rats between pure and diluted collagen-extracted formulations (100%, 5% and 10% test, respectively) and a control sample (no treatment) and commercial formulations (Carbopol, Catrix).

Maximum yields of biopolymer extracted were obtained at 0.25 mm grinding size and the use of diluted acetic acid as hydrolytic agent (24 h, room temperature), for which also a minimum swelling (better biopolymer properties) was also observed. Interestingly, samples obtained from splits exhibited a better desirability to those extracted from wet-white hides. Isolated biopolymers possessed molecular weight of ca. 300 kDa, in contrast to conventional collagen derivatives for which molecular weights are usually within 15–50 kDa for hydrolysates [82] and 50–200 kDa for gelatine [83].

Biocollagenic materials were found to be very attractive and highly useful in treatment of induced burns/wounds in mice (Fig. 11), showing in all cases improved tissue regeneration and wound healing as compared to untreated wounds and commercial formulations including Carbopol and Catrix. Even diluted formulations containing 5 wt% of the collagenic biopolymer (Fig. 11, right plot) were found to provide improved results.

FUTURE PROSPECTS AND CONCLUSIONS

Excessive disposal of food and plastic waste are deteriorating the landfill issue in many parts of the World. Waste valorization is an attractive concept that has gained increasing popularity in many countries nowadays due to the rapid increase in generation of such waste residues. Because of this, researchers are not only developing valorization strategies but also focusing on the design of greener materials utilizing a range of green technologies. One example of this could be the synthesis of magnetically separable substances [84–87]. Not only are these able to catalyze the necessary conversion, they are also economically attractive due to their simple preparation [84, 55, 85]. Also, the production of carbon-based catalysts maybe continued for research, but greener preparations (such as microwave-mediated functionalization) to lessen the energy investment, should be explored. Furthermore, the emergence of graphene as catalyst in many reactions should also be noted for valorization purposes.

As previously mentioned, an interesting valorization protocol to develop would be a photo catalytic approach. To accomplish such photo catalytic strategies, TiO_2, $Pt/CdS/TiO_2$ composite materials [88], $TiO_2/Ni(OH)_2$ [84] clusters may be used depending on the target, samples, and reaction conditions. Photo degradation has been shown to be possible toward many environmental pollutants such as chlorofluorocarbons [55], CO_2 [85], and NO [54], but whether these photoactive composites could degrade the stable polymeric structure of lignin/protein/carbohydrates is yet to be seen and perhaps understood. A recent study by Balu et al. [89], reports on the preparation of a TiO_2-guanidine-$(Ni,Co)Fe_2O_4$ photoactive material. The addition of the guanidine was made to lower the band gap of the material hence making it active under visible light. Testing the material to a model chemical reaction, using malic acid and the synthesized photo material produced simpler chemicals such as formic acid, acetic acid, and oxalic acid with a selectivity of around 80%. This study provides proof of concept that band gap engineering of semiconductors can lead to the development of photoactive materials that may be used selectively for waste valorization. A photo catalytic approach will most importantly address one of the major drawbacks of industrial valorization which is on the relatively large amounts of energy needed for processing and purification of products.

The conversion of a range of feedstock into valuable products including chemicals, biomaterials, and fuels has been demonstrated in three essentially different case studies to highlight the significant potential of advanced waste valorization strategies.

The incorporation of these and similar processes in future bio refineries for the production of value-added products and fuels will be an important contribution toward the world's highest priority target of sustainable development.

But perhaps the main and most important issue to be addressed for the sake of future generations, currently way overlooked, is society itself. The most extended perception of waste as a problem, as a residue, as something not valuable needs to give way to a general concienciation of society in waste as a valuable resource. A resource, which obviously entails a significant complexity (from its inherent diversity and variability), but one that can provide at the same time an infinite number of innovative solutions and alternatives to end products through advanced valorization strategies. These will need joint efforts from a range of disciplines from engineering to (bio) chemistry, bio(techno)logy, environmental sciences, legislation, and economics to come up with innovative alternatives that we hope to see leading the way toward a more sustainable bio-based society and economy.

ACKNOWLEDGMENTS

R. D. Arancon thanks the Department of Chemistry of the Ateneo de Manila University in the Philippines for the wonderful opportunity to learn. Also, heartfelt thanks are due to Jhon Ralph Enterina (University of Alberta, Canada) and Jurgen Sanes (Simon Fraser University, Canada) for help with some articles. Carol Sze Ki LIN acknowledges the Biomass funding from the Ability R&D Energy Research Centre (AERC) at the School of Energy and Environment in the City University of Hong Kong. The authors are also grateful to the donation from the Coffee Concept (Hong Kong) Ltd. for the "Care for Our Planet" campaign, as well as a grant from the City University of Hong Kong (Project No. 7200248). C. S. K. Lin acknowledges the Industrial Technology Funding from the Innovation and Technology Commission (ITS/323/11) in Hong Kong. R. Luque gratefully acknowledges the Spanish MICINN for financial support via the concession of a RyC contract (ref: RYC–2009–04199) and funding under project CTQ2011–28954-C02-02. Consejeria de Ciencia e Innovacion, Junta de Andalucia is also gratefully acknowledged for funding project P10-FQM-6711. R. Luque is also indebted to Guohua Chen, the Department of Chemical and Bio molecular Engineering (CBME) and HKUST for the provision of a visiting professorship as Distinguished Engineering Fellow.

REFERENCES

1. Serrano-Ruiz, J. C., R. Luque, J. M. Campelo, and A. A. Romero. 2012. Continuous-flow processes in heterogeneously catalyzed transformations of biomass derivatives into fuels and chemicals. *Challenges* 3:114–132.

2. Glasnov, T. N., and C. O. Kappe. 2011. The microwave-to-flow paradigm: translating high-temperature batch microwave chemistry to scalable continuous-flow processes. *Chem. Eur. J.* 17:11956–11968.

3. PBL Netherland Environmental Assessment Agency. Trends in global CO_2 emissions.http://edgar.jrc.ec.europa.eu/CO2REPORT2012.pdf (accessed 15 March 2013).
4. Mohan, D., C. U. Pittman, and P. H. Steele. 2006. Pyrolysis of wood/biomass for bio-oil: a critical review. *Energy Fuels* 20:848–889.
5. Heo, H. S., H. J. Park, Y.-K. Park, C. Ryu, D. J. Suh, Y.-W. Suh, et al. 2010. Bio-oil production from fast pyrolysis of waste furniture sawdust in a fluidized bed. *Bioresour. Technol.* 101:S91–S96.
6. Cho, M.-H., S.-H. Jung, and J.-S. Kim. 2009. Pyrolysis of mixed plastic wastes for the recovery of benzene, toluene, and xylene (BTX) aromatics in a fluidized bed and chlorine removal by applying various additives. *Energy Fuels* 24:1389–1395.
7. Kantarelis, E., and A. Zabaniotou. 2009. Valorization of cotton stalks by fast pyrolysis and fixed bed air gasification for syngas production as precursor of second generation biofuels and sustainable agriculture. *Bioresour. Technol.* 100:942–947.
8. Luque, R., J. A. Menendez, A. Arenillas, and J. Cot. 2012. Microwave-assisted pyrolysis of biomass feedstocks: the way forward?*Energy Environ. Sci.* 5:5481–5488.
9. Wulff, N., H. Carrer, and S. Pascholati. 2006. Expression and purification of cellulase Xf818 from *Xylella fastidiosa* in *Escherichia coli.Curr. Microbiol.* 53:198–203.
10. Atsumi, S., T. Hanai, and J. C. Liao. 2008. Non-fermentative pathways for synthesis of branched-chain higher alcohols as biofuels.*Nature* 451:86–89.
11. Gustavsson, L., P. Börjesson, B. Johansson, and P. Svenningsson. 1995. Reducing CO_2 emissions by substituting biomass for fossil fuels. *Energy* 20:1097–1113.
12. Lee, S. W., T. Herage, and B. Young. 2004. Emission reduction potential from the combustion of soy methyl ester fuel blended with petroleum distillate fuel. *Fuel* 83:1607–1613.
13. Gielen, D. J, A. J. M. Bos, M. A. R. C. de Feber, and T. Gerlagh. Biomass for greenhouse gas emission reduction.http://www.ecn.nl/docs/library/report/2000/c00001.pdf (accessed 15 March 2013).

14. Gustavsson, L., J. Holmberg, V. Dornburg, R. Sathre, T. Eggers, K. Mahapatra, et al. 2007. Using biomass for climate change mitigation and oil use reduction. *Energy Policy* 35:5671–5691.

15. Chen, K., H. Zhang, Y. Miao, M. Jiang, and J. Chen. 2010. Succinic acid production from enzymatic hydrolysate of sake lees using*Actinobacillus succinogenes* 130Z. *Enzyme Microb. Technol.* 47:236–240.

16. Oliveira, L. S., and S. F. Adriana. 2009. From solid biowastes to liquid biofuels. Agriculture Issues and Policies Series: 265. Available at:http://www.demec.ufmg.br/disciplinas/eng032 BL/solid_biowastes_liquid_biofuels.pdf (accessed May 2013).

17. Toledano, A., L. Serrano, A. M. Balu, R. Luque, A. Pineda, and J. Labidi. 2013. Fractionation of organosolv lignin from olive tree clippings and its valorization to simple phenolic compounds. *ChemSusChem* 6:529–536.

18. Du, C., J. Sabirova, W. Soetaert, and C. S. K. Lin. 2012. Polyhydroxyalkanoates production from low-cost sustainable raw materials.*Curr. Chem. Biol.* 6:14–25.

19. Balu, A. M., V. Budarin, P. S. Shuttleworth, L. A. Pfaltzgraff, K. Waldron, R. Luque, et al. 2012. Valorisation of orange peel residues: waste to biochemicals and nanoporous materials. *ChemSusChem* 5:1694–1697.

20. Au, E. 2013. Food waste management and practice in Hong Kong in Commercial and Industrial (C&I) Food Waste Recycling Seminar, 8 February 2013, Food Education Association, The Hong Kong Polytechnic University, Hong Kong.

21. Zhang, R., H. M. El-Mashad, K. Hartman, F. Wang, G. Liu, C. Choate, et al. 2006. Characterization of food waste as feedstock for anaerobic digestion. *Bioresour. Technol.* 98:929–935.

22. Russ, W., and R. Meyer-Pittroff. 2004. Utilizing waste products from the food production and processing industries. *Crit. Rev. Food Sci. Nutr.* 44:57–62.

23. Kornegay, E. T., G. W. Vander Noot, K. M. Barth, W. S. MacGrath, J. G. Welch, and E. D. Purkhiser. 1965. Nutritive value of garbage as a feed for swine. I. Chemical composition, digestibility and nitrogen utilization of various types of garbage. *J. Anim. Sci.* 24:319–324.

24. Westendorf, M. L. 1996. Pp. 24–32 *in The use of food waste as a feedstuff in swine diets.* Proceeding of Food Waste Recycling Symp. Rutgers Coop. Ext., Rutgers Univ.-Cook College, New Brunswick, NJ.

25. Grolleaud, M. 2002. *Post-harvest losses: discovering the full story.* Overview of the phenomenon of losses during the post-harvest system. FAO, Agro Industries and Post-Harvest Management Service, Rome, Italy.

26. Parfitt, J., M. Barthel, and S. Macnaughton. 2010. Food waste within food supply chains: quantification and potential for change to 2050.*Philos. Trans. R. Soc. Lond. B Biol. Sci.* 365:3065–3081.

27. Gustavsson, J., C. Cederberg, U. Sonesson, R. van Otterdijk, and A. Meybeck. 2011. *Global food losses and food waste: extent, causes and prevention.* FAO, Rome, Italy.

28. Tatsi, A., and A. Zouboulis. 2002. A field investigation of the quantity and quality of leachate from a municipal solid waste landfill in a Mediterranean climate (Thessaloniki, Greece). *Adv. Environ. Res.* 6:207–219.

29. EPD (Environmental Protection Department of HKSAR). Monitoring of solid waste in Hong Kong 2011.https://www. wastereduction.gov.hk/chi/materials/info/msw2011tc.pdf (accessed October 2012).

30. Abu-Rukah, Y., and O. Al-Kofahi. 2001. The assessment of the effect of landfill leachate on ground-water quality—a case study El-Akader landfill site-north Jordan. *J. Arid Environ.* 49:615–630.

31. Pfaltzgraff, L. A., M. De bruyn, E. C. Cooper, V. Budarin, and J. H. Clark. 2013. Food waste biomass: a resource for high-value chemicals. *Green Chem.* 15:307–314.

32. Toledano, A., L. Serrano, J. Labidi, A. Pineda, A. M. Balu, and R. Luque. 2013. Heterogeneously catalysed mild hydrogenolytic depolymerisation of lignin under microwave irradiation with hydrogen-donating solvents. *ChemCatChem* 5:977–985.

33. Toledano, A., L. Serrano, and J. Labidi. 2012. Process for olive tree pruning lignin revalorisation. *Chem. Eng. J.* 193–194:396–403.

34. Toledano, A., L. Serrano, A. Pineda, A. A. Romero, J. Labidi, and R. Luque. 2013. Microwave-assisted depolymerisation of

organosolv lignin via mild hydrogen-free hydrogenolysis: catalyst screening. *Appl. Catal. B.* doi: 10.1016/j.apcatb.2012.10.015

35. Llorach, R., J. C. Espín, F. A. Tomás-Barberán, and F. Ferreres. 2003. Valorization of cauliflower (*Brassica oleracea* L. var. botrytis) by-products as a source of antioxidant phenolics. *J. Agric. Food Chem.* 51:2181–2187.

36. González-Sáiz, J. M., C. Pizarro, I. Esteban-Díez, O. Ramírez, C. J. González-Navarro, M. J. Sáiz-Abajo, et al. 2007. Monitoring of alcoholic fermentation of onion juice by NIR spectroscopy: valorization of worthless onions. *J. Agric. Food Chem.* 55:2930–2936.

37. Dong, L.-M., X.-P. Yan, Y. Li, Y. Jiang, S.-W. Wang, and D.-Q. Jiang. 2004. On-line coupling of flow injection displacement sorption preconcentration to high-performance liquid chromatography for speciation analysis of mercury in seafood. *J. Chromatogr. A* 1036:119–125.

38. Cheng, Y., L. Fan, H. Chen, X. Chen, and Z. Hu. 2005. Method for on-line derivatization and separation of aspartic acid enantiomer in pharmaceuticals application by the coupling of flow injection with micellar electrokinetic chromatography. *J. Chromatogr. A* 1072:259–265.

39. de Boer, A. R., T. Letzel, D. A. van Elswijk, H. Lingeman, W. M. Niessen, and H. Irth. 2004. On-line coupling of high-performance liquid chromatography to a continuous-flow enzyme assay based on electrospray ionization mass spectrometry. *Anal. Chem.* 76:3155–3161.

40. Stewart, J. J., T. Akiyama, C. Chapple, J. Ralph, and S. D. Mansfield. 2009. The effects on lignin structure of overexpression of ferulate 5-hydroxylase in hybrid poplar. *Plant Physiol.* 150:621–635.

41. Sahu, R., and P. L. Dhepe. 2012. A one-pot method for the selective conversion of hemicellulose from crop waste into C5 sugars and furfural by using solid acid catalysts. *ChemSusChem* 5:751–761.

42. Chakraborty, R., S. Bepari, and A. Banerjee. 2010. Transesterification of soybean oil catalyzed by fly ash and egg shell derived solid catalysts. *Chem. Eng. J.* 165:798–805.

43. Fu, B., L. Gao, L. Niu, R. Wei, and G. Xiao. 2009. Biodiesel from

waste cooking oil via heterogeneous superacid catalyst SO_4^{2-}/ZrO_2. *Energy Fuels* 23:569–572.

44. Clark, J. H., V. Budarin, T. Dugmore, R. Luque, D. J. Macquarrie, and V. Strelko. 2008. Catalytic performance of carbonaceous materials in the esterification of succinic acid. *Catal. Commun.* 9:1709–1714.

45. Luque, R., A. Pineda, J. C. Colmenares, J. M. Campelo, A. A. Romero, J. C. Serrano-Ruiz, et al. 2012. Carbonaceous residues from biomass gasification as catalysts for biodiesel production. *J. Nat. Gas Chem.* 21:246–250.

46. Abbot, A. P., R. C. Harris, K. S. Ryder, C. D'Agostino, L. F. Gladden, and M. D. Mantle. 2011. Glycerol eutectics as sustainable solvent systems. *Green Chem.* 13:82–90.

47. Carriazo, D., M. C. Serrano, M. C. Gutierrez, M. L. Ferrer, and F. del Monte. 2012. Deep eutectic solvents playing multiple roles in the synthesis of polymers and related materials. *Chem. Soc. Rev.* 41:4996–5014.

48. Zhang, Q., K. De Oliveira Vigier, S. Royer, and F. Jerome. 2012. Deep eutectic solvents: syntheses, properties and applications. *Chem. Soc. Rev.* 41:7108.

49. Russ, C., and B. König. 2012. Low melting mixtures in organic synthesis- an alternative to ionic liquids? *Green Chem.* 14:2969–2982.

50. Serrano-Ruiz, J. C., J. M. Campelo, M. Francavilla, C. Menendez, A. B. Garcia, A. A. Romero, et al. 2012. Efficient microwave-assisted production of furfural from C5 sugars in aqueous media catalysed by Brönsted acidic ionic liquids. *Catal. Sci. Technol.* 2:1828–1832.

51. Zhang, Z., Q. Wang, H. Xie, W. Liu, and Z. K. Zhao. 2011. Catalytic conversion of carbohydrates into 5-hydroxymethylfurfural by germanium (IV) chloride in ionic liquids. *ChemSusChem* 4:131–138.

52. Colmenares, J. C., R. Luque, J. M. Campelo, F. Colmenares, Z. Karpiński, and A. A. Romero. 2009. Nanostructured photocatalysts and their applications in the photocatalytic transformation of lignocellulosic biomass: an overview. *Materials* 2:2228–2258.

53. Stillings, R. A., and R. J. V. Nostrand. 1944. The action of ultraviolet light upon cellulose. I. Irradiation effects. II. Post-irradiation effects1. *J. Am. Chem. Soc.* 66:753–760.

54. Ai, Z., W. Ho, and S. Lee. 2011. Efficient visible light photocatalytic removal of NO with BiOBr-graphene nanocomposites. *J. Phys. Chem.* 115:25330–25337.

55. Ismail, A. A., and D. W. Bahnemann. 2011. Mesostructured Pt/TiO_2 nanocomposites as highly active photocatalysts for the photooxidation of dichloroacetic acid. *J. Phys. Chem.* 115:5784–5791.

56. Bernardo, E. C. 2008. Solid-waste management practices of households in Manila, Philippines. *Ann. NY Acad. Sci.* 1140:420–424.

57. Office of the Chief Executive. 2013. Policy Address, 2013 (Office of the Chief Executive). The Hong Kong Government Special Administrative Region (HKSAR), Hong Kong. Available at http://www.policyaddress.gov.hk/2013/eng/p142.html (accessed 16 January 2013).

58. Takata, M., K. Fukushima, N. Kino-Kimata, N. Nagao, C. Niwa, and T. Toda. 2012. The effects of recycling loops in food waste management in Japan: based on the environmental and economic evaluation of food recycling. *Sci. Total Environ.* 432:309–317.

59. Bernstad, A., and J. la Cour Jansen. 2012. Separate collection of household food waste for anaerobic degradation – Comparison of different techniques from a systems perspective. *Waste Manage. (Oxford)* 32:806–815.

60. Zhang, B., L.-L. Zhang, S.-C. Zhang, H.-Z. Shi, and W.-M. Cai. 2005. The influence of pH on hydrolysis and acidogenesis of kitchen wastes in two-phase anaerobic digestion. *Environ. Technol.* 26:329–340.

61. Zhang, A. Y., Z. Sun, C. C. J. Leung, W. Han, K. Y. Lau, M. Li, et al. 2013. Valorisation of bakery waste for succinic acid production. *Green Chem.* 15:690–695.

62. Van-Thuoc, D., J. Quillaguamán, G. Mamo, and B. Mattiasson. 2008. Utilization of agricultural residues for poly(3-hydroxybutyrate) production by *Halomonas boliviensis* LC1. *J. Appl. Microbiol.* 104:420–428.

63. Du, C., S. K. C. Lin, A. Koutinas, R. Wang, P. Dorado, and C. Webb. 2008. A wheat biorefining strategy based on solid-state fermentation for fermentative production of succinic acid. *Bioresour. Technol.* 99:8310–8315.

64. Dorado, M. P., S. K. C. Lin, A. Koutinas, C. Du, R. Wang, and C. Webb. 2009. Cereal-based biorefinery development: utilisation of wheat milling by-products for the production of succinic acid. *J. Biotechnol.* 143:51–59.

65. Lin, C. S. K., R. Luque, J. H. Clark, C. Webb, and C. Du. 2012. Wheat-based biorefining strategy for fermentative production and chemical transformations of succinic acid. *Biofuels Bioprod. Biorefin.* 6:88–104.

66. Leung, C. C. J., A. S. Y. Cheung, A. Y.-Z. Zhang, K. F. Lam, and C. S. K. Lin. 2012. Utilisation of waste bread for fermentative succinic acid production. *Biochem. Eng. J.* 65:10–15.

67. García, I. L., J. A. López, M. P. Dorado, N. Kopsahelis, M. Alexandri, S. Papanikolaou, et al. 2013. Evaluation of by-products from the biodiesel industry as fermentation feedstock for poly(3-hydroxybutyrate-co-3-hydroxyvalerate) production by *Cupriavidus necator.Bioresour. Technol.* 130:16–22.

68. Delgado, R., A. J. Castro, and M. Vázquez. 2009. A kinetic assessment of the enzymatic hydrolysis of potato (*Solanum tuberosum*).*LWT Food Sci. Technol.* 42:797–804.

69. Yu, J., Z. Li, Q. Ye, Y. Yang, and S. Chen. 2010. Development of succinic acid production from corncob hydrolysate by *Actinobacillus succinogenes. J. Ind. Microbiol. Biotechnol.* 37:1033–1040.

70. Chen, K., H. Zhang, Y. Miao, P. Wei, and J. Chen. 2011. Simultaneous saccharification and fermentation of acid-pretreated rapeseed meal for succinic acid production using *Actinobacillus succinogenes. Enzyme Microb. Technol.* 48:339–344.

71. Wang, R., L. C. Godoy, S. M. Shaarani, M. Melikoglu, A. Koutinas, and C. Webb. 2009. Improving wheat flour hydrolysis by an enzyme mixture from solid state fungal fermentation. *Enzyme Microb. Technol.* 44:223–228.

72. Li, Q., J. Siles, and I. Thompson. 2010. Succinic acid production from orange peel and wheat straw by batch fermentations

of *Fibrobacter succinogenes* S85. *Appl. Microbiol. Biotechnol.* 88:671–678.

73. Quillaguamán, J., R. Hatti-Kaul, B. Mattiasson, M. T. Alvarez, and O. Delgado. 2004. *Halomonas boliviensis* sp. nov., an alkalitolerant, moderate halophile isolated from soil around a Bolivian hypersaline lake. *Int. J. Syst. Evol. Microbiol.* 54:721–725.

74. Van-Thuoc, D., H. Guzmán, J. Quillaguamán, and R. Hatti-Kaul. 2010. High productivity of ectoines by *Halomonas boliviensis* using a combined two-step fed-batch culture and milking process. *J. Biotechnol.* 147:46–51.

75. Arancon, R. A., H. R. Barros Jr., A. M. Balu, C. Vargas, and R. Luque. 2011. Valorisation of corncob residues to functionalised porous carbonaceous materials for the simultaneous esterification/transesterification of waste oils. *Green Chem.* 13:3162–3167.

76. Cabeza, L., M. M. Taylor, G. L. DiMaio, E. Brown, W. N. Marmer, R. Carrió, et al. 1998. Processing of leather waste: pilot scale studies on chrome shavings. Isolation of potentially valuable protein products and chromium. *Waste Manage. (Oxford)* 18:211–218.

77. Gelse, K., E. Pöschl, and T. Aigner. 2003. Collagens—structure, function, and biosynthesis. *Adv. Drug Deliv. Rev.* 55:1531–1546.

78. Mata, T. M., A. A. Martins, and N. S. Caetano. 2013. Valorization of waste frying oils and animal fats for biodiesel production. Pp.671–693 *in* J. W. Lee, ed. *Advanced biofuels and bioproducts*. Springer, The Netherlands.

79. Reis, R. L., N. M. Neves, J. F. Mano, M. E. Gomes, A. P. Marques, and H. S. Azevedo. 2008. *Natural based polymers for biomedical applications*. Woodhead Publishing, CRC Press, Cambridge, U.K.

80. Catalina, M., J. Cot, M. Borras, J. de Lapuente, J. González, A. M. Balu, et al. 2013. From waste to healing biopolymers: biomedical applications of bio-collagenic materials extracted from industrial leather residues in wound healing. *Materials* 6:1599–1607.

81. Catalina, M., J. Cot, M. Borras, J. de Lapuente, J. González, A. M. Balu, et al. 2013. From waste to healing biopolymers: biomedical applications of bio-collagenic materials extracted from industrial leather residues in wound healing. *Materials* 6:1599–1607.

82. Langmaier, F., P. Mokrejs, R. Karnas, M. Mládek, and K. Kolomazník. 2006. Modification of chrome-tanned leather waste hydrolysate with epichlorhydrin. *J. Soc. Leather Technol. Chem.* 90:29–34.

83. Brown, E., C. Thompson, and M. M. Taylor. 1994. Molecular size and conformation of protein recovered from chrome shavings. *J. Am. Leather Chem. Assoc.* 89:215–220.

84. Yu, J., Y. Hai, and B. Cheng. 2011. Enhanced photocatalytic H_2-production activity of TiO_2 by $Ni(OH)_2$ cluster modification. *J. Phys. Chem. C* 115:4953–4958.

85. Liang, Y. T., B. K. Vijayan, K. A. Gray, and M. C. Hersam. 2011. Minimizing graphene defects enhances titania nanocomposite-based photocatalytic reduction of CO_2 for improved solar fuel production. *Nano Lett.* 11:2865–2870.

86. Polshettiwar, V., R. Luque, A. Fihri, H. Zhu, M. Bouhrara, and J. M. Basset. 2011. Magnetically recoverable nanocatalysts. *Cheminform* 42:3036–3075.

87. Liu, J., S. Z. Qiao, Q. H. Hu, and G. Q. Lu. 2011. Magnetic nanocomposites with mesoporous structures: synthesis and applications. *Small* 7:425–443.

88. Daskalaki, V. M., M. Antoniadou, G. Li Puma, D. I. Kondarides, and P. Lianos. 2010. Solar light-responsive $Pt/CdS/TiO_2$ photocatalysts for hydrogen production and simultaneous degradation of inorganic or organic sacrificial agents in wastewater. *Environ. Sci. Technol.* 44:7200–7205.

89. Balu, A. M., B. Baruwati, E. Serrano, J. Cot, J. Garcia-Martinez, R. S. Varma, et al. 2011. Magnetically separable nanocomposites with photocatalytic activity under visible light for the selective transformation of biomass-derived platform molecules. *Green Chem.* 13:2750–2758.

Chapter 3

Sustainable Multipurpose Biorefineries for Third-Generation Biofuels and Value-Added Co-Products

Stephen R. Hughes[1], William R. Gibbons[2], Bryan R. Moser[3], and Joseph O. Rich[1]

[1]USDA, ARS, NCAUR, Renewable Product Technology Research Unit, Peoria, IL, USA

[2]Biology/Microbiology Department, South Dakota State University, Brookings, SD, USA

[3]USDA, ARS, NCAUR, Bio-Oils Research Unit, Peoria, IL, USA

INTRODUCTION

The transition to third-generation biofuels is driven by the need to integrate biomass-derived fuels more seamlessly into the existing petroleum based infrastructure. Ethanol, whether derived from corn or sugarcane in first-generation processes or biomass in second-generation facilities, has limited market access due its dissimilarity to conventional petroleum-derived fuels. Limitations include restrictions on ratios in which ethanol can be blended with gasoline, lack of compatibility with diesel and jet engines, inability to transport ethanol through existing pipeline network, and propensity to absorb water. While it is clear that biomass can provide a sustainable and renewable source of carbon to replace a significant portion of petroleum resources currently used to generate fuel, power, and chemicals [1,2], it is also obvious that technologies must be developed to convert biomass into direct replacements for petroleum products. This transition from first- and second-generation biofuels to third-generation biofuels will involve numerous facets [3], the centerpiece likely being a multipurpose bio refinery that utilizes many inputs and produces an even greater number of outputs. The first steps to incorporating each of the individual platforms into one integrated sustainable operation are well underway [4], and this transition promises to be a continuing evolution.

First-generation bio refineries use feed stocks such as corn starch or sugar cane that are renewable, but that also have feed/food uses. As production levels have increased, along with human populations, concerns about competition with food needs have arisen [5, 6]. Nevertheless, over the past 30 years these first-generation feedstocks have paved the way for production of biofuels via a more sustainable system without negative impacts on the environment or food supplies [3]. Second-generation bio refineries are based on biomass feedstocks that are more widely available and that are not directly used as food, although some are used as livestock feed. Technologies are under development to efficiently convert biomass into ethanol as well as valuable co-products. These are leading the way to sustainably meeting energy needs while also supplying materials for chemical and manufacturing industries [3]. Biomass has the unique advantage among renewable energy sources that it can be easily stored until needed and provides a liquid transportation fuel alternative for the near term. However, cellulosic ethanol can displace only the 40% of a barrel of crude oil that is used to produce light-duty gasoline. Research, development, and demonstration on a range of technologies are needed to replace the remaining 60%, which is primarily converted to diesel and jet fuel. About 15% of our current crude oil consumption is used to produce solvents, plastics, cleaners, and adhesives [7]. Thus, cost-effective technologies are needed to produce biofuels that are suitable for drop-in use in cars, trucks, and jet planes. These advanced biofuels can be sustainably produced from cellulosic and algal feedstocks. Biomass conversion technologies are also needed to produce chemical intermediates and high-value chemicals. Compatibility with the existing infrastructure will aid in process integration and increase profitability of biorefineries [7].

Biorefining has been defined as the sustainable processing of biomass into a spectrum of marketable products and energy [3, 5]. The biorefinery of the future will conduct many types of processes, including those producing advanced biofuels, commodity chemicals, biodiesel, biomateials, power, and other value-added co-products such as sweeteners and bioinsecticides. With the tools provided by molecular biology and chemical engineering, the types of co-products, chemicals and biofuels that can be derived from biomass may be almost limitless. Biorefineries combine the necessary technologies for fractionating and hydrolyzing biological raw materials with conversion steps to produce and then recover intermediates and final products. The focus is on the precursor carbohydrates, lignins, oils, and proteins, and the combination of biotechnological and chemical conversion processes of the substances [8]. Most of these processes are being developed individually, but have the potential to be more efficient and economical when combined in multi-process crossover regimens using by-products or waste materials from one process to produce advanced animal feeds, human nutritional supplements, high-value peptides, or enzymes needed in other processes [4]. Use of existing infrastructure would significantly decrease the ramp-up time for economical large-scale production of advanced biofuels [7, 9].

To fully meet the requirement for safe and sustainable energy production, third-generation biorefineries must be better integrated, more flexible, and operate with lower carbon and economic costs than second-generation facilities [5]. The main areas that must be addressed are biomass production and supply, process optimization and integration, and overall sustainability [10-12]. Technology is developing rapidly in these areas. A major task is to identify the most promising bio-based products, in particular food, feed, value-added materials, and chemicals to be co-produced with energy to optimize overall process economics and minimize overall environmental impact [13]. Challenges to achieving the promise of advanced biofuels include: overcoming biomass recalcitrance, addressing logistics of transportation of raw feedstock and finished products, providing fair prices for crops or agricultural residues, and tailoring crops and production to specific environments and cultures [14].

BIOMASS PRODUCTION AND SUPPLY LOGISTICS

Feedstock costs represent a large part of biorefinery operating costs, therefore availability of an affordable feedstock supply is crucial for the viability of every biomass processing facility. Economics of biomass production vary with location, feedstock type, political policies, current infrastructure, and environmental concerns. Biofuels may be derived from forestry (thinning and logging), agriculture (residues or dedicated biomass crops), municipal wastes, algal-based resources, and by-products or waste products from agro-industry, food industry, and food services [15]. A key factor is to identify biomass resources that are sustainable because they require minimal water, fertilizer, land use, and other inputs. Feedstock must be high in energy content, be easy to obtain in large quantities, and be amenable to the conversion processes. Intensive research is in progress on technologies to deliver high-quality, stable, and infrastructure-compatible feed stocks from diverse biomass resources to bio refineries [7].

A strategic analysis was performed in 2005 [16] and updated in 2011 [1] that identified sufficient biomass feedstock availability across the United States to meet near-term and potentially long-term bioenergy goals. The assessment took into consideration environmental sustainability and identified likely costs, assuming a farm-gate or roadside feedstock price of $40-$60 per dry ton. The study did not include additional costs for preprocessing, handling, and transporting the biomass, as these are specific to the feedstock, its condition and form, the type of handling system, and storage conditions. The analysis also did not account for feedstock density or proximity to potential processing facilities. The feedstocks evaluated were those that are currently produced from agriculture and forestry sources, including grain crops (mainly corn for ethanol, sorghum, and barley), sugarcane, sugar beets, oil crops (primarily soybeans for diesel), canola, sunflower, rapeseed, municipal solid wastes, fuelwood, mill residues, pulping liquors, and urban wood wastes, as well as potential forest and agricultural biomass and waste resources. Under conservative assumptions, the combined resources from forests and agricultural lands, assuming feedstock prices of $40-$60 per dry ton, will increase from 138-258 million dry tons in 2012 to 243-767 million dry tons by 2030 [1]. Total energy crops, including perennial grasses such as switchgrass and miscanthus, woody crops such as poplar, willow, southern pine, and eucalyptus, and annual energy crops such as high-yield sorghum, are projected to contribute significantly to this increase, going from less than 4 million dry tons in 2012 to 34-400 million dry tons in 2030. Energy crops have the potential advantages of being produced on marginal lands not used for growing food, requiring essentially no fertilizers or irrigation, and, especially if perennial, requiring little or no tilling [1].

Although sufficient biomass supply is potentially available, continued improvements in biomass feedstocks worldwide are required to achieve viable third-generation biorefineries [5]. Feedstock production improvements include maximizing yield, nutrient (N, P, and K) and water efficiency, and sustainability of production (an area with high potential for rapid gains). Screening of plant species and plant breeding is critically important to increase efficiency of biomass production while minimizing inputs, maintaining soil fertility, managing water balance, and controlling invasiveness. Techniques to estimate the biomass production potential and to evaluate the impacts and sustainability of production in a given location are required. Logistic-related improvements include increasing efficiency of harvest, addressing the issue of seasonality to provide continuous supply, and ensuring that biomass cultivation helps drive regional development. Costs in transporting biomass to the biorefinery can be reduced by using optimized harvesting equipment, appropriate preparation for shipment, and efficient collection, storage, and transfer networks, especially for multi-feedstock biorefineries [5]. Processing improvements include optimizing the composition and properties of biomass for handling and transport to meet downstream quality requirements, along with imparting traits such as greater digestibility for ease of conversion (an area where basic research has made inroads). New technologies are reducing the cost of preparing biomass for conversion. Each step of the preparation is designed to develop next-generation feedstocks. Mechanical treatments reduce the size of the feedstock, providing fractionation and separation. Thermal and chemical processes control moisture content, remove contaminants, and improve digestibility and stability to reduce fouling in process equipment. Treated or untreated biomass is typically blended or mixed in specific proportions, often with additives to improve conversion efficiency. Temperature and pressure are used to form a high-density, stable feedstock for efficient storage and transport [7].

THIRD-GENERATION BIOFUELS

Advanced biofuels were defined by the Final Rule from the United States Environmental Protection Agency (EPA) Renewable Fuel Standard (RFS) Program as being renewable fuels, other than ethanol derived from corn starch, for which lifecycle greenhouse gas emissions are at least 50% less than the gasoline or diesel fuel it displaces [17]. Advanced biofuels may include any of the following: 1) ethanol derived from cellulose, hemicellulose, or lignin; 2) ethanol derived from sugar or starch (other than corn starch); 3) ethanol derived from waste material, including crop residue, other vegetative waste material, animal waste, food waste, and yard waste; 4) biomass-based diesel; 5) biogas (including landfill gas and sewage waste treatment gas) produced through the conversion of organic matter from renewable biomass; 6) butanol or other alcohols produced through the conversion of organic matter from renewable biomass; 7) other fuel derived from cellulosic biomass. Typically, advanced biofuels are used for transportation, although some may be used in generators to produce electricity and others may eventually replace propane and heating oil's [14]. Alcohol can substitute for gasoline in spark ignition engines; biodiesel, green diesel, and dimethyl ether can be used in compression ignition engines; Fischer-Tropsch process produces a variety of hydrocarbon fuels, the main one is a diesel-like fuel for compression ignition engines [15].

Third-generation biofuels, also referred to as drop-in biofuels, are considered advanced biofuels [17]. Third-generation biofuels are direct replacements for gasoline, diesel, and jet fuels currently produced from petroleum, and are, in fact, chemically identical to their petroleum-derived counterparts. This allows third-generation biofuels to be directly substituted for petro-fuels without any alterations to pipelines and infrastructure used to deliver the fuel, nor modifications to the engines in which the fuel will be burned. These infrastructure-compatible fuels are derived from biomass or algae, typically through thermochemical processes, although some biochemical processes are being developed as well. These fuels deliver more energy per gallon than ethanol, and conversion processes also yield a range of co-products that help to enhance the economic and environmental sustainability of biorefineries. The knowledge gained and technological advances made through research on cellulosic ethanol have accelerated advances for third-generation biofuels. The previous research on cellulosic feedstock supply, pretreatment, and logistics has helped to improve feedstocks for third-generation biofuels. Similarly, technologies to break down biomass for further processing can be applied to the processing of advanced biofuels. Research on subsequent processing of intermediates and by-products into high-value biological products and chemicals is critical to improving the profitability of third-generation biorefineries [7].

BIOMASS CONVERSION TECHNOLOGIES

Lignocellulose is the least expensive and most abundant form of biomass and is cheaper than crude oil on an energy basis. Technically it is possible to convert cellulosic materials and organic wastes into biofuels. Commercialization is limited because low-cost processing technologies that efficiently convert a large fraction of the lignocellulosic biomass energy into liquid fuels have not been developed to date. Thus, it is essential to continue focused research on processes to efficiently and economically convert lignocellulosic biomass into liquid fuels. Three basic routes for this conversion are 1) gasification of biomass to syngas ($CO + H_2$) and further conversion of syngas to liquid fuels, 2) fast pyrolysis or liquefaction of biomass to produce bio-oils followed by upgrading or blending for use as fuels, and 3) hydrolysis of biomass into sugar and lignin monomer units for conversion to targeted products [18]. The conversion technologies for producing liquid biofuels from biomass are outlined inFigure 1.

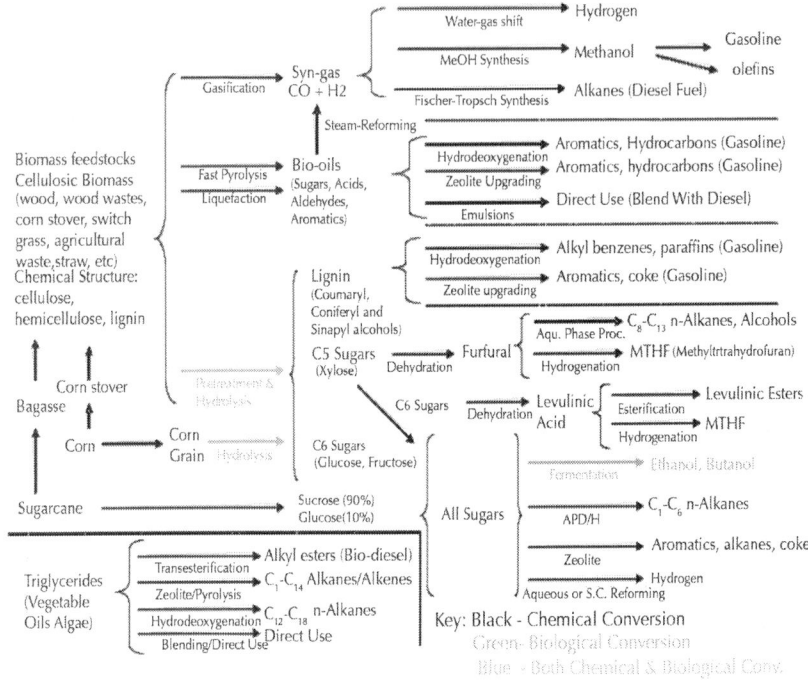

Figure 1: Production of Liquid Biofuels.

Liquid biofuel production processes use either biochemical or chemical catalysts. Biological catalysts, such as yeast to produce ethanol, are homogenous catalysts in the same liquid phase as the biomass feed. Chemical catalysts range from homogeneous acids to solid heterogeneous catalysts. (Reproduced from [9]; Adapted from [18])

Hydrolysis of Biomass and Fermentation of Sugars

Sugar streams for fermentation to biofuels can be obtained directly as sucrose from crops such as sugar cane, sweet sorghum, or energy beets. Alternatively, hydrolysis of starch crops yields glucose and hydrolysis of biomass yields glucose, xylose, and small amounts of other five carbon sugars. Lignocellulosic biomass is composed of cellulose, hemicelluloses, lignin, and proteins bound together in a complex structure that is recalcitrant to enzymatic hydrolysis by cellulase and hemicellulase enzymes. A pretreatment step is required to render the lignocellulosic biomass susceptible to the action of these hydrolytic enzymes [19]. Many factors, such as lignin content, crystallinity of cellulose, and particle size affect the digestibility of biomass. In theory, the ideal pretreatment process produces a disrupted, hydrated substrate that is easily hydrolyzed, but avoids formation of sugar degradation products and fermentation inhibitors. Various pretreatments have been proposed including comminution; delignification by white-rot fungi; chemical pretreatment with acids, alkali, organic solvents or ionic liquids; combined thermal/chemical pretreatment with steam, dilute acid, ammonia, or lime; organosolv-based fractionation treatments; and carbon dioxide or steam explosion [19 20]. Steam pretreatment, lime pretreatment, liquid hot water pretreatments and ammonia based pretreatments appear to have the most advantages for biorefinery applications. The main effects are dissolving hemicellulose and altering lignin structure, providing improved accessibility for hydrolytic enzymes [21]. Because the pretreatment process is typically the second most expensive unit cost in the conversion of lignocellulosic biomass, careful analysis and optimization of this process has the potential to significantly reduce biorefinery costs [19].

Genetic engineering of industrial microbes so they are capable of using lignocellulosic feedstocks of variable composition and without catabolite repression is crucial for development of third-generation biorefineries. Since most microbes possess carbon catabolite repression, mixed sugars derived from the lignocellulose are consumed sequentially, reducing the efficacy of the overall process. To overcome this barrier, microbes that exhibit the simultaneous consumption of mixed sugars have been isolated or developed and evaluated for the lignocellulosic biomass utilization. Specific strains of Escherichia coli, Saccharomyces cerevisiae, and Zymomonas mobilis have been engineered for simultaneous glucose and xylose utilization via mutagenesis or introduction of a xylose metabolic pathway. Other microbes, such as Lactobacillus brevis, L. buchneri, and Candida shehatae possess a less stringent carbon catabolite repression mechanism and show simultaneous consumption of glucose and xylose. Using these phenotypes, various integrated processes have been developed that incorporate both enzyme hydrolysis of lignocellulosic material and mixed sugar fermentation, thereby enabling greater productivity and fermentation efficacy [22, 23]. In addition to utilizing multiple substrates, these microbes must tolerate toxic substrate impurities such as by-products from feedstock pretreatment and hydrolysis, as well as potentially inhibitory products produced by the fermentation reaction itself.

At the present time, most recombinant strains for biorefinery applications are based on E. coli and S. cerevisiae because these organisms have been extensively studied and are relatively easy to engineer with well-developed genetic tools and established physiology [24]. However, the limited range of materials that can be fermented remains an obstacle to cost-effective bioethanol production in spite of substantial investments over the last 30 years in worldwide efforts to engineer xylose utilization in these strains [25, 26]. Although several genetically engineered strains of S. cerevisiae have been developed that will ferment xylose to ethanol [27-29], further optimization is needed. It will require the simultaneous expression at sufficiently high level of all the enzymes and proteins needed to allow industrial yeast strains to efficiently metabolize pentose as well as hexose sugars under anaerobic conditions. In addition, for cost-effective industrial ethanol production from biomass it will be necessary to express the enzymes required to saccharify the lignocellulosic feedstocks that are the source of hexose and pentose sugars. Genes considered necessary for complete fermentation of xylose and arabinose, the two major pentose sugar constituents of lignocellulosic biomass, include those encoding xylose isomerase (XI), xylulokinase (XKS), arabinose A, arabinose B, and arabinose D [27,29,30]. These genes may be obtained from microorganisms naturally capable of metabolizing these sugars. Saccharification of lignocellulosic feedstocks also requires utilization of hydrolytic enzymes including cellulases and hemicellulases after initial chemical pre-treatment [31, 32]. The cost-effectiveness of the fuel ethanol fermentation process could be further enhanced by obtaining high-value co-products and by-products from the process, such as monomers for polymer production and commercially important proteins and peptides. Genes for these proteins and peptides can be mutagenized, placed in an expression system capable of producing high levels of functional proteins or peptides, and screened in high throughput to optimize desired characteristics. [33-36].

Although extensive efforts have been made to engineer E. coli and S. cerevisiae to use both hexose and pentose sugars [22, 37-40], substrate versatility remains a significant issue. Therefore, other strains are being investigated. For example, Clostridia strains possess exceptional substrate diversity, utilizing simple and complex carbohydrates, such as cellulose, as well as CO_2/H_2 or CO. In addition, they contain a wide variety of extracellular enzymes to degrade large biological molecules (cellulose, xylans, proteins, lipids) and produce a broad spectrum of chemicals that can be used as precursors to, or directly as, biofuels and industrial chemicals [22,41-43]. Clostridia are found in virtually all anaerobic habitats containing organic matter and thus have developed the ability to ferment mono- and disaccharides as well as complex polysaccharides like cellulose and hemicellulose, which makes them ideal platforms for fermenting biomass feedstocks. They produce metabolites such as butyrate, acetate, lactate, caproate, butanol, acetone, acetoin, ethanol, and many more [24]. Clostridia are anaerobic microbes producing a large array of metabolites by utilizing simple and complex carbohydrates, such as cellulose, as well as CO_2/H_2 or CO. Efforts are underway to develop genetic and genomic tools for these microbes, and recent efforts to metabolically engineer Clostridiademonstrate their potential for biofuel and biorefinery applications. Pathway engineering to combine established substrate-utilization programs with desirable metabolic programs could lead to modular design of strains suitable for many applications. Engineering complex phenotypes--aerotolerance, abolished sporulation, and tolerance to toxic chemicals--could lead to superior bioprocessing strains [24].

Another significant challenge in using wild-type microbes to convert feedstocks into advanced biofuels is to overcome their endogenous regulation of biofuel-producing pathways that limits yields and productivities. Reconstruction of advanced biofuel pathways in specific heterologous hosts has worked, but use of data-driven and synthetic-biology approaches could further optimize both the host and the pathways to maximize biofuel production from a broader range of substrates. Research will undoubtedly lead to the creation of additional metabolic engineering techniques that can be used to improve pathway flux, and to additional synthetic-biology approaches to optimize microbial hosts for successful commercialization of third-generation biofuels [44, 45].

Gasification

Ethanol and third-generation biofuels can also be produced by a process called gasification. Gasification systems use high temperatures and a low-oxygen environment to convert biomass into synthesis gas, a mixture of hydrogen and carbon monoxide. The synthesis gas, or "syngas," can then be chemically converted into biofuels using the Fischer-Tropsch process or newer advanced catalytic processes. For example, Schmidt and co-workers [46] have combined the three reactions of older thermal gasification processes into a single, small reactor in which gasification takes place over a catalyst to directly produce third-generation biofuels. Synthesis gas can also be microbially converted into biofuels, although the low product tolerance of the microbes has been a limiting factor.

Pyrolysis

Biomass pyrolysis is the thermal depolymerization of biomass at moderate temperatures in the absence of added oxygen. The biomass is initially converted to a mixture of liquid (pyrolysis oil), solid (biochar), and gaseous fractions that can be used in the production of fuels and chemicals. Fractionation of the pyrolysis oil results in various qualities of oil needed for further upgrading into fine chemicals, automotive fuels, and energy [13]. An updated pyrolysis approach developed by Huber and co-workers uses catalysts to convert biomass into high-octane gasoline-range aromatics in a single step [47, 48]. Pyrolysis conditions can be adjusted to optimize the production or chemical composition of a given fraction [49, 50].

Transesterification of Oils and Fats to Biodiesel

Esterification and transesterification have been used for more than a decade to produce biodiesel from plant or animal-derived lipids. Any feedstock that contains free fatty acids and/or triglycerides such as vegetable oils, waste oils, animal fats, and waste greases can be converted to biodiesel. However, the product must meet stringent quality standards [51]. Consequently, fuel standards such as ASTM D6751 in the United States and EN 14214 in Europe have been implemented to ensure that only high quality biodiesel reaches the marketplace. Similar standards have been adopted elsewhere. Acquisition of refined commodity oils such as soybean oil may account for more than 80% of the cost to produce biodiesel. As a consequence, inexpensive, non-food feedstocks are critically important to improve process economics. Such low-value feedstocks often contain contaminants such as moisture and free fatty acids that render them incompatible with simple, homogeneous, alkaline-catalyzed transesterification. In such cases, alternative methods such as heterogeneous acid catalysis are needed for efficient conversion to biodiesel. An economic comparison between different conversion methods utilizing low-value feedstocks revealed that the heterogeneous acid catalyst process had the lowest total capital investment and manufacturing cost. For biodiesel to expand and mature in the market a number of key issues must be addressed, such as improving production efficiency through development of cost-effective catalysts capable of converting low-quality feedstocks into biodiesel, enhancing availability of low cost feedstocks, and managing agricultural land and water. In addition, biodiesel will require continuous improvement in producing cleaner emissions and reducing environmental impacts, although some of these issues are addressed by exhaust after-treatment technologies such as exhaust gas recirculation (EGR) and selective catalytic reduction (SCR) [52].

Technologies to Convert Carbohydrates into Mixed Hydrocarbons

Dissolved sugars can also be converted into hydrocarbons through routes that resemble petroleum processing more than fermentation. Researchers have developed several technologies in which dissolved sugars react in the presence of solid-phase catalysts under carefully controlled conditions (to avoid unwanted by-products) to produce targeted ranges of hydrocarbons for use as fuels or chemical feedstocks [53, 54]. Genetically altered microorganisms have also been developed that ferment sugars into hydrocarbons instead of alcohols [55]. Genes were isolated that, when expressed in Escherichia coli, produce alkanes, the primary hydrocarbon components of gasoline, diesel and jet fuel. If commercialized, this single step conversion of sugar to fuel-grade alkanes by a recombinant microorganism would lower the cost of producing drop-in hydrocarbon fuels that are low carbon, sustainable, and compatible with the existing fuel distribution infrastructure. The process does not require elevated temperatures, high pressure, toxic catalysts, or complex operations. The recombinantE. Coli secretes the hydrocarbons from the cell, so it is not necessary to rupture the cell. In addition, because the hydrocarbons are insoluble in water, they will form a separate organic phase and the microbes are not poisoned by the accumulating fermentation product as occurs with alcohol [56].

Renewable Diesel and Gasoline

Traditional petrochemical refinery operations such as catalytic cracking and hydro processing (HP) can be applied with modifications to biological feedstocks such as bio-oils and triglycerides to produce non-ester renewable hydrocarbon gasoline and diesel fuels [57]. Fluid catalytic cracking (FCC) may be viewed as continuous pyrolysis (400+ °C) at atmospheric pressure in the presence of heterogeneous acid catalyst and is used to produce gasoline. During FCC, long-chain hydrocarbons are cracked into smaller molecules, most of which fall within the gasoline boiling range. Among the reactions that occur (both parallel and consecutive) during FCC include protolytic cracking, dehydrogenation, decarboxylation, decarbonylation, scission, cyclization, oligomerization, coking, and hydride transfer [58]. Zeolite-based catalysts have been used for industrial FCC for over 40 years. These catalysts contain a faujasite-type zeolite as the major active component, which is embedded in a silica and/or alumina matrix. This matrix acts a binder, serves as a diluting medium, provides large mesopores for diffusion to the active zeolite crystal and facilitates heat transfer during cracking reactions [58, 59].

HP utilizes both high temperature and pressure along with hydrogen and heterogeneous catalysts to remove heteroatoms (such as oxygen, sulfur, nitrogen, and metals) and unsaturation and yields principally diesel and jet fuels [60]. Sulfur in diesel fuels is limited to 10 ppm in Europe (EN 590) and 15 ppm in the United States (ASTM D975). Consequently, an important process that occurs during HP is hydrodesulfurization (HDS), as crude oil may contain up to 2% (by weight) of sulfur [61]. During HDS, chemically bound sulfur is eliminated as H_2S [62]. A two-stage HDS unit is typically employed whereby a $Co\text{-}Mo/Al_2O_3$ catalyst is first used followed by Ni-Mo/

Al_2O_3 (or $Ni-W/Al_2O_3$). HDS over Co-Mo primarily removes sulfur from aliphatic hydrocarbons. The more active Ni-Mo facilitates hydrogenation of aromatic sulfur as well as saturation of aromatic hydrocarbons. The two-stage deep desulfurization needed to produce ultra-low sulfur (<15 ppm S) diesel (ULSD) fuel has caused changes to the chemical composition of ULSD relative to its low sulfur (<500 ppm) diesel (LSD) fuel predecessor, which was historically prepared in only one HDS stage utilizing a Co-Mo catalyst. The resulting ULSD fuel contains fewer aromatics and heteroatom-containing hydrocarbons relative to LSD [60, 63, 64]. A drawback to applying existing commercial HP catalysts to biological feedstocks is that the lack of heteroatoms (especially sulfur and nitrogen) in the biological feedstocks causes the catalysts to rapidly lose activity. Therefore, to maintain catalyst activity the feedstocks must be doped with dimethyl disulfide (DMDS) and tetrabutylamine (TBA) [65]. This is of course an undesirable solution, especially in an integrated biorefinery setting where substances such as DMDS and TBA represent non-biological inputs. Recently, zeolites such as Pt/H-ZSM5 have shown promise as HP catalysts for triglycerides such as jatropha oil to yield C15-C18 hydrocarbons directly [66].

Both FCC and HP require atmospheric distillation post-production to yield fuels with the appropriate boiling ranges. Current technology for production of biofuels from these processes involves comingling of biological feedstocks with traditional petroleum feeds to produce a hydrocarbon fuel whose carbons are primarily derived from petroleum [58, 67]. Direct production of renewable gasoline and diesel fuels from FCC and HP without comingling requires development of new catalysts with higher tolerance of biological feedstocks. If such processes are to be performed independently from the classic petroleum refinery, then process economics improvements are needed to reduce production costs. Stand-alone units or those integrated into a multi-product integrated biorefinery may become more economically competitive as scale of production increases. Demonstration facilities utilizing a patented Universal Oil Products (UOP) process for conversion of triglycerides to renewable diesel using HP have been reported [68]. Important advantages of renewable hydrocarbon gasoline and diesel fuels relative to ethanol and biodiesel are that the former are indistinguishable from their petroleum counterparts, they have greater storage and oxidative stability, they can be transported via existing pipeline infrastructure, they are not hygroscopic, and they can be blended in any proportion with conventional petroleum-derived fuels [57, 68].

Algal Biofuel Production

As a result of the interest in developing additional biomass feedstocks, research into the production of liquid transportation fuels from microalgae, is reemerging. These microorganisms use the sun's energy to combine carbon dioxide with water to create biomass more efficiently and rapidly than terrestrial plants. Oil-rich microalgae strains are capable of producing the feedstock for a number of transportation fuels—biodiesel, "green" diesel and gasoline, and jet fuel—while mitigating the effects of carbon dioxide released from sources such as power plants [56]. Research and demonstration programs are being conducted worldwide to develop the technology needed to commercialize algal lipid production. Algae store chemical energy in the form of biological oils, such as neutral lipids or triglycerides, when subjected to stresses such as nutrient deprivation [69]. The oil can be extracted from the organisms and converted into biodiesel by transesterification with short-chain alcohols such as methanol or ethanol [70] or by catalytic deoxygenation/hydrogenation of fatty acids into linear hydrocarbons [71]. Another approach is to engineer algae or cyanobacteria to directly produce fuel compounds, instead of oil [72]. These biofuel replacements for gasoline, diesel, and jet fuel will give higher fuel efficiency than ethanol and biodiesel, and will work in existing engines and fuel distribution networks.

Anaerobic Digestion

Anaerobic digestion is the use of microorganisms in oxygen-free environments to convert organic material into methane and carbon dioxide. This biogas is currently produced from crop residues, food scraps, and manure. Anaerobic digestion is also frequently used in the treatment of wastewater and to reduce emissions from landfills. When functioning well, the bacteria can convert about 90% of the biomass feedstock into biogas (containing about 55% methane), which is a readily useable energy source (combusted for thermal energy and/

or used to power electrical generators). Solid remnants of the original biomass input, which are left after the digestion process, are typically used as a fertilizer (although it should be chemically assessed for toxicity and growth-inhibiting factors). Biogas production can be part of sustainable biochemicals and biofuels-based biorefinery platform, since it can derive value from low-value by-product or waste streams. Value can be increased by optimizing methane yield and economic efficiency of biogas production [13, 73].

PLATFORM INTEGRATION

A key factor in achieving a successful biomass-based economy will be the development of biorefinery systems allowing efficient and cost-effective processing of biological feedstocks into a range of bio-based products that integrate seamlessly into the existing infrastructure [12]. Within the operation of a biorefinery, significant opportunities exist to produce commodity and high-value chemicals in conjunction with the production of bioenergy and biofuels [13]. From a technical point of view, almost all industrial materials made from petroleum resources could be replaced by their bio-based counterparts. However, the bio-based products must be no more expensive, perform at least as well, and have lower environmental impacts. Production of these materials in integrated multi-purpose biorefineries offers the most cost-effective approach to achieving this goal. In general, biofuels, both conventional and advanced, can be produced sustainably in the future only with a significant reduction in costs, which potentially can be accomplished by integrated co-production of value-added products [13]. By producing multiple products, a biorefinery can take better advantage of the intrinsic chemical complexity of biomass components and intermediates to maximize the value derived from the biomass feedstock. A biorefinery might, for example, produce one or several low-volume, but high-value, chemical products as well as a low-value, but high-volume liquid transportation fuel, while generating electricity and process heat for its own use and perhaps enough for sale of electricity. The high-value products enhance profitability, while the high-volume fuel provides economies of scale and helps meet national energy needs, and the power production reduces costs and avoids greenhouse-gas emissions [74].

The development of promising and innovative bio-based chemicals and polymers depends on the feedstock and the resulting process stream or platform [13]. These platforms include: 1) single carbon molecules such as biogas or syngas that can give rise to methanol, dimethylether, ethanol, or Fischer-Tropsch diesel, 2) six carbon carbohydrates from starch, sucrose, or cellulose and mixed streams with five and six carbon carbohydrates from hemicelluloses that can potentially produce succinic, itaconic, adipic, glutamic, and aspartic acids, and 3-hydroxypropionic acid or aldehyde, isoprene, and farnesene, plus more from the chemical processing of glucose, 3) lignin whose structure suggests it could form supramolecular materials and aromatic chemicals, 4) oils (plant-based or algal) that produce glycerol for propylene glycol, epichlorohydrin, and 1,3-propanediol and that are being developed for manufacture of polymers (polyurethanes, polyamides, and epoxy resins), 5) organic solutions from grasses such as clover or alfalfa that contain proteins, amino acids, carbohydrates, and 6) pyrolytic liquids that are expected to produce phenols, organic acids, furfural, hydroxymethyl furfural, and levoglucosan [13]. The continued growth in biobased chemicals and materials will give impetus to the cost-effective production of biofuels in a biorefinery setting. Given the expanding range of feedstocks, platform technologies, and co-products, numerous combinations for third-generation biorefineries are possible [12].

Multipurpose Biorefinery Based On Starch and Cellulosic Biofuel Platforms

Multipurpose advanced biorefineries that hydrotreat plants oils and animal fats into renewable fuels can be combined with cellulosic ethanol production via fermentation by optimized yeast strains. Concomitant production of high-value bio-based products and advanced animal feeds would be accomplished from by-products from the facility. Cellulosic n-butanol could be produced from mutant strains of Clostridium acetobutylicum and C. beijerinckii developed to tolerate high concentrations of butanol. Furthermore, engineered algae could be used for urea and ammonia production for emissions control technologies for diesel-operated trucks, for fertilizer, and for production of sucrose and algal oils. The multipurpose biorefinery would require construction of support areas, including research and pilot facilities, a strain collection building, and distillation and post-fermentation processing facilities. Unusable waste streams would be utilized as pyrolysis or biomethane feedstocks to power the biorefinery. Ideally, the biorefinery would produce third-generation biofuels that would be distributed through existing infrastructure. A high-volume animal feed station could be established for distribution to local farms. A possible arrangement of the components of a multipurpose biorefinery combining several of these platforms is shown in Figure 2.

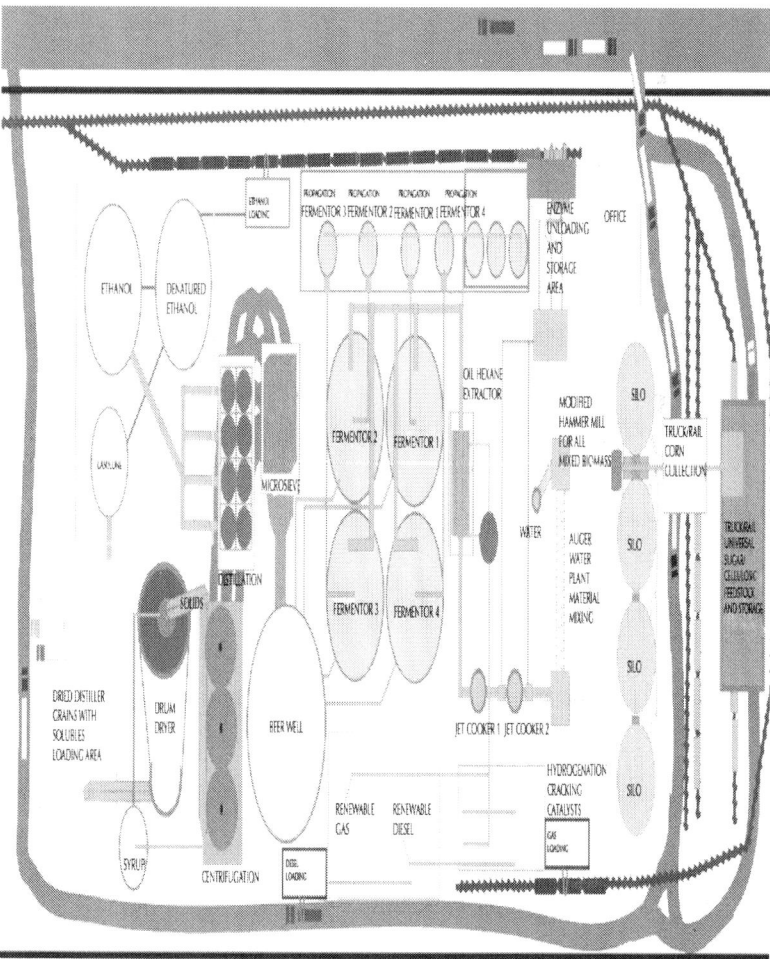

Figure 2: Multipurpose Bio refinery Combining Starch and Cellulosic Biofuel Platforms.

Cellulosic ethanol or n-butanol production by optimized yeast or Clostridium strains is combined with an existing starch ethanol production facility. Concomitant production of high-value bio-based products and advanced animal feed is also accomplished from by-products at the integrated facility.

Multipurpose Biorefinery Based on Sugar and Syngas Platforms

Another example of a multipurpose biorefinery is built on two different platforms, sugar and syngas, to promote different product slates [74]. The sugar platform is based on biochemical conversion processes and focused on fermentation of sugars extracted from biomass feedstocks. The syngas platform is based on thermochemical conversion processes and focused on gasification of biomass feedstocks and by-products from conversion processes. A diagram of this integrated biorefinery is shown in Figure 3.

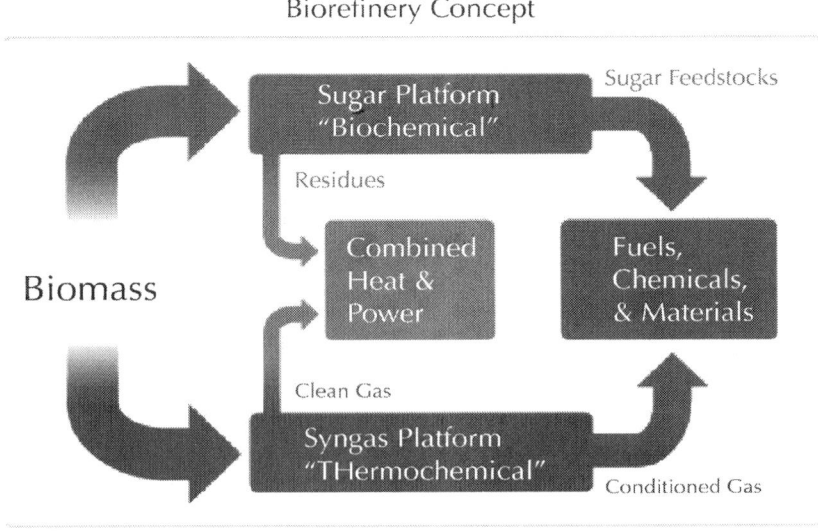

Figure 3: Integration of Sugar and Syngas Platforms.

The sugar platform uses biocatalysts such as yeast to produce liquid biofuels from fermentation of sugars. The syngas platform uses high temperatures and a low-oxygen environment to convert biomass into synthesis gas that can then be chemically converted into biofuels. (Reproduced from [74])

Projects developing this concept are focused on new technologies for integrating the production of biomass-derived fuels and other products in a single facility. The emphasis is on using new or improved processes to derive products such as ethanol, 1, 3 propanediol, polylactic acid, isosorbide, and various other chemicals. These projects include facilities to develop and validate process technology and sustainable agricultural systems to economically produce sugars and chemicals such as lactic acid and ethanol. These facilities will also develop 1) a novel biomass technology to utilize distiller's grain and corn stover blends to achieve significantly higher ethanol yields while maintaining the protein feed value, 2) a biobased technology to produce a wide variety of products based on 3-hydroxypropionic acid, produced by fermentation of carbohydrates, and 3) an integrated process for recovery of the hemicellulose, protein, and oil components from corn fiber for conversion into value-added products [74].

Conversion of Biomass Sugars to Hydrocarbon Chemicals and Fuels

Figure 4 shows an example of an integrated biorefinery that produces third-generation biofuels via chemical catalysis to convert plant-based sugars into a full range of hydrocarbon products identical to those made from petroleum, including gasoline, diesel, jet fuel, and chemicals for plastics and fibers. The biofuels are drop-in replacements that enable full utilization of existing processing, pipeline, storage, and transportation infrastructure. The process converts aqueous carbohydrate solutions into mixtures of hydrocarbons and has been demonstrated with conventional sugars obtained from existing sugar sources (corn wet mills, sugarcane mills, etc.) as well as with a wide variety of cellulosic biomass from nonfood sources. The process can accommodate a broad range of compounds derived from biomass, including C5/C6 sugars, polysaccharides, organic acids, furfurals and other degradation products generated from the deconstruction of biomass. The soluble carbohydrate streams are processed through the aqueous phase reforming (APR) step. The APR step utilizes heterogeneous catalysts at moderate temperatures and pressures to reduce the oxygen content of the carbohydrate feedstock. The reactions in the APR step include: (1) reforming to generate hydrogen, (2) dehydrogenation of alcohols / hydrogenation of carbonyls, (3) deoxygenation, (4) hydrogenolysis, and (5) cyclization [75].

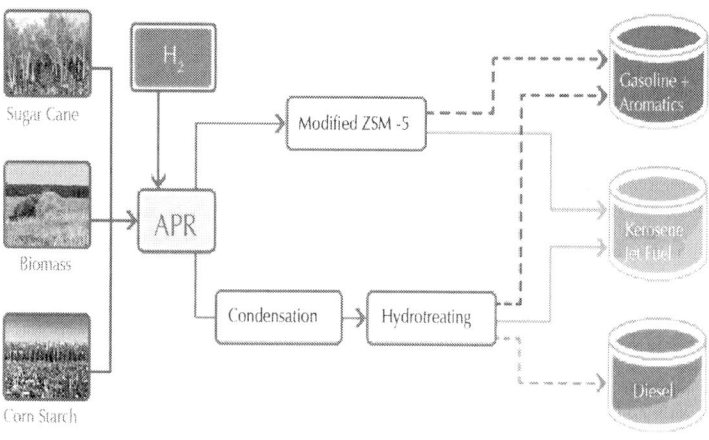

Figure 4: Biomass Conversion to Hydrocarbon Chemicals and Fuels.

Catalytic chemistry converts plant-based sugars into a full range of hydrocarbon products identical to those made from petroleum, including gasoline, diesel, jet fuel, and chemicals for plastics and fibers. (Reproduced from [75])

An advantage to this process is the ability to produce hydrogen in-situ from the carbohydrate feedstock or utilize other sources of hydrogen such as natural gas for higher yields and lower costs. The product from the APR step is a mixture of chemical intermediates including alcohols, ketones, acids, furans, paraffins and other oxygenated hydrocarbons. Once these intermediate compounds are formed they can undergo further catalytic processing to generate a cost-effective mixture of nonoxygenated hydrocarbons. A modified ZSM-5 catalyst is used to convert the chemical intermediates from the APR step to a high-octane gasoline blendstock that has a high aromatic content similar to a petroleum-derived reformate stream. The chemical intermediates from the APR step can also be converted into distillate range hydrocarbon components through a condensation step followed by conventional hydrotreating [75].

Integrated Forest Biorefinery

An integrated forest biorefinery is diagrammed in Figure 5. In ths example, a facility processing biomass to syngas to biofuels is integrated into a pulp and paper mill [76]. The biomass feedstocks for this biorefinery are forest and agricultural residuals. The biomass is dried and sized prior to gasification and then fed into the fluidized bed stream reformer through a screw feed system. It is gasified to produce syngas with the correct hydrogen to carbon ratio for gas-to-liquids processing. The syngas passes through a conventional heat recovery and gas clean-up train. The gas-to-liquids technology is the Fischer-Tropsch (FT) process, a mature technology. In the reactor the syngas, under pressure and temperature, with the FT catalyst is converted to straight chain hydrocarbons that range from light gases to heavy waxes, including gasoline, naphtha, and diesel [76].

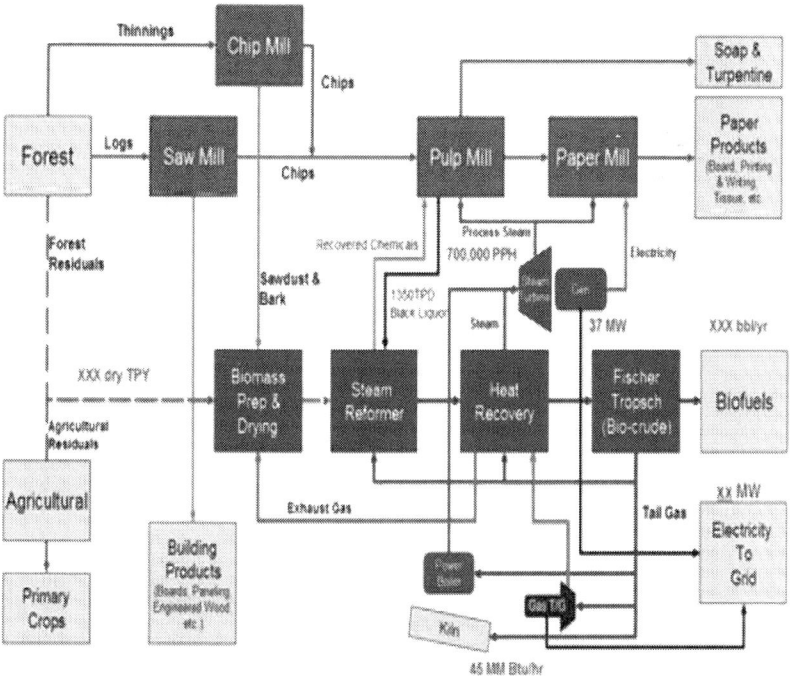

Figure 5: Integration of Paper Mill with Biomass Gasification for Biofuels.

Biomass feedstocks from an existing pulp and paper mill are used to create new revenue streams by producing high-value products such as biofuels and biochemical and at the same time improving the efficiency of the core paper-making operations (Reproduced from [76])

The gasification process is ideal for use in a forest products biorefinery because it is configured for high-performance integration with pulp and paper facilities and is capable of handling a wide variety of cellulosic feedstocks, including mill by-products (spent liquor), woodchips, forest residuals, agricultural wastes, and energy crops. The syngas can be used as a substitute for natural gas and fuel oil and as a feedstock for the production of value-added products such as biodiesel, ethanol, methanol, acetic acid, and other biochemicals [76].

CONCLUSIONS

A crucial step in developing a worldwide bio-industry is to establish integrated third-generation biorefineries that are capable of efficiently converting a broad range of biomass feedstocks into commercially viable biofuels, biopower, commodity and high-value chemicals, and other bioproducts. Integrated biorefineries are similar to conventional refineries in that they produce a range of products to optimize both the use of the feedstock and production economics. Third-generation biorefineries will use novel technologies and diverse biomass feedstocks - requiring significant investments in research, development, and deployment to reduce costs and improve performance to achieve competitiveness with petroleum fuels. These biorefineries will employ various combinations of feedstock and conversion technologies to produce a variety of products, with the main focus on producing biofuels. Co-products can include chemicals (or other materials), animal feed, and heat and power. As pretreatment, conversion, and integration technologies continue to improve, sustainable third-generation biorefineries will become a reality.

REFERENCES

1.   R. D Perlack, B. J. U. S Stokes, billion-ton update: biomass supply for a bioenergy and bioproducts industry. A study sponsored by

U. S. Department of Energy, Energy Efficiency and Renewable Energy, Office of the Biomass Program 2011

2. A Raschka, M Carus, Industrial material use of biomass: basic data for Germany, Europe and the world 2012nova-Institute for Ecology and Innovation Study Report. Nova-Institute GmbH. Germany. 28pp.

3. International Energy AgencyIEA Bioenergy Task 42. Biorefineries: Adding Value to the Sustainable Utilisation of Biomass. September 2009

4. United States Department of EnergyBiomass Program. Integrated Biorefineries.http://www1.eere.energy.gov/biomass/integrated_ biorefineries.htmlaccessed 27 September 2012

5. Strategic Research Targets for 20202020Collaboration Initiative on BiorefineriesStar-COLIBRI. European Commission's 7th Framework Programme. European Biorefinery Joint Strategic Research Roadmap for 2020.

6. C. B Field, J. E Campbell, D. B Lobell, Biomass energy: the scale of the potential resource. Trends in Ecology and Evolution 20072326572doi:j.tree.2007.12.001

7. U.S. Department of Energy, Replacing the whole barrel to reduce U.S. dependence on oil Energy Efficiency and Renewable Energy, Report DOE/EE-0762. July 2012.

8. B Kamm, M Kamm, International biorefinery systems. Pure and Applied Chemistry 2007791119831997doi:pac200779111983

9. National Science FoundationChemical Bioengineering, Environmental, and Transport Division. Breaking the chemical and engineering barriers to lignocellulosic biofuels: next generation hydrocarbon biorefineries. Ed. George W. Huber. March 2008pp.

10. G Taylor, Biofuels and the biorefinery concept. Energy Policy 2008364406

11. L. R Lynd, M. S Laser, D Bransby, B. E Dale, B Davison, R Hamilton, M Himmel, M Keller, J. D Mcmillan, J Sheehan, C. E Wyman, How biotech can transform biofuels. Nature Biotechnology 2008262169172

12. A. C Kokossis, A Yang, On the use of systems technologies and a systematic approach for the synthesis and the design of

future biorefineries. Computers and Chemical Engineering 201034913971405

13. International Energy AgencyIEA Bioenergy Task 42. Bio-based Chemicals: Value Added Products from Biorefineries. February 2012

14. Advanced Biofuels USATruly sustainable renewable future. April 17, 2012http://advancedbiofuelsusa.info/truly-sustainable-renewable-futureaccessed 29 September 2012)

15. S. N Nigam, A Singh, Production of liquid biofuels from renewable resources. Progress in Energy and Combustion Science 20113752

16. R. D Perlack, L. L Wright, A. F Turhollow, R. L Graham, B. J Stokes, D. C Erbach, Biomass as a feedstock for a bioenergy and bioproducts industry: the technical feasibility of a billion-ton annual supply. A joint study sponsored by U. S. Department of Energy and U. S. Department of Agriculture 2005

17. Federal RegisterMarch 26, 20101467014904Final Rule 40 CFR 80; 75 FR 14670; Regulation of fuels and fuel additives: changes to renewable fuels program.

18. G. W Huber, S Iborra, A Corma, Synthesis of transportation fuels from biomass: chemistry, catalysts, and engineering. Chemical Reviews 20061064044

19. V. B Agbor, N Cicek, R Sparling, A Berlin, D. B Levin, Biomass pretreatment: fundamentals toward application. Biotechnology Advances 201129675

20. D Kumar, G. S Murthy, Impact of pretreatment and downstream processing technologies on economics and energy in cellulosic ethanol production. Biotechnology and Biofuels 2011DOI:

21. A. T Hendriks, G Zeeman, Pretreatments to enhance the digestibility of lignocellulosic biomass. Bioresource Technology 200910011018DOI:j.biortech.2008.05.027

22. J. H Kim, D. E Block, D. A Mills, Simultaneous consumption of pentose and hexose sugars: an optimal microbial phenotype for efficient fermentation of lignocellulosic biomass. Applied Microbiology and Biotechnology 2010881077

23. Den Haan RMcBride JE, La Grange DC, Lynd LR, Van Zyl WH. Functional expression of cellobiohydrolases in *Saccharomyces*

cerevisiae towards one-step conversion of cellulose to ethanol. Enzyme and Microbial Technology 2007401291

24. B. P Tracy, S. W Jones, A. G Fast, D. C Indurthi, E. T Papoutsakis, Clostridia: the importance of their exceptional substrate and metabolite diversity for biofuel and biorefinery applications. Current Opinion in Biotechnology 201223364DOI:j. copbio.2011.10.008

25. A. E Farrell, R. J Plevin, B. T Turner, A. D Jones, O Hare, M Kammen, DM. Ethanol can contribute to energy and environmental goals. Science 20063115760506508

26. B. C Saha, Hemicellulose bioconversion. Journal of Industrial Microbiology and Biotechnology 2003305279291

27. K Karhumaa, Garcia Sanchez R, Hahn-Hägerdal B, Gorwa-Grauslund M-F. Comparison of the xylose reductase-xylitol dehydrogenase and the xylose isomerase pathways for xylose fermentation by recombinant *Saccharomyces cerevisiae*. Microbial Cell Factories 2007

28. M Sedlak, Ho NWY. Production of ethanol from cellulosic biomass hydrolysates using genetically engineered *Saccharomyces* yeast capable of cofermenting glucose and xylose. Applied Biochemistry and Biotechnology 2004

29. H. W Wisselink, M. J Toirkens, Berriel M del RF, Winkler AA, van Dijken JP, Pronk JT, van Maris AJA. Engineering of *Saccharomyces cerevisiae* for efficient anaerobic alcoholic fermentation of L-arabinose. Applied and Environmental Microbiology 2007731548814891

30. Y. S Jin, H Ni, J. M Laplaza, T. W Jeffries, Optimal growth and ethanol production from xylose by recombinant *Saccharomyces cerevisiae* require moderate D-xylulokinase activity. Applied and Environmental Microbiology 2003691495503

31. A Rudolf, H Baudel, G Zacchi, B Hahn-hägerdal, G Lidén, Simultaneous saccharification and fermentation of steam-pretreated bagasse using *Saccharomyces cerevisiae* TMB3400 and *Pichia stipitis* CBS6054. Biotechnology and Bioengineering 2007994783790

32. B. C Saha, L. B Iten, M. A Cotta, Y. V Wu, Dilute acid pretreatment, enzymatic saccharification, and fermentation of rice hulls to ethanol. Biotechnology Progress 2005213816822

33. S. R Hughes, S. B Riedmuller, J. A Mertens, X-L Li, K. M Bischoff, M. A Cotta, P. J Farrelly, Development of a liquid handler component for a plasmid-based functional proteomic robotic workcell. Journal of the Association for Laboratory Automation 2005105287300

34. S. R Hughes, S. B Riedmuller, J. A Mertens, X-L Li, K. M Bischoff, N Qureshi, M. A Cotta, P. J Farrelly, High-throughput screening of cellulase F mutants from multiplexed plasmid sets using an automated plate assay on a functional proteomic robotic workcell. Proteome Science 2006

35. S. R Hughes, P. F Dowd, R. E Hector, S. B Riedmuller, S Bartolett, J. A Mertens, N Qureshi, S Liu, K. M Bischoff, X-L Li, J. S Jackson, Jr., Sterner, D., Panavas, T., Rich, J.O., Farrelly, P.J., Butt, T.R., Cotta, M.A. Cost-effective high-throughput fully automated construction of a multiplex library of mutagenized open reading frames for an insecticidal peptide using a plasmid-based functional proteomic robotic workcell with an improved vacuum system. Journal of the Association for Laboratory Automation 2007124202212

36. S. R Hughes, P. F Dowd, R. E Hector, T Panavas, D. E Sterner, N Qureshi, K. M Bischoff, S. S Bang, J. A Mertens, E. T Johnson, X-L Li, Jackson JS Jr, Caughey RJ, Riedmuller SB, Bartolett S, Liu S, Rich JO, Farrelly PJ, Butt TR, LaBaer J, Cotta MA. Lycotoxin-1 insecticidal peptide optimized by amino scanning mutagenesis and expressed as a coproduct in an ethanologenic *Saccharomyces cerevisiae* strain. Journal of Peptide Science 200814910391050

37. C Laluce, Schenberg ACG, Gallardo JCM, Coradello LFC, Pombeiro-Sponchiado SR Advances and developments in strategies to improve strains of*Saccharomyces cerevisiae* and processes to obtain the lignocellulosic ethanol−a review Applied Biochemistry and Biotechnology 2012166819081926doi :s12010-012-9619-6

38. S. R Hughes, D. E Sterner, K. M Bischoff, R. E Hector, P. F Dowd, N Qureshi, S. S Bang, N Grynaviski, T Chakrabarty, E. T Johnson, B. S Dien, J. A Mertens, R. J Caughey, S Liu, T. R Butt, LaBaer J, Cotta MA, Rich JO. Engineered *Saccharomyces cerevisiae* strain for improved xylose utilization with a three-plasmid SUMO yeast expression system. Plasmid 20086112238

39. Van Maris AJAWinkler AA, Kuyper M, de Laat WTAM, van Dijken JP Pronk JT. Development of efficient xylose fermentation in *Saccharomyces cerevisiae*: xylose isomerase as a key component. Advances in Biochemical Engineering/Biotechnology 2007108179

40. B Hahn-hägerdal, K Karhumaa, C Fonseca, I Spencer-martins, M. F Gorwa-grauslund, 2007Towards industrial pentose-fermenting yeast strains. Applied Microbiology and Biotechnology 2007;745937953

41. T. C Ezeji, N Qureshi, H. P Blaschek, Bioproduction of butanol from biomass: from genes to bioreactors. Current Opinion in Biotechnology 200718220DOI:j.copbio.2007.04.002

42. N Qureshi, B. C Saha, M. A Cotta, Butanol production from wheat straw hydrolysate using *Clostridium beijerinckii*. Bioprocess and Biosystems Engineering 200730419

43. N Qureshi, X. L Li, S. R Hughes, B. C Saha, M.A Cotta, Butanol production from corn fiber xylan using *Clostridium acetobutylicum*. Biotechnology Progress 200622673

44. PP Peralta-Yahya, ,Zhzng, , Del Cardayre SB, Keasling JD. Microbial engineering for the production of advanced biofuels. Nature 2012; 488 320-328. doi:10.1038/nature11478

45. V García, J Päkkilä, H Ojamo, E Muurinen, R. L Keiski, Challenges in biobutanol production: how to improve the efficiency? Renewable and Sustainable Energy Reviews 2011152964980

46. P. J Dauenhauer, B. J Dreyer, N. J Degenstein, L. D Schmidt, Millisecond reforming of solid biomass for sustainable fuels. Angewandte Chemie- International Edition in English 2007463158645867

47. T. R Carlson, T. P Vispute, G. W Huber, Green gasoline by catalytic fast pyrolysis of solid biomass derived compounds. Chem Sus Chem 200815397400

48. G. W Huber, B. E Dale, Grassoline at the pump. Scientific American 200930115259

49. D Mohan, C. U Pittman, P. H Steele, Pyrolysis of wood/biomass for bio-oil: a critical review. Energy &Fuels 2006203848889

50. D. L Compton, M. A Jackson, D. J Mihalcik, C. A Mullen, A. A Boateng, Catalytic pyrolysis of oak via pyroprobe and bench

scale, packed bed pyrolysis reactors. Journal of Analytical and Applied Pyrolysis 201190174

51. G Knothe, Biodiesel and renewable diesel: a comparison. Progress in Energy and Combustion Science 201036364

52. J Janaun, N Ellis, Perspectives on biodiesel as a sustainable fuel. Renewable and Sustainable Energy Reviews 2010141312

53. E. L Kunkes, D. A Simonetti, R. M West, J. C Serrano-ruiz, C. A Gärtner, J. A Dumesic, Catalytic conversion of biomass to monofunctional hydrocarbons and targeted liquid-fuel classes. Science 20083225900417421DOI:10.1126/science.1159210

54. J. N Chheda, G. W Huber, J. A Dumesic, Liquid-phase catalytic processing of biomass-derived oxygenated hydrocarbons to fuels and chemicals. Angewandte Chemie- International Edition in English 2007463871647183

55. S. K Lee, H Chou, T. S Ham, T. S Lee, J. D Keasling, Metabolic engineering of microorganisms for biofuels production: from bugs to synthetic biology to fuels. Current Opinion Biotechnology 2008196556563

56. A Schirmer, M. A Rude, X Li, E Popova, Del Cardayre SB. Microbial biosynthesis of alkanes. Science 20103295991559562

57. G. W Huber, A Corma, Synergies between bio- and oil refineries for the production of fuels from biomass. Angewandte Chemie- International Edition in English 2007463871847201

58. M Al-sabawi, J Chen, S Ng, Fluid catalytic cracking of biomass-derived oils and their blends with petroleum feedstocks: a review. Energy & Fuels 2012265355

59. A.V Lavrenov, E. N Bogdanets, Y. A Chumachenko, V. A Likholobov, Catalytic processes for the production of hydrocarbon biofuels from oil and fatty raw materials: contemporary approaches. Catalysts in Industry 201133250259

60. D. C Elliot, Historical developments in hydroprocessing bio-oils. Energy & Fuels 2007211792

61. B Pawalec, R. N Navarro, J. M Campos-martin, Fierro JLG. Towards near zero-sulfur liquid fuels: a perspective review. Catalysis Science & Technology 2011112

62. A Stanislaus, A Marafi, M. S Rana, Recent advances in the science and technology of ultra-low sulfur diesel (ULSD) production. Catalysis Today 20101531

63. P. S Kulkarni, Afonso CAM. Deep desulfurization of diesel fuel using ionic liquids: current status and future challenges. Green Chemistry 2010121139

64. S. A Hanafi, M. S Mohamed, Recent trends in the cleaning of diesel fuels via desulfurization processes. Energy Sources Part A: Recovery, Utilization, and Environmental Effects 201133495

65. S Bezergianni, A Dimitriadis, A Kalogianni, K. G Knudsen, toward hydrotreating of waste cooking oil for biodiesel production. Effect of pressure, H2/oil ratio, and liquid hourly space velocity. Industrial & Engineering Chemistry Research 2011503874

66. K Murata, Y Liu, M Inaba, I Takahara, Production of synthetic diesel by hydrotreatment of jatropha oils using Pt-Re/H-ZSM-5 catalyst. Energy & Fuels 2010242404

67. J. A Melero, M. M Clavero, G Calleja, A Garcia, R Miravelles, T Galindo, Production of biofuels via the catalytic cracking of mixtures of crude vegetable oils and nonedible animal fats with vacuum gas oil. Energy & Fuels 201024707

68. T Kalnes, T Marker, D. R Shonnard, Green diesel: a second generation biofuel. International Journal of Chemical Reactor Engineering 2007Article A48.

69. R. H Wijffels, M. J Barbosa, An outlook on microalgal biofuels. Science 20103295993796799

70. Y Chisti, Biodiesel from microalgae. Biotechnology Advances 2007253294306

71. S Lestari, P Mäki-arvela, J Beltramini, G. Q Lu, D. Y Murzin, Transforming triglycerides and fatty acids into biofuels. Chem Sus Chem 200921211091119

72. R Radakovits, R. E Jinkerson, A Darzins, M. C Posewitz, Genetic engineering of algae for enhanced biofuel production. Eukaryotic Cell 201094486501doi:10.1128/EC.00364-09.PMCID: PMC2863401

73. Wisconsin Grasslands Bioenergy NetworkBioenergy conversion technologies. http://www.wgbn.wisc.edu/conversion/bioenergy-conversion-technologiesaccessed 27 September 2012

74. U.S. Department of Energy, Office of Energy Efficiency and Renewable Energy. National Renewable Energy Laboratory, Biomass Research. http://www.nrel.gov/biomass/biorefinery.html (accessed 14 October 2012)

75. VirentInc.Technology,   BioForming.   http://www.virent.
     comaccessed 14 October 2012

76. Connor,E. The integrated forest biorefinery: the pathway to our
     bio-future. Thermochem Recovery International, Inc. http://
     www.tri-inc.net/pdfs/Forest%20biorefinery%20white%20
     paper%20jan%2007%20without%20%20econ%202%20col.
     pdf (accessed 14 October 2012

Agro-Industrial Lignocellulosic Biomass a Key to Unlock the Future Bio-Energy: A Brief Review

Zahid Anwar[a, b], Gulfraz[b], and Muhammad Irshad[a, 1]

[a]Department of Biochemistry, NSMC, University of Gujrat, Pakistan
[b]PMAS Arid Agriculture University Rawalpindi, Rawalpindi, Pakistan

ABSTRACT

From the last several years, in serious consideration of the worldwide economic and environmental pollution issues there has been increasing research interest in the value of bio-sourced lignocellulosic biomass. Agro-industrial biomass comprised on lignocellulosic waste is an inexpensive, renewable, abundant and provides a unique natural resource for large-scale and cost-effective bio-energy collection. To expand the range of natural bio-resources the rapidly evolving tools of biotechnology can lower the conversion costs and also enhance target yield of the product of interest. In this background green biotechnology presents a promising approach to convert most of the solid agricultural wastes particularly lignocellulosic materials into liquid bio based energy-fuels. In fact, major advances have already been achieved to competitively position cellulosic ethanol with corn ethanol. The present summarized review work begins with an overview on the physico-chemical features and composition of agro-industrial biomass. The information is also given on the multi-step processing technologies of agro-industrial biomass to fuel ethanol followed by a brief summary of future considerations.

INTRODUCTION

Lignocellulosic materials are the most promising feedstock as natural and renewable resource essential to the functioning of modern industrial societies. A considerable amount of such materials as waste byproducts are being generated through agricultural practices mainly from various agro based industries (Pérez, Muñoz-Dorado de la Rubia, & Martínez, 2002). Sadly, much of the lignocellulosic biomass is often disposed of by burning, which is not restricted to developing countries alone. Recently lignocellulosic biomasses have gained increasing research interests and special importance because of their renewable nature (Asgher et al., 2013 and Ofori-Boateng and Lee, 2013). Therefore, the huge amounts of lignocellulosic biomass can potentially be converted into different high value products including bio-fuels, value added fine chemicals, and cheap energy sources for microbial fermentation and enzyme production (Asgher et al., 2013, Iqbal et al., 2013, Irshad et al., 2013 and Isroi et al., 2011).

PHYSICO-CHEMICAL CHARACTERISTICS OF LIGNOCELLULOSIC BIOMASS

All plant materials are mostly composed of three major units i.e., cellulose, hemicellulose and lignin. Lignocellulosic materials including agricultural wastes, forestry residues, grasses and woody materials have great potential for bio-fuel production. Typically, most of the agricultural lignocellulosic biomass is comprised of about 10–25% lignin, 20–30% hemicellulose, and 40–50% cellulose (Iqbal et al., 2011, Kumar et al., 2009 and Malherbe and Cloete, 2002). Cellulose is a major structural component of plant cell walls, which is responsible for mechanical strength while, hemicellulose macromolecules are often repeated polymers of pentoses and hexoses. Lignin contains three aromatic alcohols (coniferyl alcohol, sinapyl alcohol and p-coumaryl alcohol) produced through a biosynthetic process and forms a protective seal around the other two components i.e., cellulose and hemicelluloses (Fig. 1) (Calvo-Flores and Dobado, 2010, Jiang et al., 2010 and Menon and Rao, 2012). In general the composition of lignocellulose highly depends on its source whether it is derived from the hardwood, softwood, or grasses. Table 1 shows the typical chemical compositions of all these three components in various lignocellulosic materials that vary in composition due to the genetic variability among different sources (Bertero et al., 2012, Iqbal et al., 2013, John et al., 2006, Kumar et al., 2009, Malherbe and Cloete, 2002 and Prassad et al., 2007). To obtain a clear picture of the material, an analysis of the structure of each main component is made in the following section.

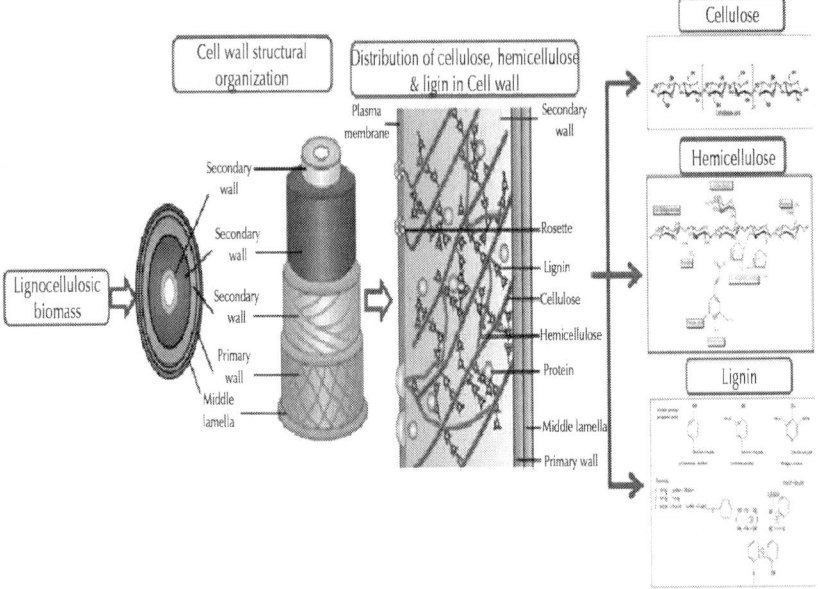

Figure 1: Diagrammatic illustration of the framework of lignocellulose; cellulose; hemicellulose and lignin.Adapted with permission from, Menon & Rao, 2012.

Table 1: Percent composition of lignocellulose components in various lignocellulosic materials

Lignocellulosic material	Lignin (%)	Hemicellulose (%)	Cellulose (%)	Reference[a]
Sugar cane bagasse	20	25	42	Kim and Day (2011)
Sweet sorghum	21	27	45	Kim and Day (2011)
Hardwood	18–25	24–40	40–55	Malherbe and Cloete (2002)
Softwood	25–35	25–35	45-50	Malherbe and Cloete (2002)
Corn cobs	15	35	45	Prassad et al. (2007)
Corn stover	19	26	38	Zhu, Lee, and Elander (2005)

Rice straw	18	24	32.1	Prassad et al. (2007)
Nut shells	30–40	25–30	25–30	Howard, Abotsi, Van Rensburg, and Howard (2003)
Newspaper	18–30	25–40	40–55	Howard et al. (2003)
Grasses	10–30	25–50	25–40	Malherbe and Cloete (2002)
Wheat straw	16–21	26–32	29–35	McKendry (2002)
Banana waste	14	14.8	13.2	John et al. (2006)
Bagasse	23.33	16.52	54.87	Guimarães, Frollini, Da Silva, Wypych, and Satyanarayana (2009)
Sponge gourd fibres	15.46	17.44	66.59	Guimarães et al. (2009)

[a]For detailed references please see Iqbal et al. (2013).

Adapted with permission from Iqbal et al., 2013.

Physical and Structural Properties of Cellulose

Cellulose is a highly stable polymer consisting of glucose and attached with linear chains up to 12,000 residues. It is majorly composed of (1, 4)-d-glucopyranose units, which are attached by -1, 4 linkages with an average molecular weight of around 100,000 (Himmel et al., 2007). Plant biomass contain 40–50% of cellulose molecules which are held together by intermolecular hydrogen bonds in native state, but they have a strong tendency to form intra-molecular and intermolecular hydrogen bonds and this tendency increases the rigidity of cellulose and make highly insoluble and highly resistant to most organic solvents. Naturally cellulose molecules are exists as bundles which aggregated together in the form of micro-fibrils order i.e., crystalline and amorphous regions (Iqbal et al., 2011 and Taherzadeh and Karimi, 2008). The chemical formula of cellulose is $(C_6H_{10}O_5)$ n and the structure of one chain of the polymer is presented inFig. 1.

Physical and Structural Properties of Hemicelluloses

Hemicellulose is the second most abundant heterogeneous polymers that mainly consist of glucuronoxylan, glucomannan and trace amounts of other polysaccharides. Grasses and straws contain arabinan, galactan and xylan, while mannan is a component of hardwood and softwood hemicellulose (Brigham, Adney, & Himmel, 1996). They are catalogued with sugar as a backbone, i.e., xylans, mannans and glucans, with xylans and mannans being the most common (Wyman et al., 2005). Galactans, arabinans and arabinogalactans are included in the hemicellulose group; however, they do not share the equatorial -1, 4 linked backbone structure. In hardwoods, glucuronoxylan (O-acetyl-4-O-methyl-glucurono- -d-xylan) is the predominant component. Xylospyranose is the backbone of the polymer and connected with -1, 4 linkages. Hemicellulosic biomass contains 25–35% of hemicellulose, with an average molecular weight of <30,000. Cellulose and hemicellulose binds tightly with non-covalent attractions to the surface of each cellulose micro-fibril. Hemicelluloses were originally believed to be intermediates in the biosynthesis of cellulose (Vercoe, Stack, Blackman, & Richardson, 2005).

Physical and Structural Properties of Lignin

Lignin is generally the most complex and smallest fraction, representing about 10–25% of the biomass by weight. It has a long-chain, heterogeneous polymer composed largely of phenyl-propane units most commonly linked by ether bonds. Lignin acts like a glue by filling the gap between and around the cellulose and hemicellulose complexion with the polymers. It is present in all plant biomass; therefore, it is considered byproduct or as a residue in bio-ethanol production process. Lignin is comprised of complex and large polymer of phenyl-propane, methoxy groups and non-carbohydrate poly phenolic substance, which bind cell walls component together (Hamelinck, Hooijdonk, & Faaij, 2005). Phenyl-propanes (3 carbons attached with 6 carbon atom rings) are main block of lignin. These phenyl-propanes denoted as 0, I, II methoxyl groups attached to rings give special structure I, II and III. These groups depend on the plant source which they are obtained. Structure I exist in plants (grasses) and structure II found in the wood (conifers) while structure III present in deciduous wood.

BIOTECHNOLOGICAL IMPORTANCE OF LIGNOCELLULOSIC BIOMASS

From the biotechnological point of view a wide variety of lignocellulosic biomass resources are available as potential candidate that are also convert able into high value bio-products like bio-ethanol/bio-fuels. From the last several years a considerable improvement from the green biotechnology related to lignocellulose biomass has appeared. The ever increasing costs of fossil fuels and their greenhouse effects are creating a core demand to explore alternative cheaper and eco-friendly bio-fuels resources as a strategy for reducing global warming (Asgher et al., 2013 and Iqbal et al., 2013). Environmental pollution, global warming, and the future of oil production are among major causes of public and private interests in natural bio based resources as an alternative or substitute for fossil fuel oil. One potential method for the low-cost production of bio-ethanol is to utilize the lignocellulosic or agro-industrial biomass because they contain carbohydrates that must be first converted into simple sugars (glucose) and then fermented into ethanol (Alonso et al., 2008, Balat and Balat, 2009 and Lin and Tanaka, 2006). Given this reality, nations around the world are investing in alternative sources of energy, including bio-ethanol. The conversion of lignocellulosic biomass into higher value added products like fine chemicals or bio-fuel production normally requires a multi-step processing that include (i) pre-treatment (mechanical, chemical, or biological etc) (ii) enzymatic hydrolysis (iii) fermentation process (Wyman, 1999 and Xiao et al., 2012). Fig. 2 illustrating a thermo-mechanical and biochemical processing of lignocellulosic biomass into various values added biotechnological products.

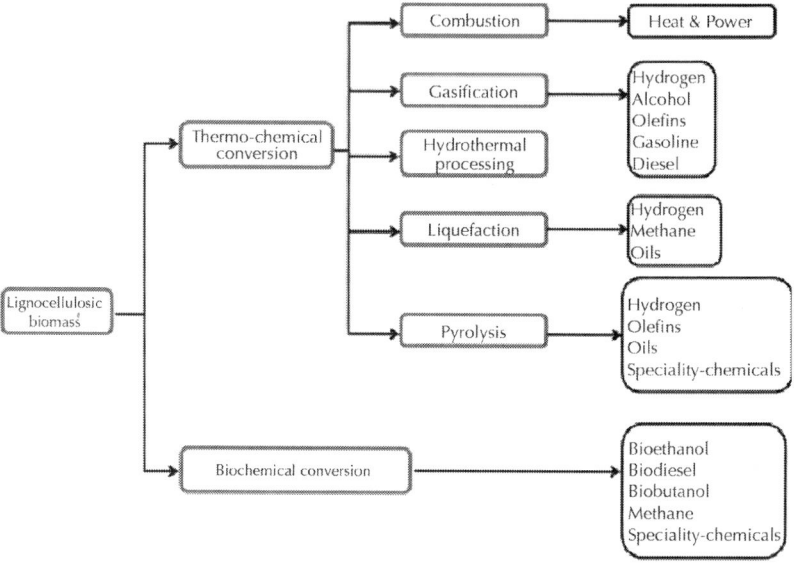

Figure 2: Thermo-mechanical and biochemical processing of lignocellulosic biomass into various values added biotechnological products. Adapted with permission from, Menon & Rao, 2012.

ROLE OF PRE-TREATMENT

Pre-treatment is an important step for the recovery of cellulosic content from lignin based biomass as compare to the starchy materials. While dealing with lignocellulosic biomasses, pre-treatment is also required to break down the lignin barrier to recover cellulose, which is further subjected to enzymatic hydrolysis to convert into fermentable sugars. During the past few decades, several pre-treatment approaches have been developed for generating cost-effective fermentable sugar from most of the agricultural cellulose and hemicellulose containing lignocellulosic materials (Yang & Wyman, 2008). In this background, there are a number of reports on pre-treatment technologies for a variety of feed-stocks. Some of the most promising pre-treatment categories have already been commercialized for the productions of bio-energy are summarized in the Table 2. An effective pre-treatment is characterized by several criteria: preserving hemicellulose fractions, to yield maximum fermentable sugar contents, limiting the loss of carbohydrate, to minimize the formation of inhibitors due to degradation products, minimizing energy input, and the process is economically efficient as well as cost-effective. While, comparing various pre-treatment options, all of the above mentioned criteria should be comprehensively considered as a basis to achieve maximal end product of interest. Hydrolysis of biomass can be done by different ways mainly including physico-chemically, chemically or biologically (Asgher et al., 2013, Hamzeh et al., 2013, Rohowsky et al., 2013, Yang et al., 2011 and Yang and Wyman, 2008).

Table 2: Most promising pre-treatment technologies

Method of pre-treatment	Sugar yield	Inhibitor formation	By-product generation	Reuse of chemicals	Applicability to different feedstock's	Equipment cost	Success at pilot scale	Advantages	Limitations & disadvantages
Mechanical	L	Nil	No	No	Yes	H	Yes	Reduce cellulose crystallinity	High power consumption than inherent biomass energy
Mineral acids	H	H	H	Yes	Yes	H	Yes	Hydrolysis of cellulose and hemicellulose. alters lignin structure	Hazardous, toxic and corrosive
Alkali	H	L	H	Yes	Yes	Nil	Yes	Removal of lignin and hemicellulose, increases accessible surface area	Long residence time, irrecoverable salts formed

134
Producing Fuels and Fine Chemicals from Biomass...

Liquid hot	H	H	L	No	–	–	Yes	Removal of hemicellulose making enzymes accessible to cellulose	Long residence time, less lignin removal
Organosolv	H	H	H	Yes	Yes	H	Yes	Hydrolyze lignin and hemicellulose	Solvents needs to drained, evaporated, condensed and reused
Wet oxidation	H or L	Nil	L	No	–	H	–	Removal of lignin, dissolves hemicellulose and causes cellulose decrystallization	–
Ozonolysis	H	L	H	No	–	H	No	Reduces lignin content, no toxic residues	Large amount of ozone required
CO_2 explosion	H	L	L	No	–	H	–	Hemicellulose removal, cellulose decrystallization, cost-effective	Does not modify lignin
Steam explosion	H	H	L	–	Yes	H	Yes	Hemicellulose removal and alteration in lignin structure	Incomplete destruction of lignine-carbohydrate matrix
AFXE	H	L	–	Yes	–	H	–	Removal of lignin and hemicellulose	Not efficient for biomass with high lignin content

Ionic liquids	H/L	L	–	Yes	Yes	–	–	Dissolution of cellulose, increased amenability to cellulase	Still in initial stages

L = low; H = high.

Adapted with permission from, Menon & Rao, 2012.

Chemical Pre-Treatments

To date chemical pre-treatment is the most studied technique among various pre-treatment categories that was originally developed and therefore has extensively been used for delignification of cellulosic materials. Chemical hydrolysis is an important treatment method for recovery of sugar monomers from cellulose and hemicellulose polymers from lignocellulosic biomass by optimizing chemical reagents. The most commonly used chemical pre-treatments include: acid and alkali based hydrolysis approaches.

Acid Based Hydrolysis

Chemical treatment of cellulosic biomass with concentrated hydrochloric acid or sulphuric acid is conventional procedure. The entire process of pre-treatment can be operating at very low temperature as compared to dilute-acid pre-treatment. On the other hand one of the possible drawbacks of this process is that it's required in higher concentration (30–70%), therefore cause high level of corrosive reaction. In this background the whole process needs further expenditure in the form of specialized non-metallic or non-corrosive material such as ceramic or carbon-brick lining. In comparison to the other pre-treatment procedures particularly dilute-acid hydrolysis the environmental hazards and high operating cost involved in concentrated-acid hydrolysis reduce the interest on industrial scale (Katzen et al., 1995 and Wyman, 1999). Dilute-acid pre-treatment has some advantages over concentrated-acid hydrolysis to solve the issues like acid recovery, toxicity, acid and special maintenance against corrosion materials (Sivers and Zacchi, 1995 and Sun and Chen,

2007). Acid pre-treatment has been applied on several biomass feed-stocks like herbaceous material (grass), hardwoods and agricultural wastes. Most of the substrates give better results by solubilizing the hemicellulose (Liao et al., 2007 and Wyman et al., 2005). Other two factors including temperature and incubation time had also important impact on alteration the structure of biomass. Major disadvantage of this process is the formation of secondary products which can lower the yield of sugars due to conversion of products in to furfural and hydroxyl-methyl furfural compounds and these compounds interfere in bio-ethanol fermentation process.

Alkali Based Hydrolysis

Alkali based pre-treatment involves the use of bases, such as sodium and ammonium hydroxide, for the pre-treatment of agricultural lignocellulosic feed-stocks. Alkaline hydrolysis causes various structural alterations inside the lignocellulosic material during treatment process such as the depletion of lignin barrier, cellulose swelling, and partial decrystallization and solvation of cellulose and hemicelluloses, respectively (Cheng et al., 2010, Ibrahim et al., 2011, Sills and Gossett, 2011 and Zhu et al., 2010). Lignocellulosic feed-stocks that have been shown to benefit from the method of alkaline pre-treatment are corn stover, switch-grass, bagasse, wheat, and rice straw (Hu et al., 2008 and Zhao et al., 2008; Zhu, Sheng, Yan, Qiao, & Lv, 2012). Zhao et al. (2008) had showed the effectiveness of sodium hydroxide pre-treatment for hardwoods, wheat straw, switch-grass, and soft-woods with less than 26% lignin content.

Biological Pre-Treatment

Biological pre-treatment employs wood degrading microorganisms, including white rot fungi (WRF), brown or soft-rot fungi, and bacteria to modify the chemical composition and/or structure of the lignocellulosic biomass. Bio-delignification is useful for pre-treatment purposes because it replaces or supplements the chemical-based pre-treatments, which include mechanical treatment with acid, alkali, and steam explosion (Iqbal et al., 2013). In spite of this biological pre-treatments are more effective, economical, eco-friendly and less health hazardous as compare to the physico-chemical or chemical-based pre-treatment approaches. Therefore, from the last few years research scientists are directing their interests towards biological delignification. Recent advances in the characterization of ligninolytic enzymes involving the degradation of lignin have given new impetus to the research in this area, which has now become amenable to biotechnological exploitation (Asgher, Iqbal, & Asad, 2012). Bio-conversions of lignocellulosic materials to useful products normally require multi-step processes that include pre-treatment, enzymatic hydrolysis, and fermentation (Xiao et al., 2012), so that the modified or pre-treated biomass is more amenable to enzyme digestion. Increasing understanding of termites and fungal systems has provided insights for developing more effective pre-treatment technologies to realize the above mentioned advantages or benefits of biological pre-treatment over some others. However, biological pre-treatment is a very slow process that also requires careful control of growth conditions and large amount of space to perform treatment. In addition to this most of the lignolytic microorganisms solubilize/consume not only lignin but also hemicellulose and cellulose. Because of these drawbacks/limitations the biological pre-treatment faces techno-economic barriers and therefore is less attractive commercially (Eggeman & Elander, 2005).

ENZYMATIC HYDROLYSIS

Enzymatic hydrolysis is an effective and economical method to achieve fermentable sugars under mild and eco-friendly reaction conditions from the pre-treated cellulosic biomass (Wyman et al., 2005). The entire process of enzymatic hydrolysis critically depends on variety of factors viz., pH, time, temperature substrates and enzyme activities, etc. Enzymatic saccharification is done separately from fermentation known as separate hydrolysis and fermentation (SHF). When cellulose hydrolysis and fermentation are carried out simultaneously the phenomenon is known as simultaneous saccharification and fermentation (SSF). Now a days this process of simultaneous saccharification of both cellulose and hemicellulose is achieved by co-fermentation of both hexoses and pentoses sugars (SSCF) with the help of genetically engineered microbes that ferment xylose and glucose in the same medium where both enzymes for cellulose and hemicelluloses are available. The major advantage of this technology is that SSF and SSCF can be performed in the same tank which makes the entire process cheap, feasible and cost-effective (reduce the capital and operational investment). Biological, physical and chemical methods have been employed for detoxification (removal of inhibitory compounds in fermentation) of lignocellulosic hydrolyzates (Olsson & Hahn-Hagerdal, 1996). Lignocellulosic materials have different degree of inhibition and tolerance levels vary according to different microbial strains. Degree of tolerance varies with different strains of Saccharomyces cerevisiae, so inhibitory compounds are detoxified by changing the substrate concentration and altering the pH of media (Linden et al., 1992 and Palmqvist et al., 1998).

FERMENTATION STRATEGY

Ethanol production from biomass is mainly categorized into three steps process (1) achieve a fermentable sugars (2) conversion of fermentable sugars into ethanol and (3) ethanol separation and purification through distillation (Asgher et al., 2014 and Demirbas, 2005). Difference between lignocelluloses or starch ethanol production is the step for obtaining sugars before fermentation. Sugar crops or starchy crops need milling and grinding for recovery of sugars by extraction and fermentation becoming a relatively simple process that requires no hydrolysis or pre-treatment steps for obtaining sugars and transformation into ethanol (Icoz, Tugrul, Saral, & Icoz, 2009). Bio-ethanol production is mainly done by fed-batch process and low ethanol produce by multi-stage continuous fermentation. Basic steps for the conversion of lignocellulosic biomass are: (1) pre-treatment process which can reduce the lignin content and render cellulose and hemicellulose content for enzymatic hydrolysis (2) steps to convert enzymatic hydrolysis to break down polysaccharide to simple sugars (3) conversions of sugars (hexoses and pentoses) for ethanol production through microorganisms (4) production of ethanol from pentose sugars. Fig. 3 illustrating a step by step procedure to convert lignocellulosic biomass into ethanol. Several fungal species belonging with genera Fusarium, Rhizopus, Monilia, Neurospora and Paecilomyces have been found potential for fermenting glucose as well as xylose (Singh, Kumar, & Schugerl, 1992). Production of bio-ethanol from cellulose is mostly conducted by using fermentative organism, but the conversion rate is very low due to byproducts formation. Filamentous fungus Fusarium oxysporum is also known for the production of bio-ethanol through SSF by direct utilizing the cellulose, but their conversion rate is low due to production of acetic acid as a byproduct (Panagiotou, Villas-Boas, Christakopoulos, Nielsen, & Olsson, 2005).

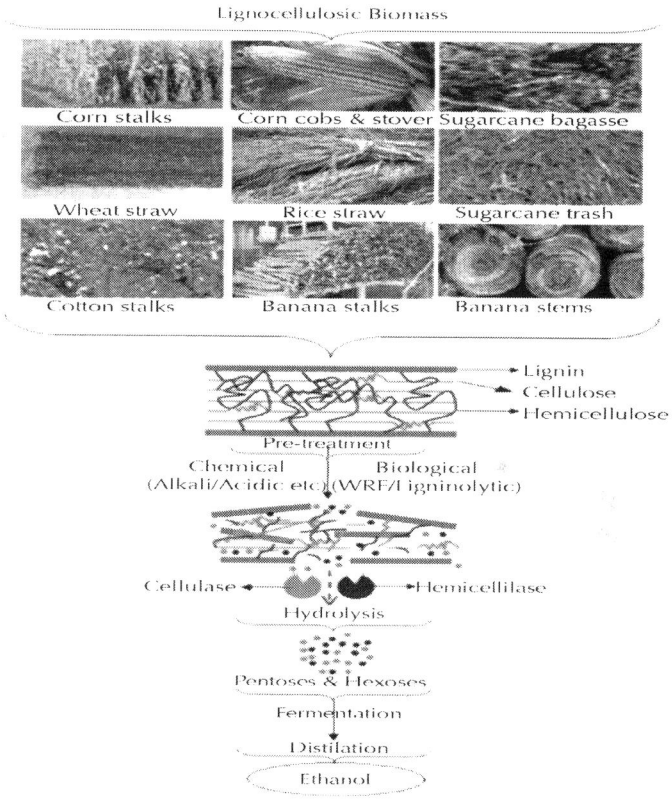

Figure 3: Generalized schematic representation of lignocellulosic materials bio-conversion into ethanol.Adapted with permission from, Asgher et al., 2014.

ENZYMES PRODUCTION

To date, the production of various ligninolytic enzymes including LiP, MnP, versatile peroxidise (VP), and laccases and other lignocellulolytic mainly endoglucanases (EC 3.2.1.4), cellobiohydrolases (EC 3.2.1.91) and -glucosidases (EC 3.2.1.21) have been widely studied in submerged and solid culture processes in the laboratory, ranging from flask shake to large scale (Elisashvili et al., 2006, Moldes et al., 2007 and Xia and Len, 1999). There are large numbers of microorganisms capable for degrading cellulose (Jungebloud et al., 2007). Trichoderma, Aspergillus, Penicillium and Fusarium species are commonly used for cellulases production (Iqbal et al., 2011). Selections of desired fungal strains depend on several factors and selection of substrate for optimizing the cellulase producing conditions (Shazia, Bajwa, & Shafique, 2007). Ligninolytic, cellulases and hemicellulases are important industrial enzymes having numerous applications and biotechnological potential for various industries including chemicals, fuel, food, brewery and wine, animal feed, textile and laundry, pulp and paper and agriculture (Asgher et al., 2011, Asgher and Iqbal, 2011, Eun et al., 2006, Iqbal and Asgher, 2013, Iqbal et al., 2011, Irshad et al., 2013, Oberoi et al., 2010, Papinutti and Forchiassin, 2007, Papinutti and Lechner, 2008, Stoilova et al., 2010 and Yoon et al., 2012). A range of different lignocellulosic materials that has successfully been adopted for the production of different enzymes having industrial importance are summarized in Table 3.

Table 3: List of various lignocellulosic materials used for the production of different microbial enzymes

Lignocellulosic material	Pre-treatment type	Microbial culture	Enzymes produced	Reference[a]
Sugar cane bagasse	Biological/chemical	P. chrysosporium;T. versicolor;Trichoderma viride;P. Sanguineus; Trichoderma viride	MnP, LiP, laccase, cellulases xylanase	El-Nasser, Helmy, and El-Gammal, 1997;El-Gammal, Kamel, Adeeb, and Helmy, 1998;Kansoh, Essam, and Zeinat 1999;Irshad et al., 2012, Irshad et al., 2012,Yoon et al., 2012 and Irshad et al., 2013
Orange peel waste	Chemical	Trichoderma viride	Endoglucanase, exoglucanase, β-glucosidase	Irshad et al., 2012 and Irshad et al., 2013

Corn cobs	Biological	Trametes versicolor; P. chrysosporium;Aspergillus niger	MnP, LiP, laccase, protease, xylanase	El-Nasser et al., 1997; Ahmed et al., 2011,Iqbal et al., 2011, Asgher and Iqbal, 2011,Asgher et al., 2012 and Asgher et al., 2012
Corn stover	Biological/chemical	P. chrysosporium;T. versicolor;Penicillium decumbens	MnP, LiP, laccase, cellulase xylanase,	El-Nasser et al., 1997; Yang et al., 2001,Iqbal et al., 2011 and Asgher et al., 2011
Rice Straw	Biological/chemical	P. chrysosporium;T. versicolor;Trichoderma reesei	MnP, LiP, laccase, cellulase	Eun et al., 2006, Iqbal et al., 2011 and Asgher et al., 2011
Penut shells	Biological	G. leucidum; P. chrysosporium	Laccase, xylanase	El-Nasser et al., 1997; Irshad, Bahadur, et al., 2012
Newspaper	Chemical	Trichoderma viride	Endoglucanase, exoglucanase, β-glucosidase	Irshad et al., 2013
Wheat straw	Biological/chemical	P. chrysosporium;T. versicolor; T. viride; F. trogii; L. edodes; P. dryinus;P. tuberregium	MnP, LiP, laccase, cellulases xylanase,	El-Nasser et al., 1997; Kachlishvili, Penninckx, Tsiklauri, and Elisashvili, 2006;Elisashvili, Kachlishvili, and Penninckx, 2008; Iqbal et al., 2011, Asgher et al., 2011 and Irshad et al., 2013
Banana stalk	Biological	S. commune; P. chrysosporium; T. versicolor; P. ostreatus	MnP, LiP, laccase, xylanase, endoglucanase	Reddy, Ravindra Babu, Komaraiah, Roy, and Kothari, 2003; Irshad, Asgher, Scheikh, and Nawaz, 2011; Iqbal, Asgher, and Bhatti, 2011; Asgher et al., 2011
Rice bran	Biological	Aspergillus niger	Protease	Ahmed et al., 2011
Wheat bran	Biological	Aspergillus niger;Morchella sculenta;F. sclerodermeus;Trametes versicolor	Protease, endoglucanase, β-glucosidase, laccase, MnP	Papinutti and Forchiassin, 2007, Papinutti and Lechner, 2008, Stoilova et al., 2010 and Ahmed et al., 2011
Apple pomace	Chemical	Trichoderma viride	Cellulases	Irshad et al., 2013

Oil palm empty fruit bunch fibre	Chemical	Thermobifida fusca;Aspergillus terreus	CMCase. FPase, β-glucosidase	Shahriarinour et al., 2011;Harun et al., 2013
Beech tree leaves	Biological	F. trogii; L. edodes;Pleurotus dryinus;P. tuberregium	Laccase, CMCase. FPase, MnP, xylanase	Kachlishvili et al., 2006;Elisashvili et al., 2008
Eucalyptus residue	Biological	Lentinula edodes	Xylanase, cellulase, MnP, laccase	Silva, Machuca, and Milagres, 2005

[a]For detailed references please see Iqbal et al. (2013).

Adapted with permission from, Iqbal et al., 2013.

CONCLUDED REMARKS AND FUTURE OUTLOOK

The energy and environmental crises which the modern world is experiencing is forcing to re-evaluate the efficient utilization or finding alternative uses for natural, renewable resources, using clean technologies. In this regard, lignocellulosic biomass holds considerable potential to meet the current energy demand of the modern world. This is also essential in order to overcome the excessive dependence on petroleum for liquid fuels. Further advanced biotechnologies are crucial for discovery, characterization of new enzymes, and production in homologous or heterologous systems and ultimately lead to low-cost conversion of lignocellulosic biomasses into bio-fuels and bio-chemicals. In current scenario future trends are being directed to lignocellulose biotechnology and genetic engineering for improved processes and products. To overcome the current energy problems it is envisaged that lignocellulosic biomass in addition of green biotechnology will be the main focus of the future research.

REFERENCES

1.   Ahmed, I., Zia, M. A., Iftikhar, T., & Iqbal, H. M. H. (2011). Characterization and detergent compatibility of purified protease

produced from Aspergillus niger by utilizing agro wastes. BioResources, 6, 4505e4522.

2. Alonso, A., Pe´ rez, P., Morcuende, R., & Martinez-Carrasco, R. (2008). Future CO2 concentrations, though not warmer temperatures, enhance wheat photosynthesis temperature responses. Physiologia Plantarum, 132, 102e112.

3. Asgher M., Ahmad, Z., & Iqbal, H. M. N. (2013). Alkali and enzymatic delignification of sugarcane bagasse to expose cellulose polymers for saccharification and bio-ethanol production. Industrial Crops and Products, 44, 488e495.

4. Asgher, M., Ahmed, N., & Iqbal, H. M. N. (2011). Hyperproductivity of extracellular enzymes from indigenous white rot fungi (P. chrysosporium IBL-03) by utilizing agro-wastes. BioResources, 6(4), 4454e4467.

5. Asgher, M., & Iqbal, H. M. N. (2011). Characterization of a novel manganese peroxidase purified from solid state culture of Trametes versicolor IBL-04. BioResources, 6, 4302e4315.

6. Asgher, M., Iqbal, H. M. N., & Asad, M. J. (2012). Kinetic characterization of purified laccase produced from Trametes versicolor IBL-04 in solid state bio-processing of corncobs. BioResources, 7, 1171e1188.

7. Asgher, M., Iqbal, H. M. N., & Irshad, M. (2012). Characterization of purified and xerogel immobilized novel lignin peroxidase produced from Trametes versicolor IBL-04 using solid state medium of corncobs. BMC Biotechnology, 12(1), 46.

8. Asgher, M., Shahid, M., Kamal, S., & Iqbal, H. M. N. (2014). Recent trends and valorization of immobilization strategies and ligninolytic enzymes by industrial biotechnology. Journal of Molecular Catalysis B: Enzymatic, 101, 56e66.

9. Balat, M., & Balat, H. (2009). Recent trends in global production and utilization of bio-ethanol fuel. Applied Energy, 86(11), 2273e2282.

10. Bertero, M., de la Puente, G., & Sedran, U. (2012). Fuels from biooils: bio-oil production from different residual sources, characterization and thermal conditioning. Fuel, 95, 263e271.

11. Brigham, J. S., Adney, W. S., & Himmel, M. E. (1996). Hemicellulases: diversity and applications. In C. E. Wyman (Ed.), Bioethanol: Production and utilization (pp. 119e141).

12. Calvo-Flores, F. G., & Dobado, J. A. (2010).Lignin as renewable raw material. ChemSusChem, 3(11), 1227e1235.

13. Cheng, Y. S., Zheng, Y., Yu, C. W., Dooley, T. M., Jenkins, B. M., & VanderGheynst, J. S. (2010). Evaluation of high solids alkaline pretreatment of rice straw. Applied Biochemistry and Biotechnology, 162, 1768e1784.

14. Demirbas, A. (2005). Bio-ethanol from cellulosic materials: a renewable motor fuel from biomass. Energy Sources, 27, 327e337.

15. Eggeman, T., & Elander, R. T. (2005). Process economic analysis of pretreatment technologies. Bioresource Technology, 96, 2019e2025.

16. El-Gammal, A. A., Kamel, Z., Adeeb, Z., & Helmy, S. M. (1998). Biodegradation of lignocellulosic substances and production of sugars and lignin degradation intermediates by four selected microbial strains. Polymer Degradation and Stability, 61(3), 535e542.

17. Elisashvili, V., Kachlishvili, E., & Penninckx, M. (2008). Effect of growth substrate, method of fermentation, and nitrogen source on lignocellulose-degrading enzymes production by white-rot basidiomycetes. Journal of Industrial Microbiology & Biotechnology, 35(11), 1531e1538.

18. Elisashvili, V., Penninckx, M., Kachlishvili, E., Asatiani, M., & Kvestiadze, G. (2006). Use of Pleurotus dryinus for lignocellulolytic enzymes production in submerged fermentation of mandarin peels and tree leaves. Enzyme and Microbial Technology, 38, 998e1004.

19. El-Nasser, N. H., Helmy, S. M., & El-Gammal, A. A. (1997). Formation of enzymes by biodegradation of agricultural wastes with white rot fungi. Polymer Degradation and Stability, 55(3), 249e255.

20. Eun, J. S., Beauchemin, K. A., Hong, S. H., & Bauer, M. W. (2006). Exogenous enzymes added to untreated or ammoniated rice straw: effects on in vitro fermentation characteristics and degradability. Animal Feed Science and Technology, 131, 87e102.

21. Guimara~es, J. L., Frollini, E., Da Silva, C. G., Wypych, F., & Satyanarayana, K. G. (2009). Characterization of banana, sugarcane bagasse and sponge gourd fibers of Brazil. Industrial Crops and Products, 30(3), 407e415.

22. Hamelinck, C. N., Hooijdonk, V., & Faaij, A. P. C. (2005). Ethanol from lignocellulosic biomass: techno-economic performance in short,middle and long-term. Biomass and Bioenergy, 28, 384e410.

23. Hamzeh, Y., Ashori, A., Khorasani, Z., Abdulkhani, A., & Abyaz, A. (2013). Pre-extraction of hemicelluloses from bagasse fibers: effects of dry-strength additives on paper properties. Industrial Crops and Products, 43, 365e371.

24. Harun, N. A. F., Samsu Baharuddin, A., Mohd Zainudin, M. H., Bahrin, E. K., Naim, M. N., & Zakaria, R. (2013). Cellulase production from treated oil palm empty fruit bunch degradation by locally isolated Thermobifida fusca. BioResources, 8(1), 676e687.

25. Himmel, M. E., Ding, S. Y., Johnson, D. K., Adney, W. S., Nimlos, M. R., Brady, J. W., et al. (2007). Biomass recalcitrance: engineering plants and enzymes for biofuel production. Science, 315, 804e807.

26. Howard, R. L., Abotsi, E., Van Rensburg, E. J., & Howard, S. (2003). Lignocellulose biotechnology: issues of bioconversion and enzyme production. African Journal of Biotechnology, 2(12), 602e619.

27. Hu, Z., Wang, Y., & Wen, Z. (2008). Alkali (NaOH) pretreatment of switchgrass by radio frequency-based dielectric heating. Applied Biochemistry and Biotechnology, 148, 71e81.

28. Ibrahim, M. M., El-Zawawy, W. K., Abdel-Fattah, Y. R., Soliman, N. A., & Agblevor, F. A. (2011). Comparison of alkaline pulping with steam explosion for glucose production from rice straw. Carbohydrate Polymers, 83, 720e726.

29. Icoz, E., Tugrul, M. K., Saral, A., & Icoz, E. (2009). Research on ethanol production and use from sugar beet in Turkey. Biomass and Bioenergy, 33, 1e7.

30. Iqbal, H. M. N., Ahmed, I., Zia, M. A., & Irfan, M. (2011). Purification and characterization of the kinetic parameters of

cellulase produced from wheat straw by Trichoderma viride under SSF and its detergent compatibility. Advances in Bioscience and Biotechnology, 2(3), 149e156.

31. Iqbal, H. M. N., & Asgher, M. (2013). Characterization and decolorization applicability of xerogel matrix immobilized manganese peroxidase produced from Trametes versicolor IBL-04. Protein & Peptide Letters, 5, 591e600.

32. Iqbal, H. M. N., Asgher, M., & Bhatti, H. N. (2011). Optimization of physical and nutritional factors for synthesis of lignin degrading enzymes by a novel strain of Trametes versicolor. BioResources, 6, 1273e1278.

33. Iqbal, H. M. N., Kyazze, G., & Keshavarz, T. (2013). Advances in valorization of lignocellulosic materials by bio-technology: an overview. BioResources, 8(2), 3157e3176.

34. Irshad, M., Anwar, Z., & Afroz, A. (2012). Characterization of Exo 1,4-b glucanase produced from Trichoderma viride through solid-state bio-processing of orange peel waste. Advances in Bioscience and Biotechnology, 3, 580e584.

35. Irshad, M., Anwar, Z., But, H. I., Afroz, A., Ikram, N., & Rashid, U. (2013). The industrial applicability of purified cellulase complex indigenously produced by Trichoderma viride through solid-state bio-processing of agro-industrial and municipal paper wastes. BioResources, 8(1), 145e157.

36. Irshad, M., Asgher, M., Scheikh, M. A., & Nawaz, H. (2011). Purification and characterization of laccase produced by Schyzophylum commune IBL-06 in solid state culture of banana stalks. BioResources, 6(3), 2861e2873.

37. Irshad, M., Bahadur, B. A., Anwar, Z., Yaqoob, M., Ijaz, A., & Iqbal, H. M. N. (2012). Decolorization applicability of solegel matrix-immobilized laccase produced from Ganoderma leucidum using agro-industrial waste. BioResources, 7(3), 4249e4261.

38. Isroi, M. R., Syamsiah, S., Niklasson, C., Cahyanto, M. N., Lundquist, K., & Taherzadeh, M. J. (2011). Biological pretreatment of lignocelluloses with white-rot fungi and its applications: a review. BioResources, 6(4), 5224e5259.

39. Jiang, G., Nowakowski, D. J., & Bridgwater, A. V. (2010). A systematic study of the kinetics of lignin pyrolysis. Thermochimica Acta, 498(1), 61e66.

40. John, F., Monsalve, G., Medina, P. I. V., & Ruiz, C. A. A. (2006). Ethanol production of banana shell and cassava starch. Dyna, Universidad Nacional de Colombia, 73, 21e27.

41. Jungebloud, A., Bohle, K., Go¨cke, Y., Cordes, C., Horn, H., & Hempel, D. C. (2007). Quantification of product-specific gene expression in biopellets of Aspergillus niger with real-time PCR. Enzyme and Microbial Technology, 40(4), 653e660.

42. Kachlishvili, E., Penninckx, M. J., Tsiklauri, N., & Elisashvili, V. (2006). Effect of nitrogen source on lignocellulolytic enzyme production by white-rot basidiomycetes under solid-state cultivation. World Journal of Microbiology and Biotechnology, 22(4), 391e397.

43. Kansoh, A. L., Essam, S. A., & Zeinat, A. N. (1999). Biodegradation and utilization of bagasse with Trichoderma reesie. Polymer Degradation and Stability, 63(2), 273e278.

44. Katzen, R., Madson, P.W., & Monceaux, D. A. (1995). Use of cellulosic feedstocks for alcohol production. In T. P. Lyons, J. E. Murtagh, & D. R. Kelsall (Eds.), The alcohols textbook (pp. 37e46).

45. Kim, M., & Day, D. F. (2011). Composition of sugar cane, energy cane, and sweet sorghum suitable for ethanol production at Louisiana sugar mills. Journal of Industrial Microbiology & Biotechnology, 38(7), 803e807.

46. Kumar, P., Barrett, D. M., Delwiche, M. J., & Stroeve, P. (2009). Methods for pretreatment of lignocellulosic biomass for efficient hydrolysis and biofuel production. Industrial & Engineering Chemistry Research, 48, 3713e3729.

47. Liao, W., Liu, Y., Wen, Z., Frear, C., & Chen, S. (2007). Studying the effects of reaction conditions on components of dairy manure and cellulose accumulation using dilute acid treatment. Bioresource Technology, 98, 1992e1999.

48. Lin, Y., & Tanaka, S. (2006). Ethanol fermentation from biomass resources: current state and prospects. Applied Microbiology and Biotechnology, 69, 627e642.

49. Linden, T., Peetre, J., & Hahn-Hagerdal, B. (1992). Isolation and characterization of acetic acid tolerant galactose-fermenting strains of Saccharomyces cerevisiae from a spent sulfite liquor plant. Applied and Environmental Microbiology, 58, 1661e1669.

50. Malherbe, S., & Cloete, T. E. (2002). Lignocellulose biodegradation: fundamentals and applications. Reviews in Environmental Science and Biotechnology, 1, 105e114.

51. McKendry, P. (2002). Energy production from biomass (part 1): overview of biomass. Bioresource Technology, 83(1), 37e46.

52. Menon, V., & Rao, M. (2012). Trends in bioconversion of lignocellulose: biofuels, platform chemicals & biorefinery concept. Progress in Energy and Combustion Science, 38(4), 522e550.

53. Moldes, A. B., Bustos, G., Torrado, A., & Dominguez, J. M. (2007). Comparison between different hydrolysis processes of vinetrimming waste to obtain hemicellulosic sugars for further lactic acid conversion. Applied Biochemistry and Biotechnology, 143, 244e256.

54. Oberoi, H. S., Chavan, Y., Bansal, S., & Dhillon, G. S. (2010). Production of cellulases through solid state fermentation using kinnow pulp as a major substrate. Food and Bioprocess Technology, 3, 528e536.

55. Ofori-Boateng, C., & Lee, K. T. (2013). Sustainable utilization of oil palm wastes for bioactive phytochemicals for the benefit of the oil palm and nutraceutical industries. Phytochemistry Reviews, 1e18.

56. Olsson, L., & Hahn-Hagerdal, B. (1996). Fermentation of lignocellulosic hydrolysates for ethanol production. Enzyme and Microbial Technology, 18, 312e331.

57. Palmqvist, E., Galbe, M., & Hahn-Hagerdal, B. (1998). Evaluation of cell recycling in continuous fermentation of enzymatic hydrolysates of spruce with Saccharomyces cerevisiae and online monitoring of glucose and ethanol. Applied Biochemistry and Biotechnology, 50, 545e551.

58. Panagiotou, G., Villas-Boas, S. G., Christakopoulos, P., Nielsen, J., & Olsson, L. (2005). Intracellular metabolite profiling of Fusarium oxysporum converting glucose to ethanol. Journal of Biotechnology, 115, 425e434.

59. Papinutti, L., & Lechner, B. (2008). Influence of the carbon source on the growth and lignocellulolytic enzyme production by

Morchella esculenta strains. Journal of Industrial Microbiology & Biotechnology, 35, 1715e1721.

60. Papinutti, V. L., & Forchiassin, F. (2007). Lignocellulolytic enzymes from Fomes sclerodermeus growing in solid-state fermentation. Journal of Food Engineering, 81, 54e59. Pe´ rez, J., Mun˜ oz-Dorado de la Rubia, T., & Martı´nez, J. (2002). Biodegradation and biological treatments of cellulose, hemicellulose and lignin: an overview. International Microbiology, 5, 53e63.

61. Prassad, S., Singh, A., & Joshi, H. C. (2007). Ethanol as an alternative fuel from agricultural, industrial and urban residues. Resources, Conservation and Recycling, 50, 1e39.

62. Reddy, G. V., Ravindra Babu, P., Komaraiah, P., Roy, K. R. R. M., & Kothari, I. L. (2003). Utilization of banana waste for the production of lignolytic and cellulolytic enzymes by solid substrate fermentation using two Pleurotus species P. ostreatus and P. sajorcaju. Process Biochemistry, 38(10), 1457e1462.

63. Rohowsky, B., Ha¨ßler, T., Gladis, A., Remmele, E., Schieder, D., & Faulstich, M. (2013). Feasibility of simultaneous saccharification and juice co-fermentation on hydrothermal pretreated sweet sorghum bagasse for ethanol production. Applied Energy, 102, 211e219.

64. Shahriarinour, M., Ramanan, R. N., Abdul Wahab, M. N., Mohamad, R., Mustafa, S., & Ariff, A. B. (2011). Improved cellulase production by Aspergillus terreus using oil palm empty fruit bunch fiber as substrate in a stirred tank bioreactor through optimization of the fermentation conditions. BioResources, 6(3), 2663e2675.

65. Shazia, S., Bajwa, R., & Shafique, S. (2007). Cellulase production potential of selected strains of Aspergilli. Pakistan Journal of Phytopathology, 19(2), 196e206.

66. Sills, D. L., & Gossett, J. M. (2011). Assessment of commercial hemicellulases for saccharification of alkaline pretreated perennial biomass. Bioresource Technology, 102, 1389e1398.

67. Silva, E. M., Machuca, A., & Milagres, A. M. F. (2005). Effect of cereal brands on Lentinula edodes growth and enzyme activities during cultivation on forestry waste. Letters in Applied Microbiology, 40(4), 283e288.

68. Singh, A., Kumar, P. K. R., & Schugerl, K. (1992). Bioconversion of cellulosic materials to ethanol by filamentous fungi. Advances in Biochemical Engineering Biotechnology, 45, 29e55.

69. Sivers, M. V., & Zacchi, G. (1995). A techno-economical comparison of three processes for the production of ethanol from pine. Bioresource Technology, 51, 43e52.

70. Stoilova, I., Krastanov, A., & Stanchev, V. (2010). Properties of crude laccase from Trametes versicolor produced by solidsubstrate fermentation. Advances in Bioscience and Biotechnology, 1, 208e215.

71. Sun, F. B., & Chen, H. Z. (2007). Evaluation of enzymatic hydrolysis of wheat straw pretreated by atmospheric glycerol autocatalysis. Journal of Chemical Technology and Biotechnology, 82(11), 1039e1044.

72. Taherzadeh,M. J., & Karimi, K. (2008). Pretreatment of lignocellulosic wastes to improve ethanol and biogas production: a review. International Journal of Molecular Sciences, 9, 1621e1651.

73. Vercoe, D., Stack, K., Blackman, A., & Richardson, D. (2005). A study of the interactions leading to wood pitch deposition. In 59th Appita annual conference and exhibition: Incorporating the 13th ISWFPC (international symposium on wood, fibre and pulping chemistry) (p. 123). Auckland, New Zealand, 16e19 May 2005: Proceedings.

74. Wyman, C. E. (1999). Biomass ethanol: technical progress, opportunities and commercial challenges. Annual Review of Energy and the Environment, 24, 189e226.

75. Wyman, C. E., Dale, B. E., Elander, R. T., Holtzapple, M., Ladisch, M. R., & Lee, Y. Y. (2005). Coordinated development of leading biomass pretreatment technologies. Bioresource Technology, 96, 1959e1966.

76. Xia, L., & Len, P. (1999). Cellulose production by solid-state fermentation on lignocellulosic waste from the xylose industry. Process Biochemistry, 34, 909e912.

77. Xiao, W., Wang, Y., Xia, S., & Ma, P. (2012). The study of factors affecting the enzymatic hydrolysis of cellulose after ionic liquid pretreatment. Carbohydrate Polymers, 87, 2019e2023.

78. Yang, B., Dai, Z., Ding, S.-Y., & Wyman, C. E. (2011). Enzymatic hydrolysis of cellulosic biomass. Biofuels, 2, 421e450.

79. Yang, B., & Wyman, C. (2008). Pre-treatment: the key to unlocking low-cost cellulosic ethanol. Biofuels, Bioproducts and Biorefining, 2, 26e40.

80. Yang, X., Chen, H., Gao, H., & Li, Z. (2001). Bioconversion of corn straw by coupling ensiling and solid-state fermentation. Bioresource Technology, 78, 277e280.

81. Yoon, L. W., Ngoh, G. C., & Chua, A. S. M. (2012). Simultaneous production of cellulase and reducing sugar from alkalipretreated sugarcane bagasse via solid state fermentation. BioResources, 7, 5319e5332.

82. Zhao, Y., Wang, Y., Zhu, J. Y., Ragauskas, A., & Deng, Y. (2008). Enhanced enzymatic hydrolysis of spruce by alkaline pretreatment at low temperature. Biotechnology and Bioengineering, 99, 1320e1328.

83. Zhu, Y., Lee, Y. Y., & Elander, R. T. (2005). Optimization of diluteacid pretreatment of corn stover using a high-solids percolation reactor. Applied Biochemistry and Biotechnology, 121e124, 1045e1054.

84. Zhu, H., Sheng, K., Yan, E., Qiao, J., & Lv, F. (2012). Extraction, purification and antibacterial activities of a polysaccharide from spent mushroom substrate. International Journal of Biological Macromolecules, 50(3), 840e843.

85. Zhu, J., Wan, C., & Li, Y. (2010). Enhanced solid-state anaerobic digestion of corn stover by alkaline pretreatment. Bioresource Technology, 101, 7523e7528.

Chapter 5

Nanotech Biofuels and Fuel Additives

Sergio C. Trindade[1]

[1]SE2T International, Ltd. and International Fuel Technology, Inc., USA

INTRODUCTION

This chapter was inspired by an invited presentation of the author at the Chemindix conference in Bahrain in October 2010. This was the 8th. International Conference & Exhibition on Chemistry in Industry, promoted by the Saudi Chapter of the American Chemical Society and Aramco. The focus is on reviewing the application of nanotechnology to biofuels production and to the utilization of fuel additives, some of which are derived from renewable materials. To introduce the topic, the broader context of petroleum fuels and biofuels is presented. A smart future of oil refining would be to increasingly utilize margins to finance a transition away from oil towards future alternative providers of mobility, in particular biofuels.

Future scenarios of liquid biofuels involve the market penetration of second and further generations of technologies and the continuous improvement of first generation processes. On the other hand, nanotechnologies are among the candidate technologies for the biofuels of the future. The nanotechnology field is vast and its applications unbound.

This is followed by a brief review of nanotechnology developments, especially as they apply to liquid particles, beyond the more common solid particle applications.

Algae growth, harvesting and conversion are presented and discussed, given the immense potential of their contribution towards an energy future where biofuels play a significant role.

Most of the current effort in second generation conversion to liquid biofuels is based on biomass cellulosic to ethanol and biodiesel. Nano processes are being pursued and will be reviewed in the chapter.

Likewise, the presently used processes to convert oils and animal fat into biodiesel are based on trans-esterification with methanol or ethanol, which inevitably generates glycerol, which must find a market or get disposed properly. Nano processes may be useful in addressing this issue.

Speculative considerations are made about the role of liquid nanoparticles of fuel additives in enhancing the performance of additized biofuel/fuel blends, in connection with surface and combustion effects.

Public concerns over the impacts of nanotechnologies on security, health and the environment are also mentioned and discussed. But, a cautionary optimistic view is presented on the huge benefits of a careful penetration of nanotechnologies in the realm of biofuels and fuel additives, and in many more applications, especially those dealing with human health.

THE FUTURE OF OIL REFINING [1] AND THE OIL TRANSITION TO ALTERNATIVES [2]

Crucial challenges to the oil industry are evolving, as the demand for energy services (mobility, lighting, rotary movement, heating and cooling) increases, which with the current technology setup, is translated into expanding demand for liquid fossil fuels. Oil producers and refiners face difficulties finding sufficient good quality crude oil in adequate amounts and reasonable costs to meet growing demand over the long run, while users and the public at large are pushing for environmental improvements, such as better air quality in the immediate future. Moreover, concerns about the impacts of climate change caused by increased greenhouse gases emissions from the production and use of liquid hydrocarbons may eventually force a transition to climate friendly energy services providing systems. This offers biofuels a market penetration opportunity in the transition towards a yet undefined new energy future.

Under this background, the oil refining industry in the USA and the European Union has been stagnant. It has been immobilized by environmental obstacles posed by an articulated public, augmented by a "not in my backyard" attitude that makes it difficult to build new refineries. In addition, declining margins for refined products may have led major players to focus more on the upstream.

On the other hand, refining is expanding in other parts of the world, such as India, China, Brazil and the Middle East, as these countries develop and the oil producers attempt to add more value to their resources. An evidence of the shift of refining towards developing economies is the fact that the largest refinery in the world is in India and belongs to Reliance.

But, all over the world, the increase in the long-term marginal cost of oil combined with environmental pressures and stricter government regulations and mandates are likely to lead to the decline of the centrality of oil in the global energy mix in favor of natural gas. This shift in dominance happened to wood and coal over the past two centuries and is now happening to oil. Oil companies are increasingly calling themselves energy companies. Some of them will leverage their current oil production margins to make a smooth transition to alternatives over time. A profitable transition to alternatives in the oil economy would require a gradual transfer of oil profits into green investments and the stretching of current oil supplies.

LIQUID BIOFUELS ISSUES [3]

The enormous global daily consumption of liquid fuels is of the order of 80 million barrels/day (e equivalent of 12.7 million m^3/day). The sugar cane area required to produce the same volume of ethanol is about 700 million hectares, assuming a yield of 6.5 m^3/ha/year of ethanol. This area is equivalent to 100 times the sugar cane cultivated area in Brazil, the second largest bio-ethanol producer in the world. Biofuels definitely face an issue of scale. In 2010, fuel ethanol and biodiesel combined displaced a mere 3% of oil in the world.

Figure 1 [4] below illustrates the scale issue by showing how much land it would take for the USA to grow its own fuel.

It appears that algae require the least area to meet the large scale demand of liquid fuels in the USA, whereas the area required by soybeans is larger than the USA's 48 continental states. The area required by corn is substantial. This suggests that the current biofuels production base of the USA would not be able to meet demand, and imports would be required to meet the colossal American energy appetite.

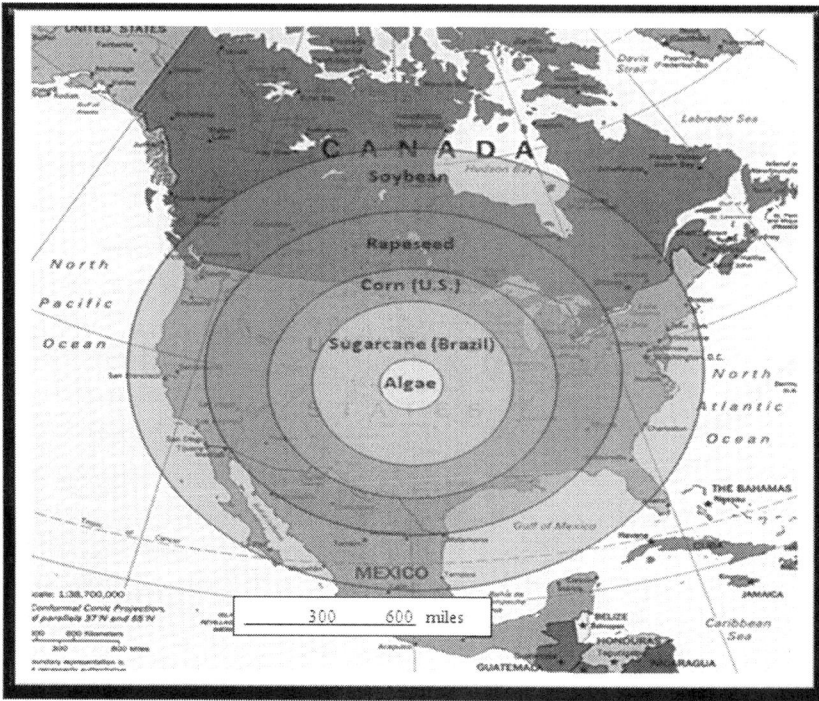

Figure 1: How much land would it take for the USA to grow its own fuel?

The scale challenge posed to biofuels relates to the labour, management, land, water and sunshine required to produce the biomass and the processing that originates them. These are scarce resources that are also needed to grow food, feed and fibre to ultimately meet various human demands. These are resources that have an opportunity cost from competing markets. To develop biofuels in the scale of commercial liquid fuels require massive financing, a resource that may have alternative uses as well. The mobilization of private capital under a perception of market and other uncertainties is another issue that biofuels have to resolve in order to thrive.

The production of biofuels is accompanied by local environmental issues that need addressing. For instance, in the case of sugar cane ethanol, stillage the liquid residue of distillation, has a high chemical and biological oxygen demand and requires appropriate processing before final disposal. From a global climate change perspective, designed and managed properly, a biofuels production system would add minimally to greenhouse gas emissions. But, in practice, many biofuel production systems in the world are contributing net GHG emissions. A bone of contention in the development of the biofuels industry is the present competition for feed stocks between the food and fuel industries. In the case of biodiesel, all commercial vegetable oils that are used in preparing food are also convertible to biodiesel. A similar situation exists with respect to fuel ethanol, especially for the starch-based feed stocks (corn and wheat). However, the hike in food prices that happened globally in 2007/8 and is happening in 2010/1 derive mostly from other causes such as droughts and other climate related phenomena, higher oil prices and market speculation.

Since the cost of biofuels is dominated by feed stocks cost, access to feed stocks in the required amounts, timing and at adequate prices is key to the success of the biofuels economy. The combination of the food versus fuel conundrum with the need to have reliable and economic access to feed stocks is shifting the industry towards non-food feed stocks and to the market penetration of second generation technologies to convert cellulosic biomass into liquid biofuels.

Concern in important consuming markets about the sustainability of biofuels producing systems is putting pressure on suppliers to abide by sustainability protocols subject to certification. The sustainability of biofuels is actually linked to freer international trade, which would tend to phase out unsustainably produced biofuels in favour of regions of the world that can meet sustainable production requirements. A valuable discussion on this matter was hosted by the Rockefeller Foundation in 2008 at its Bellagio Centre and produced a sustainable biofuels consensus. The objective was to understand the many drivers for sustainable trade, consumption and production of biofuels, and the comparative advantage of supplying regions combined with demand and technology from consuming regions [5].

However, much remains to be done to achieve free international trade of biofuels. The World Trade Organization Doha rounds have reached an impasse. Currently, biodiesel is considered an industrial product, whereas fuel ethanol is categorized as an agricultural product, which allows more protectionism. What is needed is a unified treatment of biofuels, where they are classified under environmental goods and services. But, irrespective of these drawbacks, a sign pointing to a larger role for biofuels in the future are the new biofuels technology initiatives by large oil companies, such as BP, Chevron, Exxon and Shell. The development of the international trade in biofuels is likely to distribute more evenly the production and consumption of biofuels in the world. For the time being, biofuels production is overwhelmingly concentrated in the USA, Brazil and the European Union, as shown in Fig. 2 below[6].

THE VASTNESS OF NANOTECHNOLOGY

Nanotechnology can be simply defined as the discipline of building machines/devices on the scale of molecules, a few Nanometers (10⁻⁹m) wide, way smaller than a cell. Table 1 below show some practical applications of nanotechnologies and confirms the vastness of their domain [7]:

In the practically important area of polymers, nanotechnologies originate nano-structured polymers, where applications can be found in support structures; manufacturing processes; diagnostics and therapy; pharmaceuticals; medical and dental prosthesis; and thin films for surface treatment. The main chemicals involved in nano-structured polymers are: poly-oxides; poly-acrylates; poly-vinylics; poly-saccharides; and poly-ethylenes. The main materials incorporated into polymer nano-matrices are silicon, chromium and carbon[8]

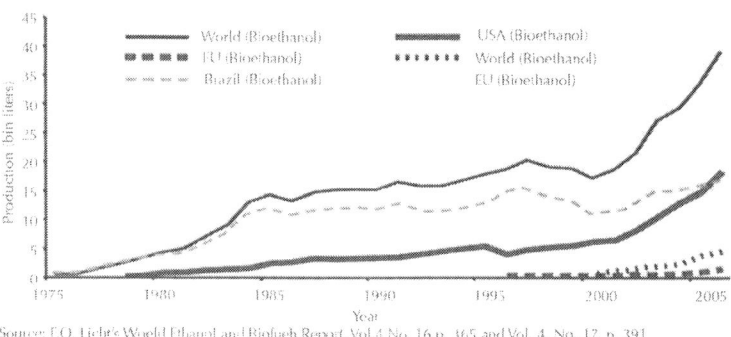

Figure 2: World's biofuels production is concentrated in the US, Brazil and the EU.

Table 1: Some applications of nanotechnology

Nanocrystalline Drugs	Oxide Nanoparticles
Nanofilm Coatings	Carbon Nanotubes
Cosmetics	Silicon Nanomaterials
Nanocatalysts	Metal nanoparticles
Ceramics	Polymer Nanocomposites
Rubbers	Inorganic Nanocrystals

TURNING ALGAE INTO BIOFUELS

As shown in Fig. 1, biofuels derived from algae offer a great potential in view of the possible high yields and smaller area requirements. In addition, algae can play a role in carbon mitigation, as one way of growing algae is to feed them carbon-dioxide (CO_2), besides water and sunlight. Algae can be fed other substrates as well, because to grow, cost-effectively, on carbon dioxide there would be a need of concentrated sources of the gas, such as found in combustion off-gases from fossil fueled power plants.

Oil can form up to 50% of the algae mass, in contrast with the best oil-bearing plants – oil palm trees – where less than 20% of the biomass is made out of foil. Algae carbohydrates can also be made into ethanol or gasified into bio-gas, or methane or hydrogen [9].

But, algae development into biofuels must overcome a number of challenges before algae can become significant sources of commercial biofuels. Since algae also need water to grow, expansion of algae production may create a dilemma of water versus fuel, similar to food versus fuel dilemma discussed previously. Another challenge is the low natural carbon dioxide concentration in the atmosphere, hence the consideration of additional sources of carbon for algal growth in a commercial biofuels system. One response to these challenges may include the use of nanotechnology to turn algae into biofuels.

As way of examples, in 2009, the company Quantum Sphere received a grant from the California Energy Commission to develop a nano-catalyzed algae bio gasification. Also in California, the Salton

Sea receives large amounts of agricultural runoff, which sometimes create large algae blooms. These algae and similar biomass have been turned experimentally into methane, hydrogen and other gases [10].

One nanotechnology relevant to algae development is the use of nanoparticles as no-harm harvesters of biofuel oils from algae, as illustrated in Fig. 3 [11]. The nano particles are shown on the left hand side of the photograph before the oil pregnant algae are added. The right hand side shows the contacting between the algae and the nano particles, which results in extracting the oil without harming the algae. Maintaining the algae alive can dramatically reduce production costs and the generation cycle.

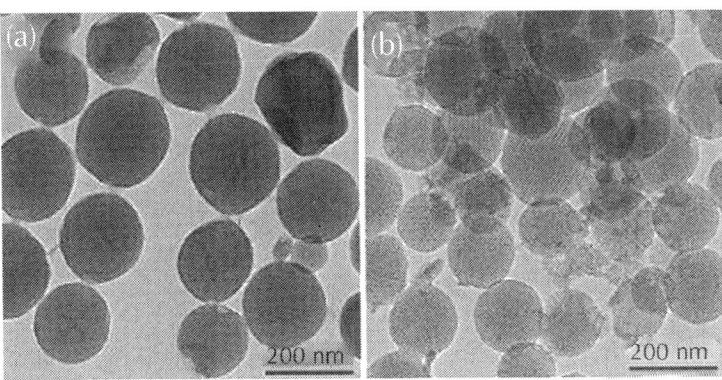

Figure 3: Nano-particles harvesting oil from algae without harming the organism.

One possible downside of the nano-harvesters is the risk that they may be released into the environment, although the spherical nano-particles are made of calcium compounds and sand [12]. The pores of the spheres are lined with chemicals, which extract algal oil without breaking the cell membrane. Nevertheless, prior to commercial market penetration of nano-harvesting, there would be a need to carry out due diligence to ensure the safety of these processes.

NANOTECHNOLOGY APPLIED TO LANDFILL FACILITIES[13]

The organic matter in landfills tend to undergo anaerobic fermentation yielding methane and CO_2 [14], which if naturally vented into the atmosphere would add to the greenhouse emissions that warm the climate. And the climate change impact of methane is 25 times larger than that of carbon dioxide for a time horizon of 100 years [15]. Thus, there is a need to sequester the carbon present in landfill methane. Nano-catalysts can crack methane into elemental carbon and hydrogen. The carbon can be produced in high-purity nano-graphite for use in aerospace, automobile, batteries, etc. This approach to handling methane can considerably improve the economics of landfills as well as of anaerobic digester plants that generate electricity from biogas fueled electricity.

NANOTECHNOLOGY TO CONVERT BIOMASS INTO BIOFUELS

Delinking biofuels production from food crops is a necessary condition to expand the scale of the market penetration of biofuels globally. Among the challenges this strategy faces is the inherent resistance of cellulosic feed stocks to conversion to simpler sugars that can be fermented into ethanol. Here, the promise lies in nano-particles used as immobilizing beds for expensive enzymes that can be used over and over again to breakdown the long chain cellulose polymers into simpler fermentable sugars [16].

The Louisiana Tech University is one among many organizations worldwide engaged in this endeavor, through the work of Dr. James Palmer, in collaborating with fellow professors Dr. Yuri Lvov, Dr. Dale Snow, and Dr. Hisham Hegab [17]. The focus is on non-edible cellulosic biomass, such as wood, grass, stalks, etc, to be converted into ethanol. This approach to produce ethanol can reduce GHG emissions by some 86% over fossils fuels.

The broader field of nanotechnology research into converting biomass into biofuels is growing fast. For example, in 2007 the oil company BP has granted a research fund of $500 million to the University of California, at Berkeley, and the University of Illinois, to explore the conversion of corn, plant material, algae and switch grass into fuel [18].

In the past, Berkeley had used nanotechnology in research for cost-effective solar panels [19]. But, the new Energy Biosciences Institute – EBI created at Berkeley will focus on fuel production with minimum environmental impacts and carbon emissions. A three pronged approach is being employed that begins with technologies for better crop production, improved feed stocks processing and development of new biofuels. The application of this approach aims at developing better feed stocks, breaking down plant material into sugars and their conversion to ethanol. Success along this pathway is expected to lead EBI to investigate the use of nanotechnology to develop other alternative fuels, such as butanol and renewable hydrocarbon fuels.

Another relevant application of nanotechnology is the use of nano-catalysts for the trans-esterification of fatty esters from vegetable oils or animal fats into biodiesel and glycerol [20]. The nano-catalyst spheres replace the commonly used sodium methoxide. The spheres are loaded with acidic catalysts to react with the free fatty acids and basic catalysts to react with the oils. This approach eliminates several production steps of the conventional process, including acid neutralization, water washes and separations. All those steps dissolve the sodium methoxide catalyst so it can't be used again. In contrast, the catalytic nano spheres can be recovered and recycled. The overall result is a cheaper, simpler and leaner process. In summary, the process claims to be economical, recyclable, to react at mild temperatures and pressures, with both low and high FFA (free fatty acid) feedstock, producing cleaner biodiesel and cleaner glycerol, greatly reducing water consumption and environmental contaminants, and can be used in existing facilities.

NANOTECH LIQUID ADDITIVES

All previous presentation and discussion referred to solid nano-particles playing a catalytic role in the obtaining biofuels from algae, landfill methane and biomass. The following segments will examine the practical opportunities that exist for liquid nano-particles or droplets [21]. Consider multifunctional surface active liquid additives, whose lubricity enhancement is achieved via the formation of a monolayer over the surfaces in contact with additized fuel. [22] The treat rate for lubricity is determined by the adsorption saturation concentration. Speculate that the improved detergency and water co-solvency is obtained by the formation of nano emulsions. Also, postulate that the more complete combustion and consequent fuel efficiency increase is the result of the behaviour of nano droplets. These nano droplets result from the surfactant action of the additive in the fuel formulation and the presence of some water in all commercial fuel systems, usually due to evening condensation.

Research by Wulff and colleagues [23] has shown that nano emulsions, which the authors call micro emulsions, with fuel (biofuel included most likely), water and surfactant are:

- Thermodynamically stable and
- Microscopically isotropic, and
- Nano-structured (thus, nano emulsions).

Their research concluded that:

- The use of these nano structures with fuel, water and surfactant is able to break the usual tradeoff between reduction of soot and NO_x emissions, by achieving them simultaneously, and
- For the same fuel consumption, higher efficiency is obtained.

Strey and collaborators filed patent applications for what they call micro-emulsions used as fuel [24]. The interpretation offered for the behavior of stable diesel (and most likely biodiesel)-water-surfactant nano emulsions is as follows:

- The surfactant components –oleic acid and nitrogen containing compounds (amines) – dissolve readily in diesel (and possibly in biodiesel) fuel and bind water to it without stirring;
- The water droplets are as small as a nanometer across, helping stabilize the emulsion
- The result is a "liquid sponge", can be stored indefinitely, like ordinary diesel fuel, without risk of phase separation
- This fuel formulation, when burned, results in the near-complete elimination of soot, and a reduction of up to 80% in nitrogen-oxide emissions
- The surfactant in the formulation also burns without creating emissions beyond water, carbon dioxide and nitrogen

PUBLIC CONCERNS OVER NANO-TECHNOLOGY: SECURITY, HEALTH AND THE ENVIRONMENT

As with all new technologies, there may be cause to concern about impacts, such as on security, health and the environment. Nanotechnologies have been the subject of many assessments seeking to anticipate possible consequences of their deployment, to humans and to the environment. For instance, the Woodrow Wilson Center carried out a Nanotechnology project [25] from 2005. The project managers said that "manipulating materials at the atomic level can have astronomic repercussions, both positive and negative. The problem is no one really knows exactly what these effects may be." This was the motivation for the Project on Emerging Nanotechnology at the Woodrow Wilson Center.

Another initiative came from the International Risk Governance Council – IRGC's Nanotechnology project [26]. Two expert workshops were held. The first in May 2005 focused on how to frame nanotechnology, its risks and its benefits. A distinction was made between the nanotechnologies of the so-called Frame One (passive or classical technology assessment) and Frame Two (active or the social desirability of innovation). The second, in January 2006, concentrated on identifying gaps in nanotechnology risk governance and developing recommendations for improved risk governance.

A symposium on the subject took place in Zurich in July 2006. A presentation by Ortwin Renn [27] discussed the policy implications of Frame One, referred to in Fig. 4. The fact is that "most people have no clear associations when it comes to nanotech. They expect economic benefits but no revolutionary technological breakthroughs. Risks are often not explicitly mentioned but there is a concern for unforeseen side effects. There is a latent concern about industry, science and politics building a coalition against public interest. And one negative incident could have a major negative impact on public attitudes."

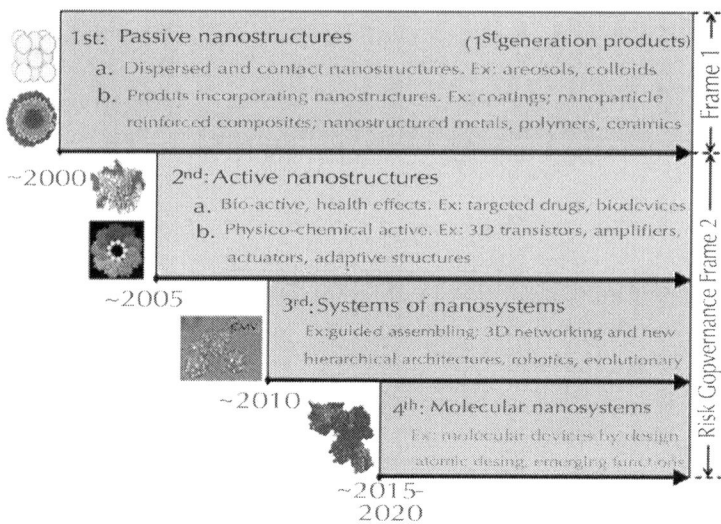

Figure 4: Frames of reference of nanotechnology generations.

The IRGC's Nanotechnology project concluded[28], among other things, that "communication about nanotechnology's benefits and risks should reflect the distinction between passive and active nano-materials and products, stressing that different approaches to managing risks are required for each. Care should also be taken to ensure that potential societal concerns about the possible impacts of Frame Two active nano-materials do not have the effect of unnecessarily increasing anxiety regarding Frame One products using only passive nanostructures." This is further expounded by Renn [29] as follows: "Frame One passive nanostructures are found, for example, in easy-to-clean surfaces, paints or in cosmetics. Frame Two refers to active nanostructures and molecular systems which could be able to interact actively or could be understood as evolutionary bio systems which change their properties in an autonomous process."

In reality, nanotechnologies are already facing challenges. Man-made nano-materials have been banned by the UK Soil Association from all its certified organic products. The 2008 annual report of the Soil Association of the UK contains the following statement [30]: "The Soil Association published the world's first standards banning nanotechnology. The risks of nanotechnology are still largely unknown, untested and unpredictable. Initial scientific studies show negative effects on living organisms, and three years ago scientists warned the Government that the release of nanoparticles should be 'avoided as far as possible'. There are many parallels with GM in the way nanotechnology is developing, particularly because commercial opportunities have run ahead of scientific understanding and regulatory control. What's more, while nano-substances are being rapidly introduced to the market, there is no official assessment process or labeling of the products – which is even worse than GM.

Health and beauty products that use nanoparticles are of concern for their potential toxicity if they get under the skin. Similar concerns exist regarding food and textiles. Definitely, more studies about health and environmental impacts are needed, to alleviate public concerns.

On the other hand, there is so much potential for nanotechnologies to do good, that Frame One and Two assessments should proceed as new applications evolve, including for instance more effective delivery of drugs to fight human and animal disease.

Fig. 5 showing a RNA nano-particle created by Peixuan Guo of Purdue University illustrates the point. Strands are spliced together from two kinds of RNA – a scaffold and a hunter to find cancer cells. This nano-structure has proven effective against cancer growth in living mice as well as lab-grown human nasopharyngeal carcinoma and breast cancer cells.

CONCLUSIONS

Increasing demand for energy services in the decades ahead will require an expanding supply of liquid fuels, despite efforts at improving energy efficiency and diversification of energy systems, including growing use of electricity in transportation. Biofuels have a key role to play in this scenario. However, the future supply of biofuels must be of such a scale that non-food feed stocks and new technologies are intensively employed. Nanotechnologies are primary candidates to play a prominent role in this energy future. They will help bring to markets liquid biofuels, including renewable hydrocarbons, from algae, carbohydrates, fatty esters and biogas. Nanotechnologies will also play a role in augmenting the efficiency of using current and future liquid fuels, especially biofuels, by providing improved

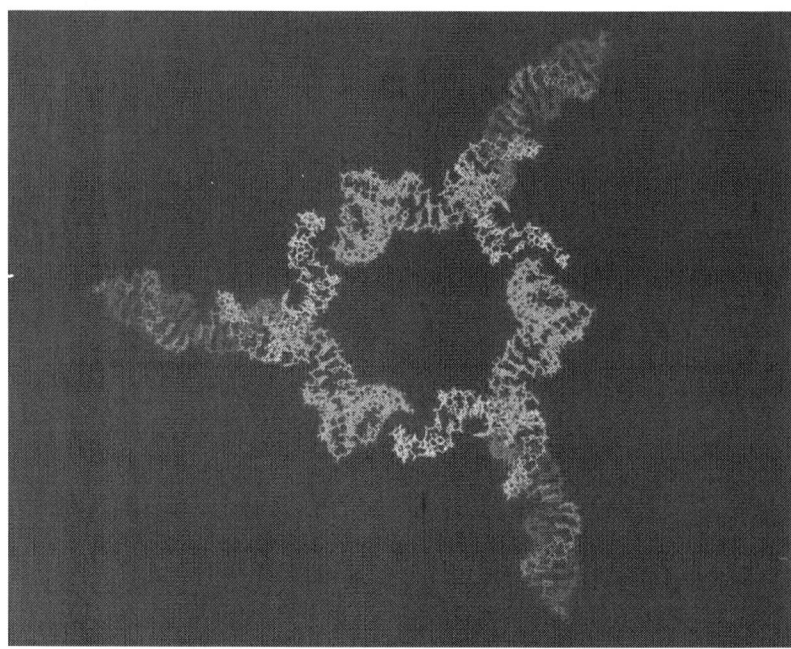

Figure 5: RNA nano-particle created by Peixuan Guo, Purdue University [31].

combustion of nanodroplets. While there are risks in each and every new technology, the world today is much better equipped to assess risks and act accordingly, that it seems possible to advance nanotechnologies applied to biofuels, without jeopardizing security, public health or the environment. But, the reach of nanotechnologies is vast and goes much beyond biofuels and offer hopes in so many areas, including importantly, human health.

REFERENCES

1.   S. C. Trindade, 2010Refining will definitely survive. Pipeline Magazine, 29 August 2010

2.   S. C. Trindade, 2010Renewable Energy Perspective- a profitable pathway from oil, Exploration and Processing, Fall 2010 [89Sep.

3.   S. C. Trindade, 2010International Biofuels Trade: Issues and Options. International Biofuels Conference_São Paulo, 2628May.

4. G. Santana, S. Quirk, 2009Growing Green: An In-Depth Look at the Emerging Algae Industry, Greener Dawn Research, 22 July, 16p.

5. A. Sustainable, Consensus. Biofuels, 2008Statement from a conference hosted by the Rockefeller Foundation Bellagio Study and Conference Center, Bellagio, Italy, 2428March 2008

6. www.defra.gsi.gov.uk2007In: F.O. Lichts's World Ethanol and Biofuels Report, 4365 and Vol4 17 p.391, Turnbridge Wells, U.K.: F.O. Licht, 2006.

7. da. Carvalho, F. Silva, P. R. da, Brum. T. N. Costa, Santos. dos, 2005Nanotechnology/Nanoscience Knowledge Managament emphasizing nanostructured polymers. Presentation, School of Chemistry, UFRJ, Brazil.

8. S. Borschiver, M. J. O. C. Guimarães, T. N. dos, F. C. Santos, Silva. P. R. C. da, Brum, 2005Patenteamento em Nanotecnologia: Estudo do Setor de Materiais Poliméricos Nanoestruturados., Polímeros: Ciência e Tecnologia, 15n° 4, 245248

9. http://ecolocalizer.com/2009/04/23/nanotechnology-to-aid-the-commercial-viability-of-algal-bio-fuel-production, 2009

10. http://www.qsinano.com/news/releases/2009_02_24.php

11. http://www.ameslab.gov/news/news-releases/nanofarming-technology-extracts-biofuel-oil-without-harming-algae

12. http://www.ameslab.gov/news/news-releases/nanofarming-technology-extracts-biofuel-oil-without-harming-algae

13. http://biomassmagazine.com/articles/2354/dudek-catalyx-nanotech-to-build-landfill-facilities

14. http://journalstar.com/news/local/article_6d5b6a34-e86f-11df-ae58-001cc4c002e0.html2010

15. http://en.wikipedia.org/wiki/Global_warming_potential

16. http://www.sciencedaily.com/releases/2009/10/091008131858.htm

17. Nanotechnology Used In Biofuel Process to Save Money, Environment Science Daily (Oct.2009

18. http://berkeley.edu/news/media/releases/2007/02/01_ebi.shtml

19. U.S. Department of Energy.Berkeley Lab Helios Project. (n.d.) Helios Solar Energy Research Center. Goals and challenges.

Retrieved December 10, 2009from http://www.lbl.gov/LBL-Programs/heliosserc/html/goals.html

20. http://www.public.iastate.edu/~nscentral/news/2007/jun/catilin.shtml

21. Cleaner diesel engines- pouring water on troubled oils, The Economist, June 3rd,201086http://www.economist.com/node/16271415

22. http://www.internationalfuel.com

23. P. Wulff, B. Lada, S. Engelskirchen, R. Strey, 2008Water-biofuel microemulsions. Institute for Physical Chemistry, University of Cologne.

24. http://strey.unikoeln.de/fileadmin/user_upload/Download/WATER___BIOFUEL_MICROEMULSIONS.pdf

25. http://strey.uni-koeln.de/333.html?&L=1

26. R. Strey, et al.2007Microemulsions and use thereof as a fuel. US Patent Application 2007/028507, Feb. 8.http://www.rexresearch.com/strey/strey.htm

27. http://www.loe.org/shows/segments.htm?programID=05-P13-00050&segmentID=3

28. http://www.irgc.org/-Nanotechnology-.html

29. http://www.yasni.ch/ext.php?url=http%3A%2F%2Fwww.irgc.org%2FIMG%2Fpdf%2FOrtwin_Renn_Nanotechnology_Frame_1_Policy_Implications_.pdf&name=Ortwin+Renn&cat=document&showads=1

30. http://www.irgc.org/Policy-Recommendations,188.html

31. http://ec.europa.eu/health/ph_risk/documents/ev_20081002_rep_en.pdf, p.14

32. http://www.soilassociation.org/LinkClick.aspx?fileticket=Moyw3Q7H%2Fp4%3Dtaid=303,22

33. http://www.eng.uc.edu/nanomedicine/Papers/1NCI.pdf

6

Producing Fuels and Fine Chemicals from Biomass Using Nanomaterials

Lei Pei[1], Markus Schmidt[1], and Wei Wei[2]

[1]Organisation for International Dialog and Conflict Management, Austria.
[2]Institute of Botany, Chinese Academy of Sciences, China.

INTRODUCTION

To build our economy on a sustainable basis, we need to find a replacement for fossil carbon as chemical industry feedstocks (Andrady and Neal, 2009). There are growing concerns about current petroleum based production, accumulation of waste in landfills and in natural habitats including the sea, physical problems for wildlife resulting from ingestion or entanglement in plastic, the leaching of chemicals from plastic products and the potential for plastics to transfer chemicals to wildlife and humans (Thompson et al., 2009). Bioplastics, derived from bio-based polymers, may provide a solution. Unlike the chemically synthesized polymers, the bio-based polymers are produced by living organisms, such as plants, fungi or bacteria. Some microorganisms are particularly capable in converting biomass into biopolymers while employing a set of catalytic enzymes. Attempts to transfer biomass to produce industrially useful polymers by traditional biotechnological approaches have obtained only very limited success, suggesting that an effective biomass-conversion requires the synergistic action of complex networks. As an interdisciplinary research field which is a unique combination of life science and engineering, synthetic biology can provide new approaches to redesign biosynthesis pathways for the synergistic actions of biomassconversion and may ultimately lead to cheap and effective processes for conversion of biomass into useful products such as biopolymers. In the following sections of this review, we will give first an introduction on bioplastics and synthetic biology (section 2). The properties of bio-based polymers for bioplastics equal to or better than their chemical synthetic counter parts will be compared. The subfields of synthetic biology related to polymer biosynthesis will be reviewed. In the Section 3, we will focus on synthetic biological approaches to improve the biological system to produce polymers for bioplastics, such as polyhydroxyalkanoates (PHAs). The fourth section then goes on to evaluate the environmental impacts of the synthetic biology derived bioplastics. In this section, we will review the current methods to measure the environmental impacts on bioplastics on greenhouse gases (GHGs) emission, direct or indiret land usage, energy consumptions and waste management, as well as the current regulation guidelines on bioplastics in Europe. And in the last section, we will summarize the perspectives of synthetic biology and bioplastics.

SYNTHETIC BIOLOGY AND BIOPLASTICS

Since industrialization, human beings have relied largely on chemicals derived from fossil carbon. Due to the unsustainable nature of petroleum and coal, it is essential to develop better alternatives to produce chemicals from renewable sources. The implementation of biological processing in industrial scale may help to reduce the undesirable damage to the environment caused by the massive scale productions from the current fossil fuel based chemical industry. Among the commodity chemicals, polymers for plastics are becoming independent for our modern lifestyle (Andrady and Neal, 2009). Plastic have transformed our everyday life and their usage is increasing. Based on the report from Plastic EU 2009, the average annual increase of plastic production and consumption is 9% globally (PlasticsEurope, 2009). The plastic demand in EU was 48.5 million tons in 2008, of which 75% was composed by mainly five high volume families-polyethylene (PE), polypropylene (PP), polyvinylchloride (PVC), and polystyrene (PS) and polyethylene terephthalate (PET). Majority of them are produced by chemical synthesis starting with petroleum feedstocks. These synthesized plastics are durable which make them resistant to biological degradation. Due to the toxic additives, for example, plasticizers like adipates and phthalate, burning plastic can release toxic pollutants. And the manufacturing of chemical industrial processes often creates environmental hazards. Due to the concerns over the climate change, limitation of the fossil carbon source and environmental mitigation, there are renewed interests in bioplastics. There are mainly three types of bioplastics in the commercial scale of production (Table 1).

- Plastics derived from fossil carbon source but biodegradable, and
- Plastics derived from polymers converted from biomass and biodegradable, and
- Plastics derived from polymers converted from biomass but not biodegradable.

In our current review, we will focus on the biodegradable, compostable and bio-based plastics. In our paper, we use the term 'bioplastics' only referring plastics made from this type of polymers.

Table 1: Polymers for bioplastics on the market (Kaeb, 2009

Polymer	Company (example)	Scale (ton/year)
Biodegradable but not bio-based		
Synthetic polyesters	Ecoflex (BASF)	14 000

Polyvinyl alcohol (PVA)	Wanwei	100 000
Biodegradable and bio-based		
Starch based materials	Novamont	60 000
Cellulose based materials	Innovia Films	30 000
Polylactide (PLA)	Natureworks	140 000
Polyhydroxyalkanoate (PHA)	Metabolix (ADM)	50 000
Biobased but non-biodegradable		
PDO from biobased glycerol	Tate & Lyle (DuPont)	45 000
PE from Bioethanol	Braskem (DOW)	200 000 (planned)
PVC from Bioethanol	Solvay	360 000 (planned)
Polyamides (PA) from oils	Arkema	6 000 (planned)

The bioplastics available in the market are made from polymers such as starch-based, polyhydroxyalkanoates (PHAs), polylactic acid (PLA) and other polymers derived from renewable sources. Polylactic acid (PLA) is a type of thermoplastic polyester resulting from the chemical polymerization of the D- and L- lactic acids obtained from fermentation (Madhavan Nampoothiri et al., 2010). Bioplastics made from PLA shows the similar properties as those made from petroleum derived polyethylene terephthalate polymer. Besides applications in plastics, PLA derived bioplastics are also used extensively in biomedical applications, such as sutures, stents, dialysis devices, drug capsules, and evaluated as a matrix for tissue engineering (Park et al., 2008; Shi et al., 2010; Shi et al., 2009; Yao et al., 2009). Polyhydroxyalkanoates (PHAs) derived plastics are considered as the best candidates to replace the current petroleum-based plastics due to their durability in use and wide spectrum of properties. The family of PHAs polymers is one of the most promising biodegradable materials to emerge in recent years. Up to date, there are more than 100 different monomers of PHA polyesters (Chen, 2009). The PHA monomers have been divided into two classes- short-chain-length hydroxyalkanoate (scl-HA, monomer of 3 to 5 carbons) and medium-chain-length HA (mcl-HA, monomer of 6 to 14 carbons) (Tsuge, 2002). Unlike other 'degradable' polymers such as those based on petrochemicals, PLA and starch polymers, PHAs have useful natural properties and, therefore, do not need to sacrifice their true biodegradability to improve their properties further. They have properties similar to those of PE and PP, ranging in properties from strong, mouldable thermoplastics to highly elastic materials to soft, sticky compositions. PHAs can be blended in the large number of copolymers which allow further engineer of polymers to the desired properties for a wide range of applications. Similar to those derived from PLA, PHAs have been used in a variety of biomedical applications such as sutures and surgical meshes. Polyhydroxybutyrate (PHB) is the simplest yet best known polyester of PHAs, which was first discovered as an intracellular biopolyester produced by the Gram-positive bacterium Bacillus megaterium (Lenz and Marchessault, 2005). PHB derived bioplastics are heat tolerant with a melting point at 175 °C. PHB is commonly used to make heat tolerant and clear packaging film. Since PHAs have many properties that are superior to those of PLA, there are more renewed interests in PHAs development (Chen, 2009). Thus, we will focus our review on the latest research on PHAs.

Bioplastics are currently only in a small portion (under 1%) of market share of plastics (EuropaBioplastics, 2011). In Europe, bioplastics on the market are defined as the plastics made from polymers according to standards (EN 13432). The polymers for bioplastics are usually produced in the biological fermentation processes using renewable and sustainable agricultural feedstocks, such as sugar, starch, oil, or lignocellulosic biomass. There are also non fermentative processes (thermochemical ones) in development to convert biological feedstocks to chemicals which have been reviewed elsewhere (Gong et al., 2008; Gray et al., 2006; Hames, 2009; Mu et al., 2010). In addition, some of the polymers can also be produced directly in planta by industrial crops constructed via genetic modification (GM) routes (Matsumoto et al., 2009; Poirier, 2001; Valentin et al., 1999).

Up to date, most of the research on the production of biopolymers is at the concept or early research stage of development, while a few have been on commercial scales (Chen, 2009). Most of these polymers are manufactured via the microbial fermentation routes. A concept of a new branch of biotechnology termed white biotechnology, has been developed, referring to the industrial development and implement strategies for chemical production based on biomass-derived carbon sources (EuropaBio, 2003; Hermann and Patel, 2007; Soetaert, 2007). Although the technologies were developed many years ago, large scale production of polymers from biomass was not feasible because those technologies were too expensive. However, in recent years, the innovations from the research sectors, particularly those on biotechnology, have made some of the biological conversions able to compete with the existing fossil based processes. To mention a few, there are productions of vitamins, antibiotics and ethanol using biotechnological approaches (DSM, 2004; EuropaBio, 2003; Hermann and Patel, 2007; Soetaert, 2007).

Table 2: Achievements in Synthetic Biology

Applications	Examles	Reference
Medical applications Fuels & Energy	Production of anti-malaria drugs precursor in engineered yeast	(Ro et al., 2006)
	Noval polio virus vaccine	(Coleman et at, 2008)
	Controlling transgene expression in subcutaneous implants	(Gitzinger et at, 2009)
	Non-fermentative pathways for synthesis of branched-chain higher alcohols as biofuels	(Atsumi et al., 2008)
	Biological system as a template for photocatalytic nanostructures	(Nam et al., 2010)
Chemicals	Synthetic protein scaffolds	(Dueber et al., 2009)
	Improved production of glucaric aicid in engineered E. coli	(Moon et at, 2010)
Biological computing	Synthetic oscillatory network of transcriptional regulator	(Elowitz and Leibler, 2000)
	Tunable synthetic mammalian oscillator	(Tigges et al., 2009)
	Synchronized quorum of genetic clocks	(Danino et al., 2010)

Organisms	Synthetic microbes	(Gibson *et al.*, 2010)
	Genetic modified insects for pest control	(Fu *et al.*, 2007)
Environmental applications	Arsenic sensor	(iGEM, 2006)
	Biosensors seek and destroy herbicide	(Sinha *et al.*, 2010)

Synthetic biology (SB) has been considered as a new way of doing biotechnology. It is an emerging science and engineering field that applies engineering principles to biology. The potential benefits of SB include the development of novel medicines, renewable chemicals and fuels (Gaisser et al., 2009). Due to the infancy of SB, a variety of definitions are circulating in the scientific community, and no consensus definition is drawn. SB-related research has been performed in several fields such as DNA synthesis (or synthetic genomics) (Carlson, 2009; Carr and Church, 2009; Gibson, 2009; Gibson et al., 2008a; Gibson et al., 2008b; Gibson et al., 2010; Gibson et al., 2009), engineering DNA-based biological circuits (Canton et al., 2008; Endy, 2005), minimal genome (or minimal cell) (2009; Luisi, 2007; Mushegian, 1999; Wegrzyn, 2001), protocells (or synthetic cells) (Bedau et al., 2009, Luisi, 2007) and xenobiology (or chemical synthetic biology) (Benner and Ricardo, 2005; Deplazes and Huppenbauer, 2009, Schmidt 2010). Among all these activities, engineering DNA-based biological circuits has contributed significantly to design advanced genetic constructs and to redesign biosynthesis pathways or fine-tuned genetic circuits for biopolymer production. One of the notable examples of SB derived biological circuits is the design of a metabolic pathway to produce a precursor of the anti-malaria compound Artemisinin, naturally found in the wormwood plant (Artemisia annua). The design and construction of this plant derived compound in engineered microbes, showed off the state of the art of enhanced metabolic engineering via SB (Hale et al., 2007; Keasling, 2008; Martin et al., 2003). A handful of successful circuits have been constructed to convert biomass to fuels and chemicals (Table 2). Besides working on practical applications, the research community also works on building so called standard biological parts (aka biobricks whereas 'biopart' is the technical expression) intending to rationalize and reduce the design complexity (Shetty et al., 2008; Smolke, 2009). By building such bioparts, it would provide a useful set of wellcharacterized, pre-fabricated, standardized and modularized genetic compounds (such as sequences of DNA) for (re)engineering biological pathways with defined functions. If succeeded, these bioparts will provide useful elements to build up systems to execute the designed functions, such as microbial fermentation to brew new biopolymers for plastics.

SYNTHETIC BIOLOGICAL APPROACHES TO CONVERT BIOMASS TO BIOPLASTICS

PHAs can be produced in almost all bacteria in the form of intracellular inclusions. In response to a certain deficient growth conditions, the biosynthesized PHAs can make up to 90% of the dry cell mass (Garcia et al., 2004; Haywood et al., 1990; Lee et al., 1999; Madison and Huisman, 1999; Yim et al., 1996). Some microbes in their wild type forms can already produce PHAs in sufficient quantity (ranging from 50% to 80% of the dry cell mass), such as Rolstonia eutropha (>80%), Alcaligenes latus (>75%), and Pesudomonas oleovorans (>60%) (Chen, 2009). There are three well known natural biosynthesis pathways of PHAs (Tsuge, 2002). The pathway I is most common which is found in many bacteria. It leads to generate 3- hydroxybutyryl (3HB) monomers from acetyl-CoA derived from sugars while a set of enzymes are involved in this process, such as PhaA (3-ketothiolase which converts acetylCoA to acetoacetly-CoA), PhaB (NADPH-dependent acetocaetly-CoA reductase, resulting in 3-HB-CoA) and PhaC (PHA synthase, polymerizing 3HB-CoA to the final monomers). The pathway II and III are more commonly found in the genus of pseudomonas. They are pathways using either sugars or fatty acids as carbon source to convert to either acetyl-CoA or acyl-CoA, resulting mainly in mcl-(R)-hydrooxyacly (3HA) monomers. The (R) - specific enoyl-CoA hydratase (PhaJ) and (R)–3-hydroxyacyl-ACP-CoA transferase (PhaG) play similar roles as PhaB in the pathway I to obtain 3HA-CoA. PHAs are truly natural polymers yet with properties similar to those of the synthetic ones. There is a renewed interest in producing PHAs in a biological process fed with sustainable sources. Several refactored microbes have been constructed to produce PHAs (Aldor and Keasling, 2003; Garcia, et al., 2004; Hofer et al., 2010; Lee, et al., 1999; Li et al., 2007; Park et al., 2002; Qiu et al., 2006; Sandoval et al., 2007; Zhang et al., 2009). One commonly used engineering strain is E. coli. Recombinant E. coli strains harbouring a set of PHAs biosysthesis genes (phaCac from Aeromonas caviae, phaCABa1 from Alcaligenes latus, or phbCAB from Ralstonia eutropha) were constructed (Taguchi et al., 1999). Sufficient amount of PHAs can be produced, while using sugars or fatty acids as carbon sources. Besides E. coli, Aeromonas hydrophila harbouring phbAB or phaPCJ were also constructed to produce PHAs using lauric acid as carbon source (Chen et al., 2001). Some of the natural PHAs producing strains have been subjected to genetic modification to enhance their productivity. For instance, double knockout mutant of P. putida was generated using suicide plasmid bearing beta-oxidation genes fadA and fadB. The mutant strains produced more PHAs (84%) than its wild type strains (50%) (Ouyang et al., 2007).

Based on these strains, a couple of biotechnology based fermentation processes have been developed for the industrial scale productions of PHAs (Chen, 2009). These fermentations are usually performed in two phases. The first phrase is a cell growth phrase to obtain high cell density and the second phrase is a PHA-production phrase of which cell growth deficient conditions are deployed in favour of the metabolic shift to PHA biosynthesis. To date, most of the PHAs production via microbial fermentations relies mainly on sugars and fatty acids as carbon sources (Chen, 2009). They are derived from the sustainable resources but with limitations while used in large scale. For instance, sugars are usually derived from starch (corn) that is also major food sources. Thus, PHAs produced by the current biotechnology processes are still far more expensive than those derived from fossil carbons. In order to make a biological based process compete with the traditional chemical processes, the PHAs will have to be produced at a higher yield preferably from a non-food biomass, more advanced PHA monomers with novel properties, less energy consumption and less greenhouse gases (GHGs) emission while full life cycle analysis (LCA) is taken into consideration. It is highly expended that the novel approaches developed by SB may help to achieve these goals. One of the contributions from SB is to develop better PHA production strains. Nearly all previous work using genetic engineering approaches has focused on reorganizing PHA biosynthesis genes derived from natural organisms. In contrast, SB will provide an integration of genetic engineering, metabolic engineering, chemistry and bioinformatics. For instance, using SB methods, microbes will not only harbour genetic circuits coding for PHA biosynthesis pathway, but also metabolic pathways which will enhance productivity. A research group of Tsinghua University has worked on such genetic circuits to enhance the production of PHAs in engineered microorganisms such as E. coli, P. putida and A. hydrophila (Jian et al., 2010; Li et al., 2010; Wei et al., 2009). In order to convert lab scale fermentation to produce PHAs into the industrial scale, cells must be engineered to be able to grow in high density. It has been proposed that the limited oxygen supply is a hurdle that cells face while growing to high density. The problem was solved by constructing synthetic pathways that are turned on in response to micro- or anaerobic condition (Jian, et al., 2010). This approach has been also applied to produce poly-3-hydroxybutyrate (PHB) (Li et al., 2009), which is currently produced from starch or glucose based feedstock by bacteria. The synthetic pathways were constructed to enhance PHB production from 29% to 48% of the cell dry weight under anaerobic conditions. One of the key obstacles to produce renewable polymers by fermentations lies on the cost intensive downstream processing steps. A novel approach has been developed to remove this obstacle by equipping engineered strains with programmed autolysis genetic circuits. An inducing lysis system has been implemented in cyanobacterium Synechocystis sp. to facilitate extracting lipids for biofuel production (Liu and Curtiss, 2009). This inducible lytic system is composed of a lytic circuit from bacteriophage-derived lysis genes and an inducible circuit of genes encoding a nickel-inducible signal transduction system. If such programmed autolysis system is implemented in PHA producing strains, the efficiency of PHA fermentation will be increased.

To maximize the benefit of SB derived PHA production, it has been suggested to design producing strains to product PHAs with desired properties for applications of high added value. There were a couple of reports on PHAs with unusual properties, such as adding functional groups side groups to the PHA monomer mediated by low specific PHA synthases (Hiraishi et al., 2006; Luengo et al., 2003; Sandoval, et al., 2007; Tsuge et al., 2005), PHA monomer of high molecular weight by mutated synthases (Zheng et al., 2006), and funtionalized

PHAs containing C-C double bonds when fed unsaturated fatty acids (methanol) (Hofer, et al., 2010).

One of the key issues hindering the large scale biopolymer production is the cost of sugars and fatty acids as feed stocks. This issue is an obstacle not only to the production of PHAs but also other commodity chemicals and fuels. World food prices reached a record high recently, according to a report from the Food and Agriculture Organization (FAO) of the United Nations. The Food Price Index reached 214.7 points in December 2010 and slightly above the previous peak of 213.5 points during food crisis in 2008 (BBC News Bussiness, 2011). It has already been in debut that increasing biofuel production from starch and sugar may post further threat to the food safety. Thus, it is a key challenge to develop biological processes on how to harness the non-food biomass. In theory, cellulose biomass may be the best sustainable carbon source which could fill the gap left in crop supplies as stocks have been diverted for commodity chemicals (mainly biofuels) production. A great deal of work has been carried out to convert cellulosic biomass into useful products. It has been suggested by Christopher E. French that the ideal microbes should be equipped with a couple of properties such as the ability to hydrolyse cellulosic material effectively, with minimal requirement for pre-processing; the ability to convert the sugars released into molecules useful as liquid fuels and/or chemical industry feedstocks; self-tolerance to these molecules at a high concentration, and suitable respects for use in industrial bioreactors (French, 2009). Since such microbes are not known to exist in nature, they can be only constructed via complex engineering, something that SB is expected to be capable of. Most of the non-edible biomass is made up by long cellulose fibres of hemi-cellulose and lignin (Lynd et al., 2002; Lynd and Zhang, 2002). With respect to their chemical composition, hemicellulose is a mixture of monomers including D-xylose, L-arabinose, D-mannose and Dgalactose, together with sugar derivatives such as 4-O-methyl-D-glucuronic acid. Lignin is a complex formed by polymerization of aromatic monomers. Such cellulose can be digested to D-glucose by enzymatic or non-enzymatic hydrolysis. A typical enzymatic hydrolysis of cellulose involves a multiple step process mediated by several enzymes. A handful of enzymes have been identified to degrade cellulose efficiently from different microbes (Lynd, et al., 2002), to mention a few, endogulcanases (enzymes able to cleave cellulose chains at random positions), exoglucanases (enzymes that move along the chain cleaving cellobiose residues from either the reducing or non-reducing end), and -glucosidases (enzymes hydrolysis cellobiose to glucose). There is research on cellulosomes which are complexes of cellulolytic enzymes produced by bacteria (Bayer et al., 1994). Cellulosomes can be used for the degradation of cellulose and hemicellulose, and they have in fact been considered as one of nature's most elaborate and highly efficient nanomachines (Fontes and Gilbert, 2010). Integration of cellulosomal components occurs via highly ordered protein-protein interactions between two proteins: cohesins and dockerins, which specificity allow the incorporation of cellulases and hemicellulases onto a molecular scaffold. Cellulosomes can be used for a range of SB applications, from clothes whitening to paper waste treatment or chemical production from lignocellulosic biomass, and the first synthetic cellulosomes have already been constructed (Mitsuzawa et al., 2009). SB will play an important role in developing cellulose degradation module along with the chemical producing module where the standardization of biological parts and interdisciplinary nature of SB enable combination of multiple modules. As it has been mentioned above, SB approaches could create artificial enzymes that are tailored for enzymatic activities that do not exist in nature. To date, degradation of lignin and cellulose is still problematic though they can be degraded by enzymes produced by very few fungi. Using SB methods may develop lignin degradation enzymes with enhanced capability and that do not exist in nature. In addition, one possible solution is to make the biomass easy in hydrolysis such as the biomass from genetic modified plants. Such attempt has been tried by lignin deficient genetically modified plants (Baucher et al., 2003). The same idea has been applied to develop genetically modified potatoes for industrial applications which have been approved to cultivate in EU (Ryffel, 2010). Thus, no doubt, biomass with properties that is more suitable for industrial scale of fermentation will be developed as well.

ENVIRONMENTAL IMPACTS OF SYNTHETIC BIOLOGICAL PRODUCTION OF BIOPLASTICS

The plastics made from fossil derived synthetic polymers have posted serious problem for environments (Thompson et al., 2009). Each year, millions tons of synthetic plastics discarded and less than one tenth has been recovered and recycled. Majority of them are end up in landfill or enter the natural habitats (waters and surface land) (Hopewell, et al., 2009). Some disposed synthetic plastics can remain in the environment for up to thousands of year (Andrady, 1994). Furthermore, the predicted depletion of fossil fuel resources, the building up landfill of plastic waste and the implementation of low-carbon environmental protection initiatives call for an intensive search for alternatives to synthetic plastics (Prieto, 2007). Bioplastics may offer benefits relative to fossil-based plastics, particularly environmental benefits. Today bioplastics have been considered as a better solution to the increasing demands for truly sustainable growth. In developing the new generation of plastic products made from renewable feedstock, the ability of biodegradability, compostability and the evaluation measured by Life Cycle Analysis (LCA) should be all taken into account to evaluate the contributions to the environmental impacts of bioplastics.

Biodegradability is the ability of organic substances and materials to be entirely converted into simple inorganic molecules such as water, carbon dioxide and methane through the biological processing mediated by microbial enzymes (Novamont, 2011a). Such a processing is part of the natural life cycle of carbon recycling of the earth. Ideally, organic waste from human activities should be removed through biodegradation to minimize their impacts to nature. It is critical to identify the key components in this processing to reach maximum efficiency. As it is known, it takes different time to biologically degrade different types of organic waste, for instance, straw and wood take longer than starch and cellulose. This implies that biodegradation is strongly influenced by the chemical nature of the substance or material to be biodegraded. To facilitate the processing, the industrial biodegradation has been used to process the urban waste. It occurs at a consistent step of composting and anaerobic digestion. Composting will produce mature compost which turns into a fertiliser, while anaerobic digestion followed by stabilisation through composting will produce biogas.

Compostability is the capacity of an organic material to be transformed into compost through the composting process (Novamont, 2011b). Composting can be conducted at a small scale such as home-hold composting bin, or at an industrial level. The production of compost and its use in agriculture represent the closure of the environmental cycle and constitute a simple way to address the problem caused by the removal of organic substances from agricultural soils, reduced soil fertility and the onset of desertification. Composting is currently applied to selected waste that only contains biodegradable organic matter where the traditional plastics are not included in composting because they resist to the biodegradation. In contrast, biodegradable plastics can be included in composting, but only if they satisfy the criteria established by the standards that define compostable materials. Incompatible materials were composted in the past in the absence of rules and in a context of unregulated definitions and test methods. This caused significant damage, not least to the trust of users and technicians responsible for composting facilities. Bioplastics that comply with these standards can play a fundamental role in the valuation and optimisation of the composting process and in the production of high quality compost.

Bioplastics, particularly those made from PHAs are truly biodegradable. PHAs can be broken down into carbon dioxide and water by hydrolysis and microbial fermentations. PHAs-degrading enzymes have been identified in some bacteria and fungi, termed depolymerises (Elbanna et al., 2004; Jendrossek and Handrick, 2002; Kim do et al., 2007; Tokiwa and Calabia, 2004). PHAs degrading microorganisms can be enriched from samples collected from various ecosystems where PHAs have been supplemented as a sole source of carbon and energy. There are mainly two types of such microbial degradation- extracellular and intracellular ones (Jendrossek et al., 1996). Besides using the native PHAs degrading strains, an advanced PHAs degrading system may be built up using the SB methods, for instances, better deploymerases or fermenting strains (Knoll et al., 2009). With more knowledge accumulating on the biodegradation of PHAs, we may expect that plastics made from polymers - which are easy to be broken down - will have more applications in the future. We also need to point out that not all bioplastics will degrade to the same extent in the anaerobic digestion step and they may perform differently in the different technologies. For this reason more research is needed on the behaviour of different bioplastics in different anaerobic systems. In addition, it has been suggested that the anaerobic fermentation on bioplastics may be combined with the production of biogas because bioplastics are carbohydrates with little or no nitrogen and therefore with a high C/N ratio, making them more suitable for energy generation during fermentation than the traditional organic waste (urban organic waste, agricultural waste, etc). SB derived degradation processes that enable the combination of degradation and energy generation will maximize the benefits of bioplastics further.

Life Cycle Analysis (LCA) has become essential in evaluating the environmental impact of a product (Novamont, 2011c). It is a method for evaluating and quantifying the energy and environmental consequences and potential impacts associated with a product/process/activity throughout the entire life cycle, involving the assessment on the entire product chain ("from cradle to grave"). Evaluations on the feedstocks, their production, use and end of life in the same context have enormous potential for many biomass derived products, particular biofuels and chemicals like bioplastics. To assess the non-renewable energy consumption and greenhouse gas emissions of PHB, the environmental performance of PHB derived from corn grain was evaluated through LCA (Kim and Dale, 2008). It showed that corn derived PHB offers environmental advantages over fossil polymers- with less non-renewable energy consumption (95% reduction) and less greenhouse gas emissions (200 % reduction) as compared to the petroleum-based plastics. Calculating the overall environmental benefits, particularly on GHGs and energy saving, attributed to lignocellulosic biomass derived products is challenging because whole production chains are complex. It can be expected that different approaches and interpretations on LCA will provoke a debate about the environmental merit of bioplastics.

A recent study on biofuel converting from lignocellulose conducted by Slade et al. showed that the most important factors affecting GHGs emission are the emissions from biomass production, the use of electricity in the conversion process; and from the potentially consequential impacts: (in)direct land-use change and fertiliser replacement (Slade et al., 2009). GHGs from land use change (the so-called "carbon debt") is one of the major contributors to the environmental impacts of biomass derived products. The GHGs of biofuel was studied, and it showed the time required for biofuels to overcome the potential carbon debt and begin providing GHGs benefits would be 100-1000 years depending on the ecosystem that was replaced. Thus, using crop biomass as feedstocks for chemical productions may take a long time to gain environmental benefits from the carbon debt. Using non-crop biomass, such as lignocelluloses, may maximize the environmental gains. Furthermore, there is ongoing research on using other inexpensive carbon sources for bioplastics productions. For example, xylose which is one of the abundant sugars in food industrial waste can be converted to lactic acid and acetic acid by an anaerobic fermentation. They can subsequently be used as feasible feedstocks for PHA production (Tsuge, 2002). Eventually, carbon dioxide would serve as an ultimate source for PHA production (Tsuge, 2002). It is known that several research groups are working using SB methods to engineer photosynthetic organisms (cyanobacteria or micro algae) that can catch carbon dioxide directly and convert it to biofuels (Fischer et al., 2008; Rosenberg et al., 2008). The same techniques can be applied to implement the PHA synthesis pathways instead of the ethanol or fatty acids synthesis pathways in the engineered microbes.

One important problem will arise: should synthetic biology be able to help solve the above technical issues, namely that more and more agricultural land will be devoted to plant energy-crops instead of food crops. In order to avoid such competition with food, we suggest also using non-food-competing biological resources such as perennial plants grown on degraded lands abandoned for agricultural use, crop residues, sustainable harvested wood and forest residues, double crops and mixed cropping systems, as well as municipal and industrial organic wastes. Besides the potential environmental benefits of bioplastics, it is worthwhile to point out another benefit of bioplastics due to the biological processing in their industrial productions. From this perspective, SB based techniques have the potential to avert the use or by-product of toxic molecules in the production process (DSM; EuropaBio, 2003; Hermann and Patel, 2007; Schmidt and Pei, 2011; Soetaert, 2007). While these productions will be at bulk chemical scale, the environmental impacts will be marked- less pollution releasing to the environment, less processing cost on industrial waste and etc. As it mentioned above, the synthetic plastics have been entered into the environment in large quantity since the last half century. And plastics are now one of the most common and persistent pollutants. Besides replacing synthetic plastics with more environmental friendly bioplastics which are degraded and compost, SB based techniques may provide new solutions for the already existing plastic waste- new approaches to break down the chemicals which are difficult to degrade and take long time. Microbes using synthetic polymers as substances might be created although there are none existed in nature. However, the consequences of the environmental leakage of the engineered plastic-eating microbes that could degrade plastic would be serious and the research should be subjected to thorough considerations (Collins, 2008).

The current legislation on bioplastics in the EU has been focused on sustainability, carbon footprint, and labelling. The European Directive 94/62 EC and Directive 2004/12/EC provide reference to European standard (EC Packaging, 2004). The EN 13432 defined the biodegradable and compostable polymers as that the plastics made from at least 90% of the organic material must be converted into CO_2 within 6 months. Directive 2004/12/EC on Packaging & Packaging Waste specifies what bio-based products are. Furthermore, the EU environmental regulations played a role in providing incentives for the emergence of the bio-based product market. There are different rationales in supporting the R&D in bioplastics. For the industrial stakeholders, bioplastics have high market potential. For the agricultural stakeholders, bioplastics will be a new market potential for agriculture derived raw materials. For the environmental protective agencies, bioplastics might contribute to reduce the so-called carbon footprint. At the national level, some countries (for instance, Germany) set specific targets, e.g. that the more than 75% of the packaging should be from renewable sources (the Federal Environment Ministry Germany, 2009).

CONCLUSIONS

Bioplastics, defined as plastics derived from renewable carbon sources that are also biodegradeable, have been considered as good candidates for sustainable development as well as eco-friendly environment. The R&D on SB approaches is highly expected to deliver feasible approaches to bulk production of bioplastics. There are many aspects where SB can contribute to obtain this goal (Figure 1). Both SB and bioplastics are in their early stage of development. Most of the SB related research on bioplastics is focused on improving production of PHAs. More matured processes for large-scale conversion of biomass to polymer by SB design approaches could significantly influence the bulk production of bioplastics. However, adoption by industry on a new biological processing is expected to be slow, even though there are clear benefits on environmental perspectives over the longterm. Regarding to the market potential of bioplastics, the incentives and subsidy from the legislation body are currently the main driving force of the R&D of bioplastics and have high influence to the marketing of the end products. Bioplastics derived from lignocellulosic sources hold promise for emission reductions yet the techniques in harnessing lignocellulosic biomass is at the development stages. If the techniques become feasible, it will make the feedstock price of bio-based products lower and competed with those of fossil sources. Designer polymers with new properties will broaden the application of bioplasticsnot only in packaging (currently), but also in other applications (medicine, textile, electronics etc). Improvements in biodegradation of bioplastics particularly combining with energy generation will make the benefits maximized.

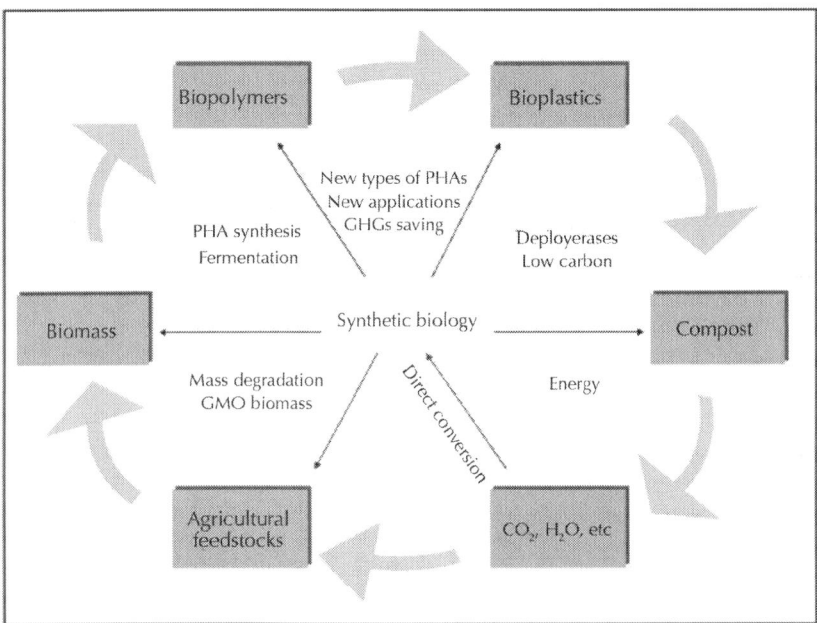

Figure 1: Synthetic biology, Bioplastics and Environment.

ACKNOWLEDGMENTS

This work was supported by a grant by the FWF (Austrian Science Fund) project "Investigating the biosafety and risk assessment needs of synthetic biology in Austria (Europe) and China," project number I215-B17; the EC-FP7 KBBE-2009 project TARPOL Targeting environmental pollution with engineered microbial systems à la carte (Grant agreement no.: 212894); and National Natural Science Foundation of China (NSFC 30811130544). Some of this work is based in part on the TARPOL project report, authored by Schmidt M, Mahmutoglu I, Porcar M, Armstrong R, Bedau MA, Morange M, Pei L, Danchin A, Schachter V, and Chanal A., titled Assessing economic, environmental and ethical implications of synthetic biology applications in environmental biotechnology.

REFERENCES

1. Aldor, I. S. and Keasling, J. D. (2003). Process design for microbial plastic factories: metabolic engineering of polyhydroxyalkanoates. Curr Opin Biotechnol 14, 475-83.

2. Andrady, A. L. (1994). Assessment of environmental biodegradation of synthetic polymers. Polymer Reviews 34, 25-76.

3. Andrady, A. L. and Neal, M. A. (2009). Applications and societal benefits of plastics. Philos Trans R Soc Lond B Biol Sci 364, 1977-84.

4. Atsumi, S., Hanai, T. and Liao, J. C. (2008). Non-fermentative pathways for synthesis of branched-chain higher alcohols as biofuels. Nature 451, 86-9.

5. Baucher, M., Halpin, C., Petit-Conil, M. and Boerjan, W. (2003). Lignin: genetic engineering and impact on pulping. Crit Rev Biochem Mol Biol 38, 305-50.

6. Bayer, E. A., Morag, E. and Lamed, R. (1994). The cellulosome--a treasure-trove for biotechnology. Trends Biotechnol 12, 379-86.

7. BBC news (2011). World food prices at fresh high, says UN. BBC. Available from: www.bbc.co.uk/news/business-12119539.

8. Bedau, M. A., Parke, E. C., Tangen, U. and Hantsche-Tangen, B. (2009). Social and ethical checkpoints for bottom-up synthetic biology, or protocells. Syst Synth Biol 3, 65-75.

9. Benner, S. A. and Ricardo, A. (2005). Planetary systems biology. Mol Cell 17, 471-2.

10. Canton, B., Labno, A. and Endy, D. (2008). Refinement and standardization of synthetic biological parts and devices. Nat Biotechnol 26, 787-793.

11. Carlson, R. (2009). The changing economics of DNA synthesis. Nat Biotechnol 27, 1091-4.

12. Carr, P. A. and Church, G. M. (2009). Genome engineering. Nat Biotechnol 27, 1151-62.

13. Chen, G. Q. (2009). A microbial polyhydroxyalkanoates (PHA) based bio- and materials industry. Chem Soc Rev 38, 2434-46.

14. Chen, G. Q., Zhang, G., Park, S. J. and Lee, S. Y. (2001). Industrial scale production of poly (3- hydroxybutyrate-co-3-hydroxyhexanoate). Appl Microbiol Biotechnol 57, 50-5.

15. Coleman, J. R., Papamichail, D., Skiena, S., Futcher, B., Wimmer, E. and Mueller, S. (2008). Virus attenuation by genome-scale changes in codon pair bias. Science 320, 1784-7.

16. Collins, J. J. (2008). Collins: Boston University University Professor Lecture titled Biology by Design. Available from: wyss.harvard.edu/viewpage/145/collins-bostonuniversity-university-professor-lecture.

17. Danino, T., Mondragón-Palomino, O., Tsimring, L. and Hasty, J. (2010). A synchronized quorum of genetic clocks. Nature 463, 326-330.

18. Deplazes, A. and Huppenbauer, M. (2009). Synthetic organisms and living machines: Positioning the products of synthetic biology at the borderline between living and non-living matter. Syst Synth Biol 3, 55-63.

19. DSM (2004). Industrial (white) biotechnology: An effective route to increase EU innovation and sustainable growth.

20. Dueber, J. E., Wu, G. C., Malmirchegini, G. R., Moon, T. S., Petzold, C. J., Ullal, A. V., Prather, K. L. and Keasling, J. D. (2009). Synthetic protein scaffolds provide modular control over metabolic flux. Nat Biotechnol 27, 753-9.

21. EC Packaging (2004). Directive 2004/12/EC of the European Parliament and of the Council of 11 February 2004 amending Directive 94/62/EC on packaging and packaging waste Available from: eurlex.europa.eu/LexUriServ/LexUriServ.do?uri=CONSLEG:1994L0062:20050405:EN: HTML.

22. Editorial NBT (2009). Unbottling the genes. Nat Biotechnol 27, 1059-1059.

23. Elbanna, K., Lutke-Eversloh, T., Jendrossek, D., Luftmann, H. and Steinbuchel, A. (2004). Studies on the biodegradability of polythioester copolymers and homopolymers by polyhydroxyalkanoate (PHA)-degrading bacteria and PHA depolymerases. Arch Microbiol 182, 212-25.

24. EN 13432 Standard EN 13432 and EN 14995 – Proof of compostability of plastic products. Available from: www.european-bioplastics.org/index.php?id=158.

25. Endy, D. (2005). Foundations for engineering biology. Nature 438, 449-53.

26. Elowitz, M. B. and Leibler, S. (2000). A synthetic oscillatory network of transcriptional regulators. Nature 403, 335-8.

27. EuropaBio (2003). White Biotechnology: Gateway to a More Sustainable Future.

28. EuropaBioplastics (2011). Bioplastics at a Glance. Available from: www.europeanbioplastics.org/index.php?id=182.

29. Fischer, C. R., Klein-Marcuschamer, D. and Stephanopoulos, G. (2008). Selection and optimization of microbial hosts for biofuels production. Metab Eng 10, 295-304.

30. Fontes, C. M. and Gilbert, H. J. (2010). Cellulosomes: highly efficient nanomachines designed to deconstruct plant cell wall complex carbohydrates. Annu Rev Biochem 79, 655-81.

31. French, C. E. (2009). Synthetic biology and biomass conversion: a match made in heaven? Journal of the Royal Society Interface 6, S547-S558.

32. Fu, G., Condon, K. C., Epton, M. J., Gong, P., Jin, L., Condon, G. C., Morrison, N. I., Dafa'alla, T. H. and Alphey, L. (2007). Female-specific insect lethality engineered using alternative splicing. Nat Biotechnol 25, 353-7.

33. Gaisser, S., Reiss, T., Lunkes, A., Muller, K. and Bernauer, H. (2009). Making the most of synthetic biology. Strategies for synthetic biology development in Europe. EMBO reports 10 Suppl 1, S5-8.

34. Garcia, B., Olivera, E. R., Sandoval, A., Arias-Barrau, E., Arias, S., Naharro, G. and Luengo, J. M. (2004). Strategy for cloning

large gene assemblages as illustrated using the phenylacetate and polyhydroxyalkanoate gene clusters. Appl Environ Microbiol 70, 5019-25.

35.   Gibson, D. G., Benders, G. A., Axelrod, K. C., Zaveri, J., Algire, M. A., Moodie, M., Montague, M. G., Venter, J. C., Smith, H. O. and Hutchison, C. A., 3rd (2008b). Onestep assembly in yeast of 25 overlapping DNA fragments to form a complete synthetic Mycoplasma genitalium genome. Proc Natl Acad Sci U S A 105, 20404-9.

36.   Gibson, D. G. (2009). Synthesis of DNA fragments in yeast by one-step assembly of overlapping oligonucleotides. Nucleic Acids Res 37, 6984-90.

37.   Gibson, D. G., Young, L., Chuang, R. Y., Venter, J. C., Hutchison, C. A., 3rd and Smith, H. O. (2009). Enzymatic assembly of DNA molecules up to several hundred kilobases. Nat Methods 6, 343-5.

38.   Gibson, D. G., Glass, J. I., Lartigue, C., Noskov, V. N., Chuang, R. Y., Algire, M. A., Benders, G. A., Montague, M. G., Ma, L., Moodie, M. M., Merryman, C., Vashee, S., Krishnakumar, R., Assad-Garcia, N., Andrews-Pfannkoch, C., Denisova, E. A., Young, L., Qi, Z. Q., Segall-Shapiro, T. H., Calvey, C. H., Parmar, P. P., Hutchison, C. A., Smith, H. O. and Venter, J. C. (2010). Creation of a Bacterial Cell Controlled by a Chemically Synthesized Genome. Science 329, 52-56.

39.   Gitzinger, M., Kemmer, C., El-Baba, M. D., Weber, W. and Fussenegger, M. (2009). Controlling transgene expression in subcutaneous implants using a skin lotion containing the apple metabolite phloretin. Proc Natl Acad Sci USA 106, 10638-10643Gibson, D. G., Benders, G. A., Andrews-Pfannkoch, C., Denisova, E. A., Baden-Tillson, H., Zaveri, J., Stockwell, T. B., Brownley, A., Thomas, D. W., Algire, M. A., Merryman, C., Young, L., Noskov, V. N., Glass, J. I., Venter, J. C., Hutchison, C. A., 3rd and Smith, H. O. (2008a). Complete chemical synthesis, assembly, and cloning of a Mycoplasma genitalium genome. Science 319, 1215-20.

40.   Gong, R., Zhong, K., Hu, Y., Chen, J. and Zhu, G. (2008). Thermochemical esterifying citric acid onto lignocellulose for enhancing methylene blue sorption capacity of rice straw. J Environ Manage 88, 875-80.

41. Gray, K. A., Zhao, L. and Emptage, M. (2006). Bioethanol. Curr Opin Chem Biol 10, 141-6.

42. Hale, V., Keasling, J. D., Renninger, N. and Diagana, T. T. (2007). Microbially derived artemisinin: a biotechnology solution to the global problem of access to affordable antimalarial drugs. Am J Trop Med Hyg 77, 198-202.

43. Hames, B. R. (2009). Biomass compositional analysis for energy applications. Methods Mol Biol 581, 145-67.

44. Haywood, G. W., Anderson, A. J., Ewing, D. F. and Dawes, E. A. (1990). Accumulation of a Polyhydroxyalkanoate Containing Primarily 3-Hydroxydecanoate from Simple Carbohydrate Substrates by Pseudomonas sp. Strain NCIMB 40135. Appl Environ Microbiol 56, 3354-9.

45. Hermann, B. G. and Patel, M. K. (2007). Today's and Tomorrow's Bio-Based Bulk Chemicals From White Biotechnology. Appl Biochem Biotechnol 136, 361-88.

46. Hiraishi, T., Hirahara, Y., Doi, Y., Maeda, M. and Taguchi, S. (2006). Effects of mutations in the substrate-binding domain of poly[(R)-3-hydroxybutyrate] (PHB) depolymerase from Ralstonia pickettii T1 on PHB degradation. Appl Environ Microbiol 72, 7331-8.

47. Hofer, P., Choi, Y. J., Osborne, M. J., Miguez, C. B., Vermette, P. and Groleau, D. (2010). Production of functionalized polyhydroxyalkanoates by genetically modified Methylobacterium extorquens strains. Microb Cell Fact 9, 70.

48. Hopewell, J., Dvorak, R. and Kosior, E. (2009). Plastics recycling: challenges and opportunities. Philos Trans R Soc Lond B Biol Sci 364, 2115-26.

49. iGEM (2006). Arsenic biosendor. Available from: parts.mit.edu/wiki/index.php/Arsenic_Biosensor.

50. Jendrossek, D. and Handrick, R. (2002). Microbial degradation of polyhydroxyalkanoates. Annu Rev Microbiol 56, 403-32.

51. Jendrossek, D., Schirmer, A. and Schlegel, H. G. (1996). Biodegradation of polyhydroxyalkanoic acids. Appl Microbiol Biotechnol 46, 451-63.

52. Jian, J., Li, Z. J., Ye, H. M., Yuan, M. Q. and Chen, G. Q. (2010). Metabolic engineering for microbial production of

polyhydroxyalkanoates consisting of high 3- hydroxyhexanoate content by recombinant Aeromonas hydrophila. Bioresour Technol 101, 6096-102.

53. Kaeb, H. (Year). Bioplastics: Technology, Markets, Policies. Available from: www.ecoembes.com/es/envase/prevencion/.../ HaraldKaeb.pdf.

54. Keasling, J. D. (2008). Synthetic biology for synthetic chemistry. ACS Chem Biol 3, 64-76.

55. Kim do, Y., Kim, H. W., Chung, M. G. and Rhee, Y. H. (2007). Biosynthesis, modification, and biodegradation of bacterial medium-chain-length polyhydroxyalkanoates. J Microbiol 45, 87-97.

56. Kim, S. and Dale, B. E. (2008). Energy and greenhouse gas profiles of polyhydroxybutyrates derived from corn grain: a life cycle perspective. Environ Sci Technol 42, 7690-5.

57. Knoll, M., Hamm, T. M., Wagner, F., Martinez, V. and Pleiss, J. (2009). The PHA Depolymerase Engineering Database: A systematic analysis tool for the diverse family of polyhydroxyalkanoate (PHA) depolymerases. BMC Bioinformatics 10, 89.

58. Lee, S. Y., Choi, J. and Wong, H. H. (1999). Recent advances in polyhydroxyalkanoate production by bacterial fermentation: mini-review. Int J Biol Macromol 25, 31-6.

59. Lenz, R. W. and Marchessault, R. H. (2005). Bacterial polyesters: biosynthesis, biodegradable plastics and biotechnology. Biomacromolecules 6, 1-8.

60. Li, Q., He, Y. C., Xian, M., Jun, G., Xu, X., Yang, J. M. and Li, L. Z. (2009). Improving enzymatic hydrolysis of wheat straw using ionic liquid 1-ethyl-3-methyl imidazolium diethyl phosphate pretreatment. Bioresour Technol 100, 3570-5.

61. Li, R., Zhang, H. and Qi, Q. (2007). The production of polyhydroxyalkanoates in recombinant Escherichia coli. Bioresour Technol 98, 2313-20.

62. Li, Z. J., Jian, J., Wei, X. X., Shen, X. W. and Chen, G. Q. (2010). Microbial production of meso-2, 3-butanediol by metabolically engineered Escherichia coli under low oxygen condition. Appl Microbiol Biotechnol 87, 2001-9.

63. Liu, X. and Curtiss, R., 3rd (2009). Nickel-inducible lysis system

in Synechocystis sp. PCC 6803. Proc Natl Acad Sci U S A 106, 21550-4.

64. Luengo, J. M., Garcia, B., Sandoval, A., Naharro, G. and Olivera, E. R. (2003). Bioplastics from microorganisms. Curr Opin Microbiol 6, 251-60.

65. Luisi, P. L. (2007). Chemical aspects of synthetic biology. Chem Biodivers 4, 603-21.

66. Lynd, L. R., Weimer, P. J., van Zyl, W. H. and Pretorius, I. S. (2002). Microbial cellulose utilization: fundamentals and biotechnology. Microbiol Mol Biol Rev 66, 506-77, table of contents.

67. Lynd, L. R. and Zhang, Y. (2002). Quantitative determination of cellulase concentration as distinct from cell concentration in studies of microbial cellulose utilization: analytical framework and methodological approach. Biotechnol Bioeng 77, 467-75.

68. Madhavan Nampoothiri, K., Nair, N. R. and John, R. P. (2010). An overview of the recent developments in polylactide (PLA) research. Bioresour Technol 101, 8493-501.

69. Madison, L. L. and Huisman, G. W. (1999). Metabolic engineering of poly (3- hydroxyalkanoates): from DNA to plastic. Microbiol Mol Biol Rev 63, 21-53.

70. Martin, V. J. J., Pitera, D. J., Withers, S. T., Newman, J. D. and Keasling, J. D. (2003). Engineering a mevalonate pathway in Escherichia coli for production of terpenoids. Nat Biotechnol 21, 796-802.

71. Matsumoto, K., Murata, T., Nagao, R., Nomura, C. T., Arai, S., Arai, Y., Takase, K., Nakashita, H., Taguchi, S. and Shimada, H. (2009). Production of short-chainlength/medium-chain-length polyhydroxyalkanoate (PHA) copolymer in the plastid of Arabidopsis thaliana using an engineered 3-ketoacyl-acyl carrier protein synthase III. Biomacromolecules 10, 686-90.

72. Mitsuzawa, S., Kagawa, H., Li, Y., Chan, S. L., Paavola, C. D. and Trent, J. D. (2009). The rosettazyme: a synthetic cellulosome. J Biotechnol 143, 139-44.

73. Moon, T. S., Dueber, J. E., Shiue, E. and Prather, K. L. (2010). Use of modular, synthetic scaffolds for improved production of glucaric acid in engineered E. coli. Metab Eng 12, 298-305.

74. Mu, D., Seager, T., Rao, P. S. and Zhao, F. (2010). Comparative life cycle assessment of lignocellulosic ethanol production: biochemical versus thermochemical conversion. Environ Manage 46, 565-78.

75. Mushegian, A. (1999). The minimal genome concept. Curr Opin Genet Dev 9, 709-14. Nam, Y. S., Magyar, A. P., Lee, D., Kim, J. W., Yun, D. S., Park, H., Pollom, T. S., Jr., Weitz, D. A. and Belcher, A. M. (2010). Biologically templated photocatalytic nanostructures for sustained light-driven water oxidation. Nat Nanotechnol 5, 340-4.

76. Novamont (2011a).BIODEGRADABILITY. Available from: www.novamont.com/default.asp?id=491.

77. Novamont (2011b). COMPOSTABILITY. Available from: www.novamont.com/default.asp?id=493.

78. Novamont (2011c). Life Cycle Assessment Available from: www.novamont.com/default.asp?id=958.

79. Ouyang, S. P., Luo, R. C., Chen, S. S., Liu, Q., Chung, A., Wu, Q. and Chen, G. Q. (2007). Production of polyhydroxyalkanoates with high 3-hydroxydodecanoate monomer content by fadB and fadA knockout mutant of Pseudomonas putida KT2442. Biomacromolecules 8, 2504-11.

80. Park, S. J., Park, J. P. and Lee, S. Y. (2002). Metabolic engineering of Escherichia coli for the production of medium-chain-length polyhydroxyalkanoates rich in specific monomers. FEMS Microbiol Lett 214, 217-22.

81. Park, Y. M., Shin, B. A. and Oh, I. J. (2008). Poly (L-lactic acid)/polyethylenimine nanoparticles as plasmid DNA carriers. Arch Pharm Res 31, 96-102.

82. PlasticsEurope (2009). The Compelling Facts about Plastics 2009. Available from: www.plasticsconverters.eu/docs/Brochure_FactsFigures_Final_2009.pdf.

83. Poirier, Y. (2001). Production of polyesters in transgenic plants. Adv Biochem Eng Biotechnol 71, 209-40.

84. Prieto, M. A. (2007). From oil to bioplastics, a dream come true? J Bacteriol 189, 289-90.

85. Qiu, Y. Z., Han, J. and Chen, G. Q. (2006). Metabolic engineering of Aeromonas hydrophila for the enhanced production of poly(3-

hydroxybutyrate-co-3-hydroxyhexanoate). Appl Microbiol Biotechnol 69, 537-42.

86. Ro, D.-K., Paradise, E. M., Ouellet, M., Fisher, K. J., Newman, K. L., Ndungu, J. M., Ho, K. A., Eachus, R. A., Ham, T. S., Kirby, J., Chang, M. C. Y., Withers, S. T., Shiba, Y., Sarpong, R. and Keasling, J. D. (2006). Production of the antimalarial drug precursor artemisinic acid in engineered yeast. Nature 440, 940-943.

87. Rosenberg, J. N., Oyler, G. A., Wilkinson, L. and Betenbaugh, M. J. (2008). A green light for engineered algae: redirecting metabolism to fuel a biotechnology revolution. Curr Opin Biotechnol 19, 430-6.

88. Ryffel, G. U. (2010). Making the most of GM potatoes. Nat Biotechnol 28, 318.

89. Sandoval, A., Arias-Barrau, E., Arcos, M., Naharro, G., Olivera, E. R. and Luengo, J. M. (2007). Genetic and ultrastructural analysis of different mutants of Pseudomonas putida affected in the poly-3-hydroxy-n-alkanoate gene cluster. Environ Microbiol 9, 737-51.

90. Schmidt, M. and Pei, L. (2011). Synthetic Toxicology: Where engineering meets biology and toxicology. Toxicol Sci. 120, Suppl 1:S204-24.

91. Shetty, R. P., Endy, D. and Knight, T. F., Jr. (2008). Engineering BioBrick vectors from BioBrick parts. J Biol Eng 2, 5.

92. Shi, X., Jiang, J., Sun, L. and Gan, Z. (2010). Hydrolysis and biomineralization of porous PLA microspheres and their influence on cell growth. Colloids Surf B Biointerfaces.

93. Shi, X., Sun, L., Jiang, J., Zhang, X., Ding, W. and Gan, Z. (2009). Biodegradable polymeric microcarriers with controllable porous structure for tissue engineering. Macromol Biosci 9, 1211-8.

94. Sinha, J., Reyes, S. J. and Gallivan, J. P. (2010). Reprogramming bacteria to seek and destroy an herbicide. Nat Chem Biol 6, 464-70.

95. Slade, R., Bauen, A. and Shah, N. (2009). The greenhouse gas emissions performance of cellulosic ethanol supply chains in Europe. Biotechnol Biofuels 2, 15.

96. Smolke, C. D. (2009). Building outside of the box: iGEM and the BioBricks Foundation. Nat Biotechnol 27, 1099-102.

97. Soetaert, W. (2007). White biotechnology: A key technology for building the biobased economy.

98. Taguchi, K., Aoyagi, Y., Matsusaki, H., Fukui, T. and Doi, Y. (1999). Co-expression of 3- ketoacyl-ACP reductase and polyhydroxyalkanoate synthase genes induces PHA production in Escherichia coli HB101 strain. FEMS Microbiol Lett 176, 183-90.

99. The Federal Environment Ministry (Germany) (2009). Ordinance on the Avoidance and Recovery of Packaging Wastes Available from: www.bmu.de/english/waste_management/downloads/doc/37115.php.

100. Tigges, M., Marquez-Lago, T. T., Stelling, J. and Fussenegger, M. (2009). A tunable synthetic mammalian oscillator. Nature 457, 309-12.

101. Thompson, R. C., Moore, C. J., vom Saal, F. S. and Swan, S. H. (2009). Plastics, the environment and human health: current consensus and future trends. Philos Trans R Soc Lond B Biol Sci 364, 2153-66.

102. Tokiwa, Y. and Calabia, B. P. (2004). Degradation of microbial polyesters. Biotechnol Lett 26, 1181-9.

103. Tsuge, T. (2002). Metabolic improvements and use of inexpensive carbon sources in microbial production of polyhydroxyalkanoates. J Biosci Bioeng 94, 579-84.

104. Tsuge, T., Yano, K., Imazu, S., Numata, K., Kikkawa, Y., Abe, H., Taguchi, S. and Doi, Y. (2005). Biosynthesis of polyhydroxyalkanoate (PHA) copolymer from fructose using wild-type and laboratory-evolved PHA synthases. Macromol Biosci 5, 112-7.

105. Valentin, H. E., Broyles, D. L., Casagrande, L. A., Colburn, S. M., Creely, W. L., DeLaquil, P. A., Felton, H. M., Gonzalez, K. A., Houmiel, K. L., Lutke, K., Mahadeo, D. A., Mitsky, T. A., Padgette, S. R., Reiser, S. E., Slater, S., Stark, D. M., Stock, R. T., Stone, D. A., Taylor, N. B., Thorne, G. M., Tran, M. and Gruys, K. J. (1999). PHA production, from bacteria to plants. Int J Biol Macromol 25, 303-6.

106. Wegrzyn, G. (2001). The minimal genome paradox. J Appl Genet 42, 385-92.

107. Wei, X. X., Shi, Z. Y., Yuan, M. Q. and Chen, G. Q. (2009). Effect of anaerobic promoters on the microaerobic production of polyhydroxybutyrate (PHB) in recombinant Escherichia coli. Appl Microbiol Biotechnol 82, 703-12.

108. Yao, D., Zhang, W. and Zhou, J. G. (2009). Controllable growth of gradient porous structures. Biomacromolecules 10, 1282-6.

109. Yim, K. S., Lee, S. Y. and Chang, H. N. (1996). Synthesis of poly-(3-hydroxybutyrate-co-3- hydroxyvalerate) by recombinant Escherichia coli. Biotechnol Bioeng 49, 495-503.

110. Zhang, X., Luo, R., Wang, Z., Deng, Y. and Chen, G. Q. (2009). Application of (R)-3- hydroxyalkanoate methyl esters derived from microbial polyhydroxyalkanoates as novel biofuels. Biomacromolecules 10, 707-11.

111. Zheng, Z., Li, M., Xue, X. J., Tian, H. L., Li, Z. and Chen, G. Q. (2006). Mutation on Nterminus of polyhydroxybutyrate synthase of Ralstonia eutropha enhanced PHB accumulation. Appl Microbiol Biotechnol 72, 896-905.

Chapter 7

Functionalized Activated Carbon Derived from Biomass for Photocatalysis Applications Perspective

Samira Bagheri, Nurhidayatullaili Muhd Julkapli, and
Sharifah Bee Abd Hamid

Nanotechnology & Catalysis Research Centre (NANOCAT), University
of Malaya, IPS Building, 50603 Kuala Lumpur, Malaysia

ABSTRACT

This review highlighted the developments of safe, effective, economic, and environmental friendly catalytic technologies to transform lignocellulosic biomass into the activated carbon (AC). In the photocatalysis applications, this AC can further be used as a support material. The limits of AC productions raised by energy assumption and product selectivity have been uplifted to develop sustainable carbon of the synthesis process, where catalytic conversion is accounted. The catalytic treatment corresponding to mild condition provided a bulk, mesoporous, and nanostructure AC materials. These characteristics of AC materials are necessary for the low energy and efficient photocatalytic system. Due to the excellent oxidizing characteristics, cheapness, and long-term stability, semiconductor materials have been used immensely in photocatalytic reactors. However, in practical, such conductors lead to problems with the separation steps and loss of photocatalytic activity. Therefore, proper attention has been given to develop supported semiconductor catalysts and certain matrixes of carbon materials such as carbon nanotubes, carbon microspheres, carbon nanofibers, carbon black, and activated carbons have been recently considered and reported. AC has been reported as a potential support in photocatalytic systems because it improves the transfer rate of the interface charge and lowers the recombination rate of holes and electrons.

INTRODUCTION

Increasing environmental problems and the need for competitive and cost-effective products are becoming two major principles in modern material research [1–4]. Former developed routes to get a periodic porous carbon network were successful, but they did not take into account any criteria of sustainability [5–7]. For the last 20 years, countless laboratories of research institution have done some indepth researches on the conversion of biomass to carbon based materials without using the catalysts [5, 8, 9]. Studies of such researches have covered many model carbon material compounds, for example, methanol, lignin, glucose, cellulose, and some real biomass compounds [10–12]. As prosperous demonstrations were accumulated, kinetics, thermodynamics, and detailed reaction mechanism have created a solid base for subsequent researches [13–15]. However, in order to enhance the selectivity of carbon material manufactures, immense activation energy is necessary for the reaction without the use of catalyst. The excessive cost of tools and operations has undoubtedly become the biggest hindrance to the formulation of this technology [15]. Hence, the main problem of carbon synthesis under sustainable conditions was currently revisited and executed by several terms, where catalytic treatment of biomass through either heterogeneous or homogeneous catalytic approach corresponding in mild condition provided a bulk, mesoporous, and nanostructure carbon materials [16–18]. Catalytic effects of homogeneous catalyst, especially ionic liquid on the biomass conversion, have been established by many of the open literature. The main characteristics of this catalytic technique are to have a conversion system with minimal energy to confirm the high yield of carbon materials [18–20]. If we compare homogeneous catalysts with heterogeneous catalysts, heterogeneous catalysts have the advantages of being highly selective, recyclable, and environmentally friendly [19, 21]. Heterogeneous catalysts have the advantages of being highly selective; therefore, heterogeneous catalyst with the ample range of solid acid, ion exchange resin, metal oxide, and zeolite has become a research hotspot in this field [21]. Both techniques have been explored substantially, with the need for creating cheap and sustainable ways to obtain chemicals and carbon from raw materials other than natural gas or crude oil that could lead to a reexploration of this area. In addition, the implementation of low-priced pathway to recycle the products of farmed biomass would furthermore represent a way to sequester particular amount of CO_2 creating a material advantaged at the same time [22, 23].

This paper will not debate on the preparation of the catalyst, but instead it would focus primarily on the use of the catalyst to bring out carbon based materials through the conversion process of biomass, which can be additionally used as a photocatalyst.

Various published papers and examined articles have indicated the theory and environmental supplication of heterogeneous photocatalysis by the employment of semiconductors [24, 25]. One of the biggest drawbacks of using these semiconductors is the power necessity because of the utilization of ultraviolet light [26]. However, upgrading the photocatalytic reactors may upgrade reaction rate and hence downgrade the time of residence and minimize consumption of energy per unit of volume being treated. It is accepted that this might be accomplished by depositing the photocatalyst on a high surface product that will particularly absorb the polluted molecules and will fixate them around the photocatalyst [26–28]. Lamentably, active absorption of pollutant lowers the diffusion rate into semiconductor powder, which may thus hinder the activeness of the photocatalyst. Such drawback has assisted various researchers to find any worthy initiative of semiconductor for the operation of some particular pollutants or try to formulate the latest techniques of deposition [29, 30]. The pH of the solution, the support, and the kind of pollutant play a significant role in the accomplishment of photocatalytic process.

Various efforts have announced utilization of activated carbon (AC) as a platform for the semiconductor but it has been cautioned that effective absorption of pollutants into the absorbent area may hinder diffusion into the catalyst and thus may affect the entire process [31, 32]. AC acts as a brace for the titanium oxide (TiO_2) which could give tremendous results over the other mediums [32]. These consist its potential to swiftly absorb pollutants and also its high absorption ability because of its surface area and porosity [31]. As indicated, medium pores AC will make easier the diffusion of pollutants and product discharging from the surface [32,33]. Furthermore, high ability to absorb fluid of AC may reduce the penetration of ultraviolet lights into small areas, and it may cause confining of pollutants within the pores without getting able to diffuse into the outer surface for further reaction with the OH radical of pollutants [34–36]. Additionally, some types of pollutants, for example, phenol, may go through polymerization on the AC's carbon surface, which causes unchangeable absorption [37, 38]. The general processes involved in semiconductor particles upon bandgap excitation are illustrated in Figure 1.

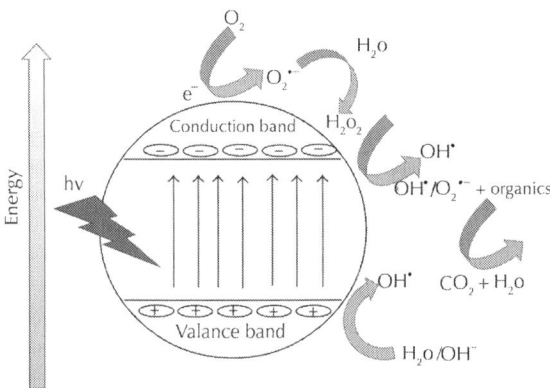

Figure 1: Processes involved in semiconductor particles upon bandgap excitation [39].

ACTIVATED CARBON

Properties of Carbon Materials

Carbon materials technology has made extraordinary progress in current years because of its diversity of physicochemical properties, such as tunable porosity, lightweight, exciting electronic properties, electrical conductivity, chemical and thermal balancing, and the potential to acquire an immense range of morphologies [40–43]. Hence, carbon materials have found a large number of applications in different domains, varying from environmental science [44], absorbent [45], drug delivery [46], catalyst [44], electrode materials [47], stationary phase in the chromatography system [48], energy storage [49], and many others according to its structure, morphology, and chemical properties (Figure 2).

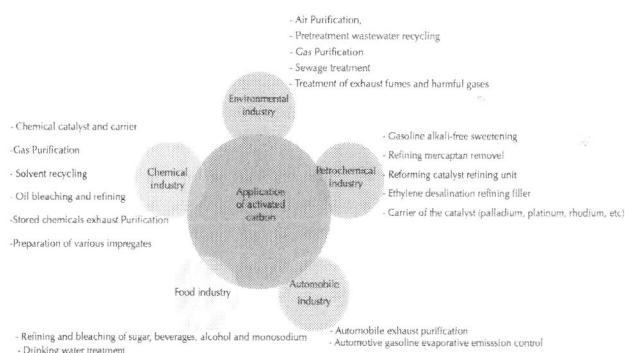

Figure 2: General prospective on application of AC.

However, for some particular applications, functionalization is essential at controllable size and shape [48, 49]. Nevertheless, the production of such materials usually requires very harsh conditions and has several limits such as extreme temperature of the carbonization process in the first steps up to >800°C and followed by chemical or physical activation to transform carbon materials into activated carbon [43–49]. Furthermore, it is significantly important here to explore economical and sustainable ways to get carbon materials from raw materials other than crude oil or natural gas which leads to a reexploration of this field (Table 1; Figure 3).

Table 1: Classification of AC materials

Criteria	Particle size	Properties	Applications	References
Powdered AC	<1.0 mm Diameter: 0.15– 0.25 mm	Higher surface area, extraordinary volume per gram, and greater purity	As an additive in vessel, waste water treatment, classifiers, and gravity filter	[50]
Granule AC	0.42 to 0.84 mm	Suitable for many organic chemicals, able to improve taste/odor, and removes chlorine	Vapor and liquid adsorption, water treatment, deodorization, separation of components	[51]
Extruded AC	0.8 to 130 mm	Low pressure drop, high mechanical strength, and low dust content	Gas phase applications	[52]
Beads AC	0.35 to 0.8 mm	Low pressure drop, high mechanical strength, and low dust content	Adsorbent for waste water	[53]
Impregnated AC	0.8 to 200 mm	Porous carbon impregnated with inorganic materials (iodine, silver, cations) and antimicrobial antiseptic	Pollution control and purification of domestic water	[54, 55]

Figure 3: Classification of AC materials.

Conventional Conversion of Biomass to Carbon

The use of biomass extracted products or biomass is becoming vitally significant for the enhancement of effective and environmentally friendly technology and together it solves the issues of agricultural and forestry waste use [5–9]. Different synthetic methods, such as carbonization [56], high-voltage-arc electricity [57], laser ablation [58], or hydrothermal carbonization [56], have been disclosed for the preparation of carbonaceous, amorphous, crystalline carbon materials or porous with different sizes, chemical composition, and shape of the biomass feedstocks [56–59].

Moreover, the application of a low-priced pathway to recycle by-products of biomass formed would furthermore represent a way to separate significant amount of CO_2 and at the same time a material advantage would also be created.

Biomass conversion is very significant, yet, working with this complex biomass feedstock is challenging, and approaches based on the creation of simpler and more six balanced intermediate derivatives, known as platform molecules, have been shown to be active for efficient biomass conversion into chemicals and fuels [55], catalytic routes consisting deoxygenation, and reaction combined with –C–C– coupling processes [57]. Biomass conversion to carbon materials through catalytic approach normally begins with hydrolysis of dehydration and cellulose chains [57, 59] and is divided into monomer's soluble products that come from the hydrolysis of cellulose [58], condensation or polymerization of the soluble products, aromatization of the polymers hence formed, and growth of the nuclei so created by linkage and diffusion of species from the solution to the nucleus surface and finally appearance of short burst of nucleation [56–59] (Figure 4).

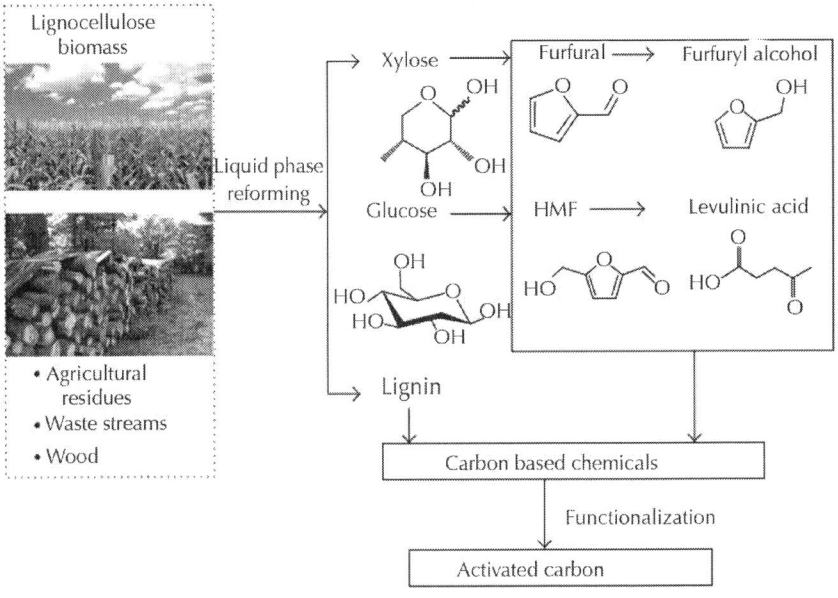

Figure 4: The route of AC derived from biomass.

Catalytic Conversion of Biomass to Carbon

Green chemistry, now days, is becoming more and more connected with the catalytic process on biomass conversion to carbon materials. Table 2 shows the list of catalysts corresponding to their category and advantages towards the conversion process of cellulose to carbon based materials.

Table 2: Catalytic approach on conversion of cellulose to carbon based materials

Categories	Types	Advantages
Homogeneous	Ionic liquid (e.g.: [BMIM]Cl; [EMIM]Cl and [EMIM]BF$_4$)	(i) Low melting point (ii) Appears as crystal in normal condition (iii) Acts as template on production of porous carbon structure (iv) Solvent reaction media (v) High yield of carbon production
Heterogeneous	Solid acid catalyst (i) Bronsted/Lewis solid acid (e.g.: ZrP; SiO$_2$-Al$_2$O$_3$, WOX/ZrO$_2$, c-Al$_2$O$_3$,) (ii) HPA (e.g.: H$_3$PW$_{12}$O$_{46}$; H$_4$SiW$_{12}$O$_{40}$; Cs$_{2.5}$H$_{0.5}$PW; Cs$_{2.5}$H$_{0.5}$PW$_2$O$_{40}$)	(i) High catalyst selectivity (ii) Good separation process (iii) High promotion on depolymerization of cellulose (iv) Low formation of soluble oligomer (v) Low cellulose self-hydrolysis (vi) Favors direct formation of lactic acid (vii) High stability
Heterogeneous	Ion exchange resin (e.g.: Amberlyst; McM-41; HnbM$_0$O$_6$; mixed oxides; niobic acid; silica-niobic; niobium phosphate)	(i) High accessibility of saccharides (ii) Satisfactory reaction rate

Heterogeneous	Zeolite (e.g.: ZSM-5; Beta; Mordenite; Ferrient; FCC; Al-MCM-42; SBA-15; Al-MSU-F; MOR)	(i) High selectivity to adsorb molecules (ii) Good separation process (iii) Good thermal and hydrothermal stabilities (iv) Production of porous carbon
Heterogeneous	Metal ions (e.g.: Cr^{3+}; Mn^{2+}; Fe^{3+}; Fe^{2+}; Co^{2+})	(i) Good catalyst for ring opening and hydrogenation cellulose (ii) Require short reaction time (iii) High yield of glucose (iv) High turnover amount of catalyst (v) High activity and selectivity (vi) Good recyclability (vii) Easily separable
Heterogeneous	Metal oxide (e.g.: $HnbM_0O_6$; Al_2O_3)	(i) Green process (ii) Nonvolatility (iii) Highly stable (iv) Nontoxic (v) Reusability (vi) Low cost

Homogeneous Catalyst

Convention of sulfuric acid solution and catalyst cellulose hydrolysis into glucose is a time-consuming and well-formulated process [56–58, 60–65]. Many large scale segments have been developed, but there are rigid conditions including the treatment and the recycling of the waste sulfuric solutions of acids, which also suffer from the complex separation of products from the solution, the lack of glucose selectivity, toxicity, and high prices which take this process away from the original approach of sustainability [56–58, 60–63].

Thus, the solvents of catalyst as ionic liquids have received enough attention because of their low vapor pressure, stability, and recyclability [62, 62–73]. The kinds of novel green solvent are ionic liquids with relatively less melting point and appear as a crystal in general conditions [64]. Cellulose is balanced via inter- and intramolecular bonds of hydrogen, so that rigid bundles could be created, which makes it difficult to solubilise with common organic solvents and water [16–18, 74–79]. It is important to make solvents for cellulose so as to initiate a system of minimal efficiency to confirm greater yield of carbon materials in the conversion of biomass [17–20]. Ionic liquids have significant roles; it acts both as a soft template to formulate the characterized pore structuring for the development of a hierarchical porous carbon structure and as a catalyst which results in enhanced ionothermal carbon yields [64–66, 71–79]. Thus, it has been demonstrated that in the presence of an acid catalyst, the utilization of ionic liquids, can embrace the efficiency of the hydrolysis of glucose cellulose [66, 67, 80–84].

Heterogeneous Catalyst

The utilization of heterogeneous catalysis along with an immense range of designs is less expensive and extraordinarily stable at high temperature [85–88]. They are believed to upgrade catalyst characteristics and process conditions to get high yields of hydrocarbons while minimizing coke development in the wide range of reaction conditions [86–88]. This method has the benefit of being very economical and mild if we follow some rules of green chemistry since it does not add organic solvent [85, 86] with resulting carbon, which is spherically shaped and the surface is decorated with oxygenated functional groups [86]. This method also involves simple reaction mechanisms for the creation of carbon, which involves the dehydration of carbohydrate into a furan like molecules, mainly 5-(hydroxymethyl)-2-furaldehyde as an initial step and further polymerization and carbonization as the next step [89–92]. This reaction possesses high potential for the catalytic improvement biomass since –C–C– coupling takes place with consistent oxygen removal (the reaction involves the dispatchment of CO_2 and H_2O) from carboxylic acids [89, 90], and the latter of which are mutual intermediates in the process of biomass conversion [91, 92]. If we compare homogeneous catalysis, carbon materials obtained through a heterogeneous catalytic process consist of an aromatic core containing polyfuran-type units which is surrounded by oxygen rich polar functional groups such as COOH, –OH, –C–, and –O–, which makes the materials more hydroscopic, hydrophilic, and have a lesser degree of graphitisations [93, 94]. These functional surface groups could act as a premier binder and depositor to promote and stabilize the carbon to form nanocables [95–98], a novel carbon-encapsulated core-shell composite, and hybrids. Moreover, heterogeneous catalysts are easy to recover and reuse [96, 97] (Figure 5).

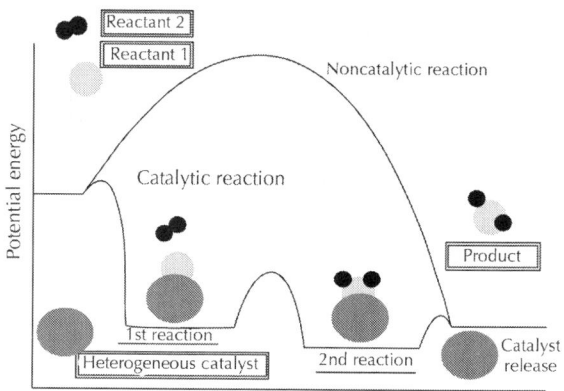

Figure 5: Heterogeneous catalytic routes in chemical reaction.

Nevertheless, the effective exploitation of cellulose is a main challenge in heterogeneous catalysis application, since cellulose itself has a tough, mainly crystalline, chemically stable, and water insoluble characteristics, which are induced from the intra/intermolecular hydrogen bonds [99–101]. The most commonly used heterogeneous catalysts for the conversion of biomass are alkali salts [76, 102, 103], metals (Ni, Rh, PT, Ru, Pd, Ir, and Ag) [104–106], metal oxides (CeO_2, ZrO_3 TiO_2, and Al_2O_3) [107, 108] usually on supports, and metal oxide catalysts at medium temperatures (300 to 425°C) which involves deoxygenation reaction combined with –C–C– coupling processes [109].

(1) Solid Acid Catalyst: Transformation of cellulose to water soluble sugars via solid acid catalysts has received much attention in these years as a solution to remove mineral acids in the formulation of furfural which is easily recovered from the reaction mixture, reused without losing the activity, and minimized the posttreatment cost [75, 110–113]. Solid acid catalysts are basically a Bronsted or Lewis acid [113, 114] and it is of various types such as Vanadyl phosphate, ZrO_2, zeolites, inorganic oxides, and ion exchange resins.

(1.1) Bronsted or Lewis Solid Acid Catalyst: These series of catalysts include WO_x/ZrO_2, $C-Al_2O_3$, Zr–P, $SiO_2-Al_2O_3$, and HY Zeolite which are used for mainly aqueous phase dehydration of xylose [115–119]. Lewis acid sites lower furfural selectivity through catalyzing a side reaction. In the hot water (190°C), solid Lewis acids promoted the cellulose depolymerization and lowered the creation of soluble oligomers and polymers as in connection to the cellulose self-hydrolysis which occurred in the familiar conditions [117–124]. By contrast, in normal conditions, strong Bronsted solid acid has not improved the extent of the cellulose depolymerization but has affected the product distribution [98, 122, 123]. By comparison, Lewis solid acids were not only potential to upgrade the extent of cellulose depolymerization but also favorable for direct formation of lactic acid, which gives high yield, approximately 30% [96, 125]. This comparison represents highly potential ways to optimize the conversion of cellulose and stabilize Bronsted acid catalyst which helps to understand the reaction pathways [111].

This also brought an idea for various researchers to study the effect of Bronsted acidities of water soluble heteropoly acid catalyst, known as HPA [112]. This heterogeneous HPA catalyst which is micellar, clean, economical, facile, and environmentally friendly could be recycled via centrifuge [112, 126]. The exhibited activity for the hydrolysis of polysaccharides comprising disaccharides, starch, and cellulose, is known to be HPA [121, 126]. Tungstophosphoric acids ($H_3PW_{12}O_{40}$) [119] and tungstosilicious acids ($H_4SiW_{12}O_{40}$) [109] are the vigorous acids used in the series of HPA catalysts in reaction of hydrocarbons and have been employed, for example, in alkylation, acylation, esterification, dehydration, and isomerization of the ethanol process [127–129]. Salt of acidic cesium $H_3PW_{12}O_{40}$ and $Cs_{2.5}H_{0.5}PW_{12}O_{40}$ ($Cs_{2.5}H_{0.5}PW$) with intense acidity is insoluble in organic solvents and water and has meso- and micropores with relatively high surface area

$(130 \, m^2 gk)$ [119]. This $Cs_{2.5}H_{0.5}PW$ has been described as a solid acid catalyst with prudent and environment friendly specifications and is recommendable in industrial process like hydration of olefins or ester and liquid phase dehydrated alcohol [130, 131]. Till now, cellulose hydrolysis into saccharides employing a range of HPA catalyst has been researched for capable applications, in contrast to conventional acid-catalyzed reaction, whereas the yield of glucose is less [130–132]. It is because of the insolubility of cellulose in any of the solvents and problems of solid to solid mass transport [77, 133]. In a few cases, the acidity of mineral acid is lower than that of HPA. For example, (H_0) of $H_5BW_{12}O_{40}$ (0.7 M at 100°C), the Hammet acidity function, is -2.1 lower than that of HCl (0.7 M) and H_2SO_4. Hence, HPAs have been anticipated to be reusable and active catalyst for the hydrolysis. Along with that theory, amorphous carbon bearing SO_3H, COOH, and OH groups had been described to show hydrolysis activity (10% of glucose yield) for 3 hours at 100°C. Meantime, Yabushita et al. demonstrated that the cellobiose hydrolysis and cellulose in water assisted by HPAs such as $Sn_{0.75}PW_{12}O_{40}$, $H_3PW_{12}O_{40}$, and $H_4SiW_{12}O_{40}$ record total yields of decreasing sugar and it was around 40% at 200°C for 16 hours of reaction [134]. It has been found by Hara that $H_5BW_{12}O_{40}$ exhibited a good performance for the transformation of crystalline cellulose to give glucose 77% yield and various types of HPAs like $H_3PW_{12}O_{40}$ (glucose yields 8%) and $H_4SiW_{12}O_{40}$ (37%) which are less active than $H_5BW_{12}O_{40}$ [135]. Other research groups also administered the screening of counter cations of $PW_{12}O_{40}$ for the cellobiose hydrolysis and they found that there was a volcano-type correlation between Lewis acidities [136–138] and TOFs for glucose formation. They also observed that the acidity and the role of decreasing CrI were the significant factors for the catalytic activity of HPAs [139]. In reality, the H_0 role was $H_3PW_{12}O_{40} < H_4SiW_{12}O_{40} < H_5BW_{12}O_{40}$ in a similar order of catalytic activity. The H_0 roles were corresponding to the concentration of acids and the number of anions. Hence, HPAs involving highly negatively charged ions are desirable. The anions were dissociated from hydrogen bonding between cellulose molecules to lower the CrI. Moreover, HPAs protons have also exhausted the bonds of hydrogen from cellulose and greater concentration of protons was effective in this role [135, 140]. Therefore, a strong catalyst for the cellulose hydrolysis is $H_5BW_{12}O_{40}$, which was recovered through the extraction and it was recycled for around 10 times [141].

(2) Ion Exchange Resins: Sulfonic acid functionalities of ion exchange resin with solid acid catalyst and sulfonic acid functioned materials resulted in high yield at 63% and 76% in pure dimethyl sulfoxide (DMSO) solvent, correspondingly [142, 143]. Solid acids such as amorphous carbon materials consisting of SO_3H groups layered transition ($HNbMoO_6$) metal oxide and resin sulfated have been tested for the cellulose hydrolysis, but the yield of glucose is still comparatively less [144, 145]. It has been hypothesized that side reactions were abolished employing aprotic or organic solvents [146]. Vigorous acidic resins (Amberlite IR-200 and IR-120), niobic acid, mixed oxides (silica-zirconia and silica-alumina), silica-niobia, and niobium phosphate created a strong acidity which was protonic and accessibility of simple saccharides to the most effective sites on the surface catalyst allows satisfactory reaction rates to be obtained [147–150].

Sulfonic resins which are acidic are represented in different literatures of active catalyst system for the hydrolysis of starch, cellulose, and disaccharides [151–153]. Generally, the rigid conditions in terms of temperature (>120°C temperature in the water and in critical condition) and high concentration of saccharide (>100 g) were employed to push the catalytic actions of solid acid towards the achievement of high transformations [154–156]. Sulfonated activated-carbon could transform cellulose of amorphous into glucose with 41% of yield for 24 hours at a temperature of 200°C [157]. Such things have been founded by Sun and Zhang demonstrated that p-toluenesulfonic acid can catalyze cellulose hydrolysis in ionic [EMIM][Cl] system, giving THF yields of 28% and 13% and a yield of mono- and disaccharides 10% and 3%, respectively [158]. Natural bamboo can be converted by a sulfonated biomass char with cotton and starch around 20% yield of glucose under microwave assistance [159–161]. The process of hydrolysis consisting of starch and cellulose was obtained via layered transition metal oxide, despite the fact that the yield from glucose was less than cellulose. Sulfonated carbon with a mesoporous like structure was used by the groups of Vyver for cellulose hydrolysis getting the yield of glucose around 75%, which is considered to be the highest recorded yield via solid-acid catalyst [162, 163]. Similarly, some interest has been shown in cellulose depolymerization in water because solid supported acid catalysis was used [164–166]. Current reports describe the cellulose hydrolysis by solid catalysts such as layered transition metal oxides, Amberlyst resin acid modified amorphous carbon, and sulfonated silica or carbon nanocomposites [167–169]. Also, the depolymerization of cellulose was also considered under catalysis of both $FeCl_3$ and Nafion supported on amorphous silica to be tested on a continuous flow reactor, given the residual that unreacted cellulose can be easily eliminated from the system [75, 170]. For the conversion of glucose, these surface species which are acidic were quite active [169]. Certainly, we are aware that the hydrolysis rate of cellulose depends on the acid strength. As an expansion on the previous reports on the usage of Nafion as a solid assisted acid catalyst for the transformation of cellulose into glucose and levulinic acid, many researchers have incorporated the reaction with alkali metal salts to embrace the reaction's yield [171, 172]. Namchot et al. and Klamrassamee et al. have recently formulated carbon-based solid acid with immense density of Bronsted acid sites (SO_3H and COOH) to pyrolytically carbonize sugar, such as cellulose, sucrose, or glucose, and subsequently sulfonate the prepared carbons [173, 174]. Interestingly, these sulfonated carbon materials are very strong for the microcrystalline cellulose hydrolysis to produce water soluble saccharides with low reaction temperature (100°C) with the conventional and strong Bronsted acid catalyst such as H-mordenite, niobic acid, and others. The particular surface area of the sulfonated carbon was around $2\,m^2g^{-1}$ but the soluble saccharides yield reached nearly 70%. Catalytic performance of soluble saccharides was applied to its intrinsic ability to adsorb -1,4-glucan, which is not absorbed on the other solid acids. Thus, it can be concluded that heterogeneous catalysis are more active and environmentally benign, mainly because of a hasty product separation and also catalysts recovery. Jule and Schoonover described that acid resins with considerably big pores could actively depolymerize cellulose in ionic liquid, but the main products were cellooligomers which failed to be dissolved in water [175].

(3) Zeolite Catalyst: The workhorse of the petroleum industry is zeolite catalyst that efficiently converts petroleum based feedstocks into the targeted chemicals and fuels and chemicals [176–178]. Crystalline microporous solids are an important part of zeolites because of their widespread application in absorption, separation, and catalysis [178]. Its importance stems from its unique structures of pores, which makes it highly particular to absorb molecules for separation reasons or towards

product molecules in catalysis [175, 179, 180]. Furthermore, during heterogenous catalytic reaction, zeolite shows good hydrothermal and thermal stabilities. However, from the last twenty years, there have been many studies that focused on the catalytic transformation of biomass and its derived feedstocks with a variety of zeolite catalysts, including Ferrient, FCC, Al-MCM-41, ZSM-5, beta zeolite, Y zeolite, SSZ-20, IM-5, TNU-9, MOR mordenite, SBA-15, Al-MSU-F FER, ZSM-23, MCM-22, and MFI [181–186]. ZSM-11 and ZSM-5 among these series had the lowest amount of coke and the highest yield of aromatic because of its pore space and steric effects [181]. Manufacturization of ordered porous carbon material has been obtained previously through replication of ordered zeolite inorganic and nanocasting [187, 188]. Inside zeolite many reactions have occurred which include decarbonylation, dehydration, isomerisation, and decarboxylation and with that removing oxygen as carbon dioxide, water, and carbon monoxide and conversion of carbon and hydrogen into aromatics and olefins [181–188]. In these catalysts of zeolite, ZSM-5 has exhibited the highest olefin and aromatic yields from biomass of lignocellulosic. With a pore size of around 5.5 to 5.6 A, ZSM-5 has a three-dimensional pore system. This small size of pore, internal volume, and internal structure has made it problematic for greater aromatic coke antecedents to form inside the pores [189]. It has been demonstrated by Zapata et al. that tin consisting zeolite is a highly efficient catalyst for the isomerisation of glucose in water [190]. Nevertheless, other studies have mentioned that, at low conversions in aqueous environments, faujasite and mordenite both resulted in great furfural selectivities up to 80% and 90% in 200°C as the selectivity lowers with high conversion and the final yield was relative [191, 192].

(4) Metal Ion Catalyst: Furfural and HMF are the two main and significant intermediates which are derived from biomass. They were directly manufactured from the microcrystalline cellulose hydrolysis with metal ions in ionic liquids such as Fe^{3+}, Fe^{2+}, Cr^{3+}, Co^{2+}, and Mn^{2+} as a catalyst under mild conditions [193–195]. Metal ions as an acidic support are a nice catalyst for hydrogenation and ring opening of cellulose polymers.

Many reports are there on carbon production through cellulose hydrolysis at a moderate temperature up to 250 and 300°C metal catalyst in a very less reaction time to control the deep exploitation of the formed glucose [196]. Some researchers even used Ru that is a ruthenium catalyst to enhance the transformation of oligosaccharides and increased glucose yield (almost 30%) with the TON that is the turnover number of the catalyst was immense (145 based on bulk Ru) in contrast to those of the sulfonated catalysts of carbon [197,198]. Wang et al. found that the glucose yield was increased around 31% by upgrading the Ru loading to 10 wt% along with the recyclable number of catalyst up to five times without losing the activity or Ru leaching [199]. Various reports on the efficiency of the hydrolyzate of cellulose formulated by Ru catalysis as a source of carbon for the bacterial PHA production found that under the aqueous solution the reaction which occurs is desirable for delivering microbes which thereupon make it easily disconnected from yielded sugar via facilitating the race of the catalyst and filtration [200–203]. Such reports have accepted that Ru species were in fact in an oxidized state and acted as the real and effective site for the oligosaccharides hydrolysis. Prior, Ru/Co_3O_4 catalyst has showed considerate selectivity and activity and good recyclability in the biomass conversion to carbon [204].

Researchers have even exclaimed that iron oxide nanoparticles and iron ions could adequately catalyze the hydrothermal carbonization of rice grains and starch beneath mild conditions (<200°C) and had a powerful influence on the creation of nanomaterials of carbon with different shapes [205, 206]. Catalysts of magnetite with sulfonic groups like a mesoporous silica composite, Fe_3O_4, SBA-15 treated by sulfonated $CoFe_2O_4$ embedded silica, and sulfuric acid were also employed for the cellulose hydrolysis [207, 208]. After the catalytic reactions they were easily separable via magnet. Fe_3O_4-SBA-SO_3H catalyst managed glucose in 26% yield from microcrystalline cellulose at a temperature of 200°C for 3 hours, though levulinic acid becomes an important product giving 42% yield by delaying the reaction time up to 12 hours [209].

(5) Metal Oxide Catalyst: In many catalytic processes metal oxides play an important role [210, 211]. For example, metal oxide nanostructures are important components commercially available for synthesis of methanol [212–215]. There are many more benefits of these catalysts which make them efficient candidates for green processes. The characteristics of such catalysts are stable, nontoxic, low cost, availability, nonvolatility, and reusability [214]. Several studies have observed the processes of decomposition connected with

formic acid on nanometal oxide surface, methanol, and formaldehyde [210–215]. It is strongly expected that metal oxide nanostructures would have a better catalytic activity in developing the conversion of cellulose to the value added products in hydrothermal media with an efficient separation from the reaction matrix in regard to the increased surface area of the nanomaterials [214]. Layered transition metal oxides containing niobium were found to be specifically active in the hydrolysis of disaccharide, suggesting the importance to investigate niobium containing catalyst as an energy inefficient factor for the conversion of biomass [213–215].

The employment of transition metal oxide like $HNbMoO_6$ was also reported as an efficient solid catalyst to generate glucose from cellulose. Similarly, in the presence of noble metal consisting catalyst, for example, Pt/Al_2O_3, Felica et al. have found the creation of sugar alcohols in yield up to 31% from cellulose in hydrothermal conditions. On the other hand, the researchers proclaimed that Pt free catalysts have generated only poor glucose hexitols amount [216].

ACTIVATED CARBON IN PHOTOCATALYST SYSTEM

Activated carbon (AC), a carbonaceous material structured on plant-based material, is a porous, amorphous solid carbon [31, 32]. Well-developed porous surface, high pore volume, and extended surface area make AC the most commonly used technique for controlling pollution [35]. Well-developed pores over the surface are one of the main uses of AC as the photocatalyst. Nontoxic, chemical stability and being economical are the main reasons that in the past decades the heterogeneous technology has attracted the attention of many researchers [217, 218]. As the organic pollutants can be mineralized into neutral by-products such as H_2O, CO_2, and mineral acids as one of the main properties of heterogeneous AC, photocatalysis methods include the destruction of the wider category of organic compounds. Promoting solar radiation and working on the low temperature, eventually saving a lot of energy, make it very economical [35–39]. One of the advantages of the AC includes the regeneration of spending absorbent and demolishing of absorbed organic material on the site converting the loss of absorbents to burn them concluding that thermal regeneration is efficient [218, 219]. AC being the strong light absorbing compound has been successfully used as photoactive species [38]. Determining the band gap of the AC (band gap less than 4 eV) resulted in a semiconductor and therefore, a photoelectric material in the presence of ultraviolet radiations [218]. Recent reports suggest the abnormal reaction towards the aqueous environment by directing ultraviolet irradiation of the sample in the presence of the AC provided that no other photoactive materials are present [220] showing that the AC improved the photooxidation of phenol, beyond the degradation of photolysis in comparison to bare or unmoved TiO_2 [220,221]. To find the difference whether this reaction is only shown by the AC or also by other carbon compounds, some researchers have worked on different porous AC materials obtained from different sites, procedures, and reactions and examined their behavior to the exposed ultraviolet radiations [222, 223].

The final solution we want to reach is that to remove any vagueness in the aqueous medium, ultraviolet radiations and absence of semiconductor AC are able to demolish the organic materials in the respective conditions [220]. Regardless of the type, AC acts as a catalyst during the removal of diatrizoate. Gamma radiations based AC is more efficient in a way that it has a higher proportion of C atoms and contains sp^2 hybridized [224]. With more than 53% of synergistic effect in diatrizoate in the first minute of reaction, commercial carbon is produced from the ultraviolet/AC system [222, 223]. Providing more oxygen, the synergistic act of AC is boosted up. The reutilized AC is quite similar to the original C; presence of O in the sample increases the rate of removing diatrizoate by the ultraviolet/AC system, but the ultraviolet inclusion of system results in some modifications in AC chemically. There is a very vague relationship between the textual properties and the synergistic contribution of the AC. Gamma radiation involvement with AC reduces the band gap which results in a more efficient removal of organic compounds [225, 226]. One of the actions proposed include that AC is the photocatalyst which will promote the electrons in the valence band to the conducting band, resulting in enhanced generation of OH free compounds present in the polluting medium.

Activated Carbon Surface Properties in Photocatalyst System

For commercial use, textures and surface properties can alter, depending on the crude material used, activation conditions and carbonization procedure which may potentially result in well-defined photocatalytic performance [39]. Many researchers and authors reported many surface and textual features of AC [227, 228]. It is clear that basic AC in the presence of low oxygen in this medium has much more potential than the phenol adsorption [225]. An increase in the physical absorption and surface polymerization of phenols can be done by ensuring the deprivation of the acidic categories on the surface of the AC. Irreversible absorption and catalyzing the oxidative coupling of phenolic compounds can be enhanced by the O_2 containing basic categories over the face of AC [221]. Boosting up the interactions between ϖ-ϖ electrons in AC and phenols is due to the graphene layers of an activated carbon which increases ϖ electron density [222]. In carbon of basic nature, phenol is considered to be higher regardless of textual propertie and absorption capacity [228]. Retention of phenol is supposed to become less in the acidic carbon as compared to the basic carbon; in fact there is no clear relationship between the two.

On of AC limitations is that the waste organic materials are not really destroyed but are transformed from one phase to another and in result the used AC is transformed into a dangerous product [226]. So, regenerating AC becomes necessary for its reusability, which makes an economical process. The AC is carrying some limitations like its adsorption capacity, which is a function of inert concentration that results in low quality products [229]. Already used AC is to be disposed of as it is hazardous material or is regenerated to be used again. Thermal regeneration contains many disadvantages because of its off-site regeneration that hardly converts the pollutant from one phase to another. This may result due to depletion of carbon and may cause damage to the structure of activated carbon.

The process, which could produce high yield efficiencies, could be a chemical regeneration of spent AC but it has some drawbacks regarding chemical consumption nondestructive pollutant elimination and creation of unimportant steams of waste [230, 231]. However, techniques like ultrasonic regeneration, microwave regeneration, and electrochemical regeneration are also being proposed as alternative techniques for the process of chemical regeneration of spent AC [230]. Chemical consumption, having increased footprints of carbon or having expensive facility requirement, is through the bench scale studies in which it has been proved to be effective but in industrial applications it has a limited appeal. Due to the limitations of the present technique, there is a need to develop another technique, which is more economical and environmentally friendly. Thus, to make one hybrid system, there is a need to merge semiconductor with AC [230, 231]. Organic pollutants, issues of destruction, and other hazardous problems are expected to arise from the oxidation semiconductor element. The reason lies in the generation of radical species like $O^{\bullet}$ and $OH^{\bullet}$ from the catalyst particles of the semiconductor, which causes oxidation of such species [230]. Another technique that has been studied for AC is the combination of heterogeneous photocatalysis and the Fenton reaction with the catalytic process, which is an oxidation based process.

Studies on this spent AC carried out previously were mostly about granular AC. As an example, it was demonstrated that there was an improved efficiency in herbicide removal from the water when granular AC adsorption photocatalyst hybrid system was used in comparison with a photocatalysis system [227, 228]. Similarly, it was also reported that the combination of photocatalytic and rotating adsorbent showed better efficiency in removing formaldehyde in comparison with adsorption [232].

Transition Metal Oxide Hybrids: AC in Photocatalysis System

In future, it anticipated that the coming photocatalyst generation would have improved internal efficiency regarding separation and also would be placed in contact with molecules of external pollutants [222–224]. So to improve the photocatalytic efficiency and separating catalyst from aqueous solution, the hybrid photocatalyst was designed by not moving the metal oxides having a large surface area, to condense pollutants which are diluted [233, 234]. In environmental purification field, heterogeneous photocatalysis with metal oxides of semiconductor has been applied as an efficient method [235].

By impregnation and adsorption along with various methods, metal oxides are expected to be impregnated into the surface of carbon; the same is applied for complex experimental procedures and process, which operate at high temperature. Considering ideal conditions, a photocatalyst should be inexpensive, highly proactive, nontoxic, and stable [233]. One more criterion that plays a role in the degradation of organic compounds is the potential of the redox residing in the band gap of the semiconductor. There are many semiconductors with the band gap energies ample to catalyze several chemical reactions, which include WO_3, Fe_2O_3, $SrTiO_3$, ZnS, TiO_2, and ZnO [233–236]. The metal sulfate group with insufficient stability in the process of catalysis is kids, PbS, or CdS. This compound undergoes photoanodic corrosion readily and is toxic. For example, Fe_3O_4 undergoes photoanodic corrosion readily so they are not suitable [233]. ZnO (3.2 eV) is unstable with Zn $(OH)_2$ in water; also it has band gap similar to anatase, so it results in the deactivation of the catalyst [228].

TiO_2 : AC Photocatalyst System

(1) TiO_2 Photocatalyst System: The most promising semiconductor for photocatalytic destruction organic pollutants is titanium dioxide (TiO_2) [229]. It provides the most excellent agreement in aqueous media between stability and catalytic performance. Since it is nontoxic, cheap, biocompatible, and stable in sunlight, so it is of immense importance and that is why it is also considered usable in cleaning environmental operations [230, 231]. Consequently, the electron pairs and positive holes are created at the surface of TiO_2. Once it has been irradiated with the UV light of the wavelength of 380 nm, TiO_2 would form reactive oxidants, such as OH radicals, hydrogen peroxide, superoxide anions, and other reactive species of oxygen and reactions that are reductive, to contribute in the organic compound decomposes which are adsorbed on the surface of TiO_2 [235–237] (Figure 6). The highest photocatalytic detoxification of TiO_2 is anatase phase. Deep studies have shown that photodegeneration of components like herbicides, phenols, dyes, pesticides, surfactants, and organic components (e.g: salicylic acid and sulfosalicylic acid) has been possessed by TiO_2 that is present in water wastes [238, 239].

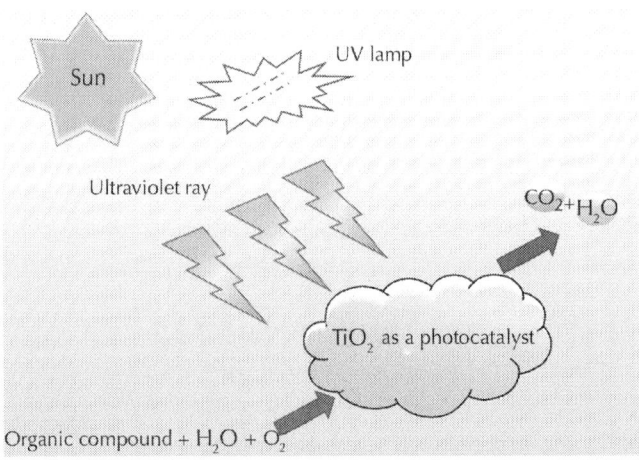

Figure 6: Basic principle on TiO_2 photocatalyst.

TiO_2 powders contributed to some drawbacks in separating phase in photocatalysis, with the purpose of its emission in the atmosphere because of their small particle size and recovery, the loss of photocatalyst if the separation is not promising, the need of fluidization of the powder in gaseous phase with cost and energy, and the scaling difficulties involved [237, 240]. Also, since radiation from the light compromises 47% of visible light, 48% infrared radiation, and 5% UV light, so TiO_2 acts as a benchmark of UV photocatalysis that it goes deactivated under visible light because of its wide band gap [241]. Moving on, the holes and photogenerated electrons present in an excited state play a vital role in the degeneration of pollutant and are unstable and without any effort they can recombine, they lead to low order efficiency which results in photocatalysis activities [231]. It is clear that the use of high potential solar photocatalysis cannot be made by TiO_2.

(2) Activated Carbon Supported TiO_2 in Photocatalyst System: For easy manipulation in a process of total photocatalytic operation and quick decomposition of organic pollutants, it might speed up the process to load photocatalysis to suitable adsorbents to increase the strength of pollutants around the photocatalysis system [237, 238]. Therefore, researchers had made attempts to support TiO_2 on different matrixes as silica gel, clay, carbon materials, alumina, and zeolites which can be nanotubes, carbon microspheres, carbon black, carbon nanofibres, and AC [242, 243]. TiO_2 particles are hydrophilic when

exposed to direct UV light whereas organic pollutants are hydrophobic [229]. The use of AC as a reference will provide help to molecules of pollutant closer to the active site of TiO_2 for a quick and effective photodegeneration process (Figure 7). The AC in comparison to organic pollutants that are approaching, in which secondary degradation takes place intermediately in situ, can generate new adsorption centers.

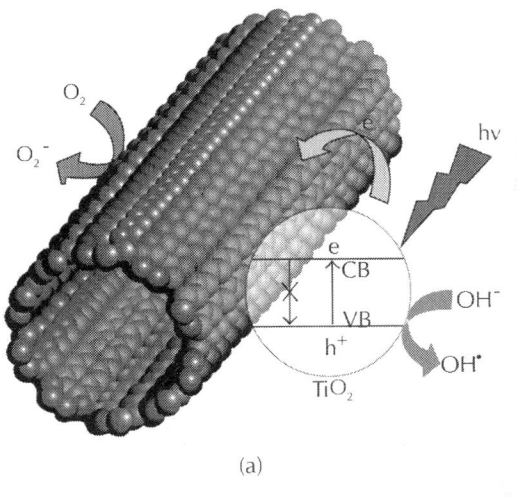

(a)

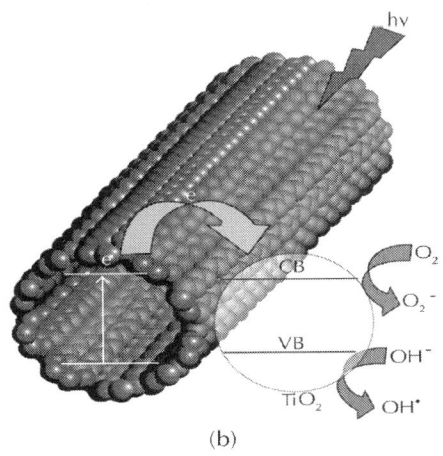

(b)

Figure 7: General mechanism of adsorption and photodegradation of TiO_2 supported CNTs photocatalyst [243].

In gas and water remediation for support purposes, AC is used widely because of its high porosity, good adsorption, supported TiO_2, and low cost that has marked the effects on disappearance of pollutants kinetics, with each pollutant being more quickly degenerated [244]. For example, the TiO_2 surface becomes static over glass surface; it has the benefits of high photodegradation productivity. The major limitation is the adhesion force in TiO_2 membrane and glass is poor, so TiO_2 is easy to decrease, which causes the decrease of the photodegradation productivity [231].

Hybrid of TiO_2 with AC support, as a sensitizer which is able to absorb light, was proven to be the best approach to developed photo-responding photocatalyst with great activity. The formation of heterojunction between TiO_2 with a small band gap and negatively charged AC may result in the inoculation of conducting band electrons from AC to TiO_2 and it is very useful for electrons and hole division [245, 246]. At the same time, the immobilization of TiO_2 onto the AC support can compromise for the loss of photocatalytic ability of TiO_2 because of the difficulty to effectively disperse in water for complete interaction with pollutants. So, many researchers reported that there is an optimum used amount of TiO_2 and AC pore formation for attaining the higher photocatalytic productivity than TiO_2 [245–247]. The 3D relation between the particles of AC, TiO_2 photocatalyst small particles, and the molecules of organic toxic is reported representationally in the absence of light and in the presence of ultraviolet enlightenment [245, 248]. The organic pollutants are supposed to be small to be adsorbed in microspores. In most of the AC amounts, a large group of micropores exists over the broad surface of the substantial pores, mesopores, and macropores; a large amount of toxic particles is well balanced due to physical adsorption [247–250]. Instead, only a little number of pollutant particles are adsorbed on the surface of TiO_2. By depositing TiO_2 molecules onto AC particles, some mesopores and micropores become closer to their openings and this causes a marked lessening in the uncovered surface area [245]. The ultraviolet enlightenment over such TiO_2 used AC molecules; oxidative OH free radicals take birth on the TiO_2 and can destroy toxic molecules by oxidation (Figure 8).

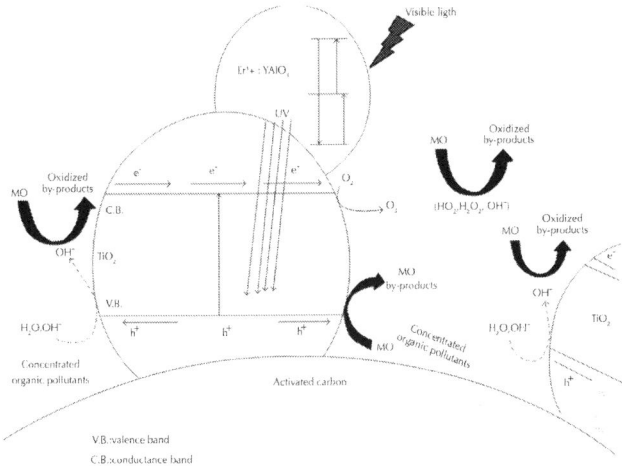

Figure 8: Mechanism on photodegradation of MO with TiO$_2$/AC under visible light irradiation [245].

Though these radicals will take birth over the surface of TiO$_2$ molecules and they are in the access of ultraviolet radiations and do not locate on the surface in the same radicals, they cannot diffuse through long distances and are limited to an area close to the active centers in TiO$_2$ [231–235, 251] (Figure 9). Adsorbed pollutant particles are photocatalytically demoted; they have to move along the surface of AC and TiO$_2$ molecules with the surface are not necessarily interacted to ultraviolet radiations [249].

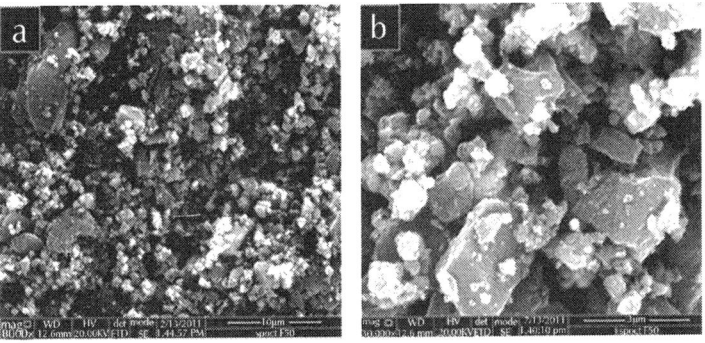

Figure 9: SEM images of TiO$_2$/AC powder. (a) 10 µm and (b) 3 µm [250].

The moving force of migration is actually the saturated gradient between organic toxic particles over the enlightenment of TiO_2 interface and on the other side some are over the surface of changing pore sizes of the AC molecules. The molecules diffused inside micropores of AC migrate with greater retaliation towards TiO_2 molecules resided on the interface of the AC particles. Thus, highly microporous AC particles are not usually advantageous for the TiO_2 : AC to have preferred the photocatalytic response [247, 248]. The effect of the substrate pore skeleton has been observed using AC surface area (770–1150 m^2g^{-1}) and a dip-hydrothermal process of photocatalyst preparation. Improved photocatalytic demoting of methyl orange (MO) has been attained with TiO_2 : AC than with a simple mixture of TiO_2 and AC [255] (Figure 10) and (Figure 11).

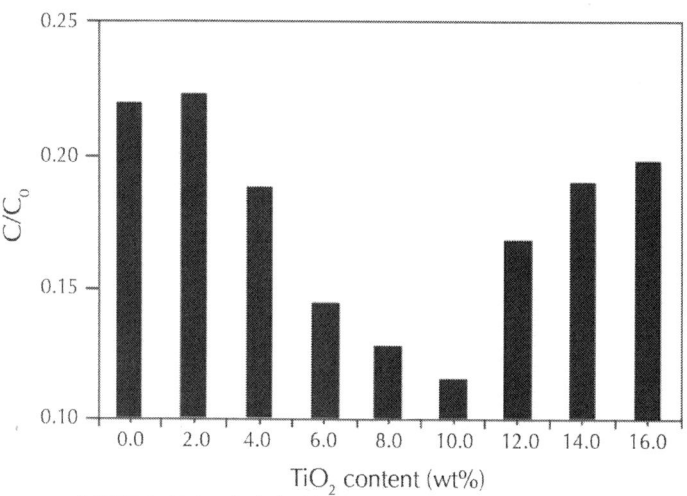

Figure 10: Effect on TiO_2/AC photocatalyst at different contents of TiO_2 on MO degradation [250].

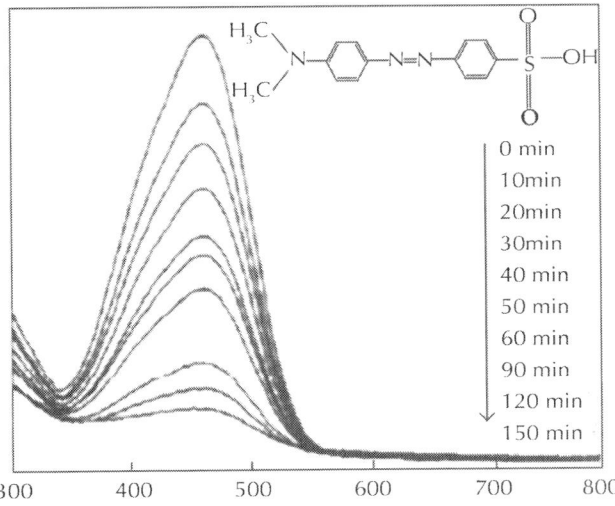

Figure 11: UV-Vis spectra of MO degradation with TiO_2/AC under visible light [245].

This got into the design of highly effective TiO_2:AC hybrid heterojunction photocatalyst; also the demand of commonsensible crosscheck capacity of band potentials among hybrid modules, the spatially and flat accessible transmission of holes and electron at the exposed surface, and the hole and electron movement of the hybrid system are important to enhance the photocatalytic action. It has been reported that the surface chemistry and map of AC revealed major effect on the collection of TiO_2 particles and photocatalytic deprivation of 4-chlorophenol [256, 257]. Adding to this, some researchers have constructed TiO_2 with AC microspheres to both maintain spreading and speeding up separation due to the AC microsphere that can be balanced with airy bubbles and it can be speedily settled in the reactor base with the help of some gravity due to the air bubbles [241]. In addition, some authors have fabricated TiO_2 with AC microspheres to both sustain spreading and acceleration division, because AC microsphere can be balanced with air bubbles and be able to speedily settle on top of the reactor base by gravity due to air bubbly. To overcome this shortcoming, the use of cobinder upholds the expansion of another method to obtain a fresh form of the TiO_2:AC photocatalyst for such a high action and better division performance.

Jamil et al. found that TiO_2 supported with AC sample heated at almost 500°C, which mainly consisted of rutile phase, showed the greatest photoactivity for deprivation and elimination of methyl orange from aqueous medium [255]. Therefore, most samples which were cooked at higher temperature were very detrimental to photoactivity. Also, using different types of AC revealed the connected effects between TiO_2 and AC during the 4-chlorophenol photodegradation and found a clear enhancement of photoactivity due to an increase in electron density of the AC support [258, 259]. It is researched that attendance of AC in interaction by TiO_2 is beneficial due to its burly adsorption capability. In the same way it advances the relocation rate of the interfacial change and lessens the rejoining rate of the holes and electrons [260, 261]. This synergistic effect of the interaction of AC and TiO_2 has been previously been stated for deprivation of some organic compounds in the photocatalytic process. It has been credited to a common contact between the different solid phases, in which AC acts as an efficient adsorption agent for the organic pollutants [257] (Figure 12).

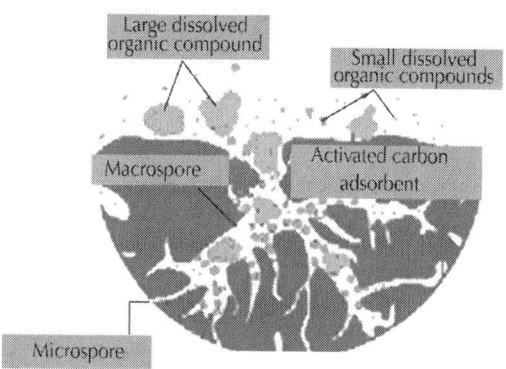

Figure 12: Adsorption mechanism of AC towards organic compounds.

The organic more efficiently moved to the TiO_2 surface, where it is immediately photocatalytically degraded by a mass movement of the photoactivated TiO_2 [258]. Thus, the organic burning rate observed on TiO_2:AC is like heading both with surface diffusion toxin particles with the photocatalytic process rate; because adsorption occurs gradually, the variation in relative pollutant proportion with irradiation time depends on both adsorption and photodecomposition, mainly at the start of ultraviolet treatment. It is stated that the variance of the proportion of phenol (as model of organic pollutants) remaining in the solution by ultraviolet treatment time is compared for TiO_2:AC, which were prepared by hydrolysis of tetraisopropyl orthotitanate and heat treatment at 650 to 900°C. For example, we noted that, in the duration of first 1 hour, adsorption of pH occurs in native AC and after 3 hours, in the presence of UV irradiations saturation achieved [261]. Adsorption as well as the photodecomposition of pH takes place simultaneously, but on the other side the former is supposed to become the dominate method in the beginning, being similarly latter in the next stage [221, 222]. The noted trend was based on two linear processes; the change of one process to the other takes place approximately in 1 hour of irradiation.

(3) Synthesis of TiO_2: AC Photocatalyst: A number of methods are available for the composition of TiO_2:AC catalyst, such as precipitation, chemical vapor deposition (CVD), hydrothermal, aerosol pyrolysis, hydrolysis, dip coating, and sol-gel [235, 262, 263]. However, selection process which is used for the selection of a suitable impregnation method depends only on the support used in it and the pollutant which is degraded [262]. It is clear that those physicochemical properties of TiO_2:AC catalyst have a heavy impact on the structure of the supported catalyst and they depend basically on the preparation method used, for example, thermal treatments. The main advantages of using physical methods are simple, low-cost and the use of commercially is present for photocatalyst with the wanted functionalities. By using the common wet methods, TiO_2:AC hybrid was also being prepared, the mismatches in the level of lattice among two hybrid components lessen the required efficiency of separation and transmission of photogenerated carriers (electron and hole) [264, 265]. Adherence of AC surface to TiO_2 particles appears significant for increment of photocatalytic action and as well for useful applications of hybrid system. To increase the anchorage of TiO_2 on AC, the wet process of synthesis is warranted [264]. This is only because the physical stable TiO_2:AC hybrid is in disagreement with hydrodynamic shearing method, surface chemistry plus AC pore structure can have a sufficient impact on spreading of TiO_2 over the synthesis; it results in the different photocatalytic presentations of TiO_2:AC. Nevertheless, TiO_2 photocatalyst usually has low precise surface region. Due to

the crystalline particle development to happens in heat operation like that. To prevent TiO_2 sintering, or else to make it precise surface region better, AC can be examined like a better help for photocatalysis material [266]. They found that the TiO_2 particles calculated at 450°C can collect and go through the great pores of the activated carbon substrate, including a very burly contact among carbon matrix and TiO_2. The contact between them leads to visible synergy to increase photocatalytic capacity for the degradation of the chromotrope 2R [267]. The prepared nanocrystal anatase TiO_2 particles are installed on activated carbon at a fewer temperature with the hydrolysis of the titanium but oxide in the acidic aqueous solution [268]. It is noticed that phenol toxin was absorbed by AC and after that drifted constantly over TiO_2, which consequently accelerate photocatalytic oxidation. On the other side, for the chemical vapor, deposition method is used for nanosized TiO_2 particles which were exposed to stick to activated carbon and tetra-butyl titanate and to offer large activities on the behalf of the photodecomposition of methylene orange in the water. Adding of water in titanium tetra-isopropoxide vapors was described to make possible CVD method at a higher deposition rate and lower temperature [269]. Introduction of H_2O vapors for the duration of CVD method and adsorption on the AC in prior were announced to be critical to get hold of anatase type TiO_2 nanoparticles at AC surface [270]. Investigations have also made it clear that the HNO_3 treatment results in more orderly TiO_2 loading by CVD, in comparison with other oxidation treatment. TiO_2 : AC ratio has been formed as a result of using dropping the support in solution produce with the alkoxide hydrolysis as well [269, 270] and subjected to heat operation at 300 to 500°C. In an alternative process, it can be produced by adding $TiCl_4$ drop by drop in aqueous suspensions of AC, come after by heat operation at 500°C in N_2 atmosphere [271]. Load the TiO_2 powder exactly over AC has been achieved as a result of combining TiO_2 in AC aqueous suspension with stirring. On the other side, loading over AC filter was produced as a result of gluing granular AC over the glass cloth and it was formed by the water suspension of five mass % TiO_2 and the conclusion TiO_2 : AC particles inside CCl_4 solution of pitch, come after by heat operation at 750°C [263]. In resultant hybrid system, TiO_2 particles over AC were expected to be layered with carbon, composed of the pitch at the stage of the heat operation; it may function as a mean repair TiO_2 particle over the AC surfaces. TiO_2 particles also were able to load over

AC by spray-desiccation procedure, with a little modification in the pore structure of the AC [272, 273]. In the other study, loading of TiO_2 over the AC surface was taking place via dipping the AC particles in a peroxotitanate solution; after that heating at 180°C in a Teflon lined stainless-steel vessel came after in calcinations at 300 to 800°C [274, 275]. With the help of the AC particles of 0.16 to 0.26 millimeter, the disjunction of the particles from the solution was not too much harder, and photocatalytic action on behalf of the decomposition of MO almost remains identical for 5 cycles. TiO_2 has also been achieved by plugging the pore of AC by paraffin [275]. After loading TiO_2, by removing the paraffin at 250°C in the air, the high surface region of pristine AC, it could be recuperated and the high photocatalytic action was procured basically for the decay of methylene blue (MB) [276]. TiO_2:AC has also been composed by mixing TiO_2 particles with some liquid or solid state carbon precursor. By hydrolysis of tetraisopropyl orthotitanate, TiO_2 was caused in the exterior region of the poly vinyl butyral (PVB) and TiO_2 overloaded PVB was carbonized at a high temperature in the flow of CO_2. TiO_2 loaded carbon microspheres with 25 μm diameter have been prepared from the TiO_2 loaded cellulose microspheres, composed with one step stage division by using the sodium polyacrylate aqueous solutions and cellulose xanthate with the isolated TiO_2 powder [277–279].

(4) Performance of TiO_2: AC Photocatalyst: Although it is a hard to understand how light could it penetrate the carbon particles to reach the inoperative photocatalyst, TiO_2:AC composites have quite clear high efficiencies for the photodegradation of a variety of pollutants [238]. In such a case, the presence of the AC seems to change the photocatalytic activity of TiO_2 towards the abasement of organic pollutants beyond the so-called "synergistic" effect [227]. The harmonious aftereffect of the adsorption with AC and photocatalytic disintegration by TiO_2 has been noticed during the deprivation of many kinds of organic toxin.

The basic principle of photocatalysis over illuminated TiO_2:AC system. But model of a TiO_2:AC photocatalytic process can be more of a complex issue, it starts the photointensity to the classical aspects of the heterogeneous catalytic system; for example, temporal variations in concentration of iminoctadine triacetate (IT) that is frequently used in excess plus orchard fields like an insecticide and in the water path of the fields are exposed for pristine AC and three TiO_2:AC [280, 281]. The hybrid systems were kept without any light on them for 200h to saturate IT adsorption and after that were exposed to ultraviolet irradiation. 800h behind this [280], the sample was separated from given test solution and spread once more in the virgin 1.87×10^{-4} mol/L solution and again kept without any light in it for approximately 200h and then exposed to ultraviolet radiations. As far as pristine AC is concerned, concentration of pollutants was initiated to survive approximately steady without any light and to spread to some extent beneath ultraviolet irradiation. The bottom line is that the photocatalyst system of TiO_2:AC could have similar photocatalytic results without any light and under ultraviolet radiations [282]. Before mentioned data was supported by the one who stated that the enhancement of photocatalytic activity for the organic burning of pollutants via TiO_2: AC enzyme. Stated that is, the concentration of 4-chlorophenol solution demotes rapidly for the TiO_2:AC than TiO_2, saying that it enhances photoorganic burning approximate by a pseudo-first-order equation with a linear relationship between time and concentration change. The same harmonious result was noticed in the organic burning of pH and 2,4-dichlorophenoxyacetic acid applying similar AC and TiO_2 sample [283]. Furthermore, the pH disintegration is found to be dependent on the mass ratio of TiO_2 to AC (5/10 to 75/10) [231, 284, 285]. The harmonious result is thought to be attributable to the fast movement of pH molecules at the start adsorbed on the AC on the outer surface of TiO_2; the motivating force of that movement is most probably the differentiation in the surface concentration of pH between AC and TiO_2. Defined proportion of pH was found to remain at AC, even after the proportion in sample becomes negligibly small.

ZnO:AC Photocatalyst System

After TiO_2, ZnO supported AC finds broader attraction of use, due to some of the excellent behaviors of ZnO, such as wider availability, stability, and suitable band gap of energy [228]. Problems associated with the usage of ZnO alone as photocatalyst are as well partitioned, especially the complexity in unraveling the powder with the sample after the reaction is over; gathering of particles in delay, particularly at high loading and complexity in purpose to the consistent flow of the domain system has been approved by its surface properties [37, 235] (Figure 13).

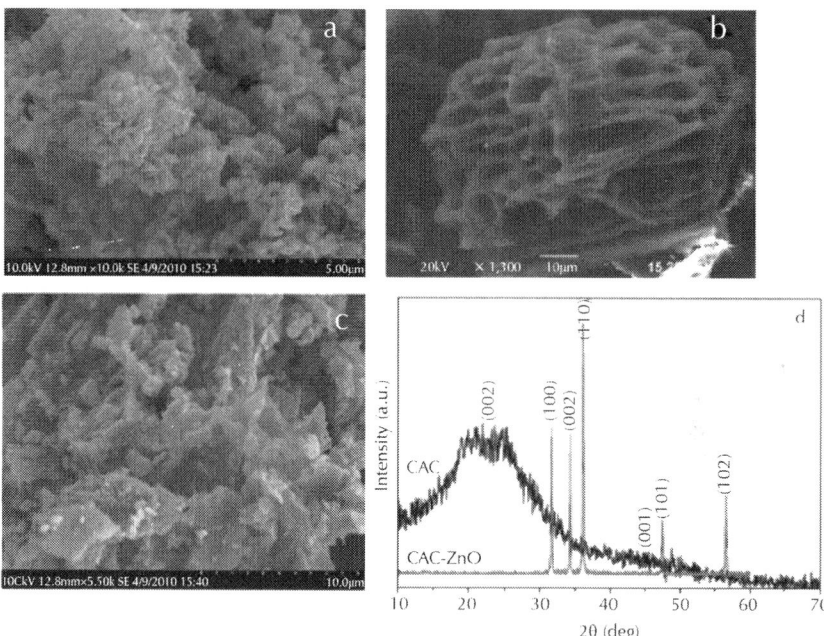

Figure 13: Surface morphology of the (a) pure ZnO, (b) pure AC, and (c) AC–ZnO mixture. XRD spectrum of pure AC and AC–TiO_2 mixture [37].

The problem is AC has been declared good as for the support of the ZnO photocatalysis system. Replying on using of dioxygen, photo-, and entirely mineralized organic as well as inorganic substances and particularly biorecalcitrant, make the technique is environmentally friendly for toxic waste reduction schemes [225]. Spherical AC particles having ZnS and ZnO were formed from a caption-exchange resin (polystyrene with sulfonate groups and cross-linked by divinylbenzene) and $ZnCl_2$ aqueous solution, followed by carbonization at almost 500 to 900°C [286]. It has been reported that ZnO is an appropriate substitute to TiO_2 for the photodegradation of Acid Red 14, an azo dye, because it is photodegradation process that has been proved to be similar to TiO_2 [37, 287] (Figure 14).

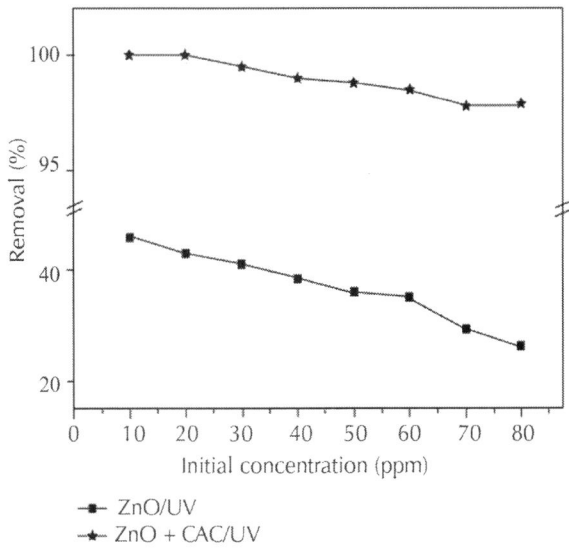

Figure 14: Effect of initial concentration of dye in the presence and absence of AC in ZnO/AC photocatalyst system (contact time: 90 min; dose of the catalyst: 200 mg; dose of the CAC: 40 mg) [37].

Recently it was reported that the instantaneous destruction of inorganic toxin like Cr(VI) and organic toxin, like 4-CP, can be recognized in ZnO : AC photocatalytic reaction system [286–289]. Hence, this technique can be functional over the broader level for aqueous waste reduction.

ADVANCE ACTIVATED CARBON PHOTOCATALYTIC SYSTEM

Granular and Spherical Activated Carbon Photocatalytic Systems

In all kinds of AC, the spherical AC has benefits, such as its frictionless surface, high quality fluidity, and good strength on the powdered and rough AC. Carrying this forward, many proposals have lately examined the sustain of TiO_2 to spherical AC [290, 291]. But there are some statements on the granular AC that supported TiO_2 photocatalyst which could enhance the demolishing efficiency of many organic compounds, regarding many environmentally related conditions [252, 292] (Figure 15).

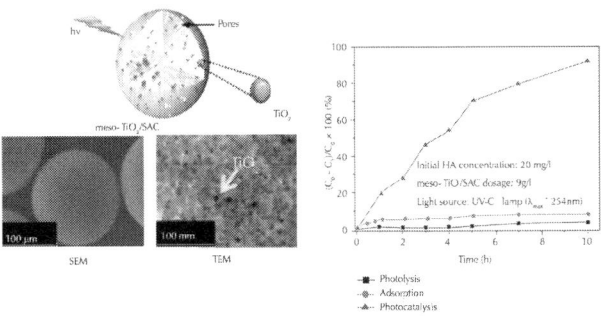

Figure 15: Effect of granular and spherical AC on adsorption and photocatalytic activity [252].

Granule AC supports TiO_2 powdered by adding more pollutants and alternatives around the TiO_2. The pollutants and alternatives can diffuse to the surface of TiO_2. Granular AC also reduces TiO_2 jelling, which lessens its surface, thereby reducing its enzymatic activities [292]. This was brought by the theory that the absorptivity of AC depending strongly on the molecular morphology and size of the pollutant particles to the TiO_2 surface happens straightforwardly from the solution, and not through the AC surface [293, 294]. The TiO_2 can destroy the pollutants, leading again to generation of granular AC in this situation [294]. Most of the porous AC still in granules, and the problem of separation and recovering of the photocatalyst from the reaction environment is already present.

ACF Photocatalytic Systems

AC is a freshly developed type of photocatalyst supporting materials consisting of nanographites known as AC fiber ACF. In comparison to the granular AC, ACF has a larger surface area, having greater pore volume, more uniform micropores size distribution, a greater rate of adsorption and desorption, and a rapid attainment of adsorption equilibrium with ACF in the form of fleet are preferable to the handling than granular supports [245, 253] (Figure 16).

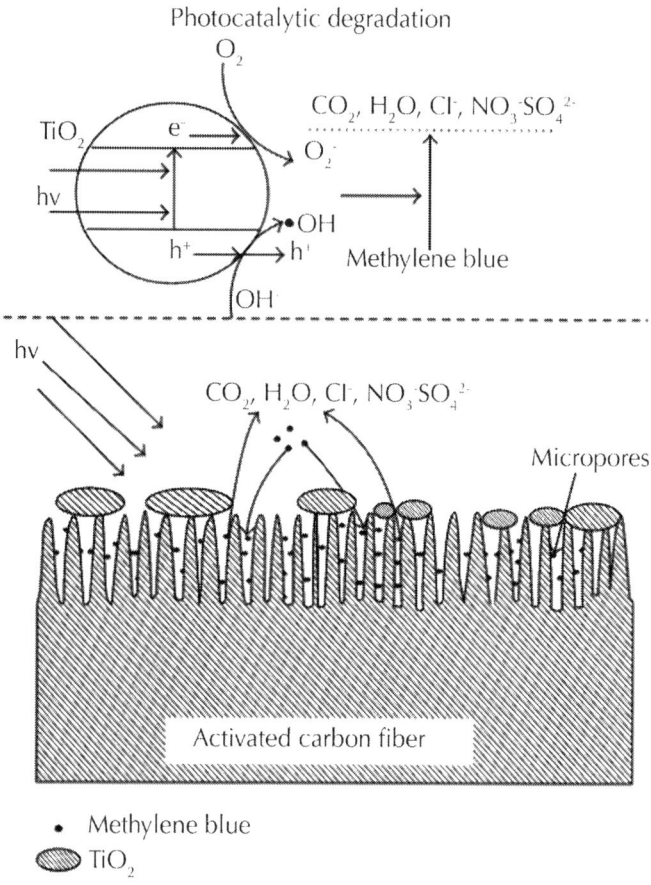

Figure 16: Schematic diagram for the adsorption and photocatalytic degradation of methylene blue on the TiO_2/ACFs [253].

The surface-area characteristics of the ACF are identified to depend powerfully on the creative processes, affecting the load of TiO_2 and eventually the adsorption of pollutant particles [255]. Nevertheless, the ACF supported TiO_2 photocatalyst has sometimes been used for the removal of gas phase pollutants in the environment (Figure 17).

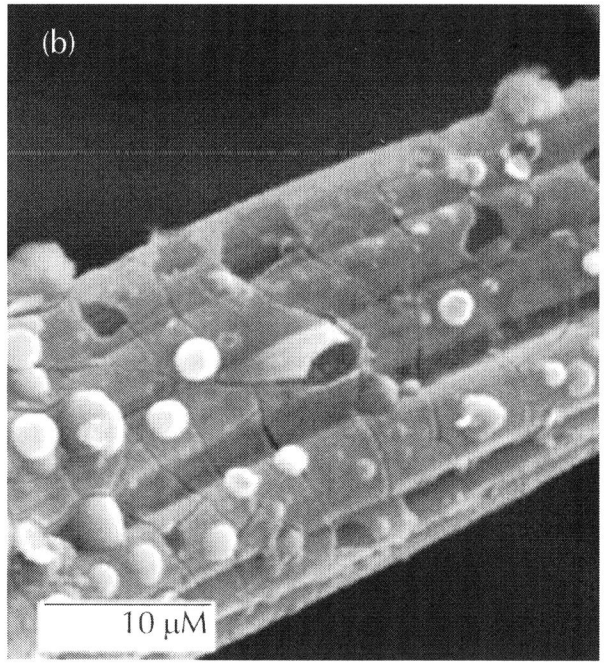

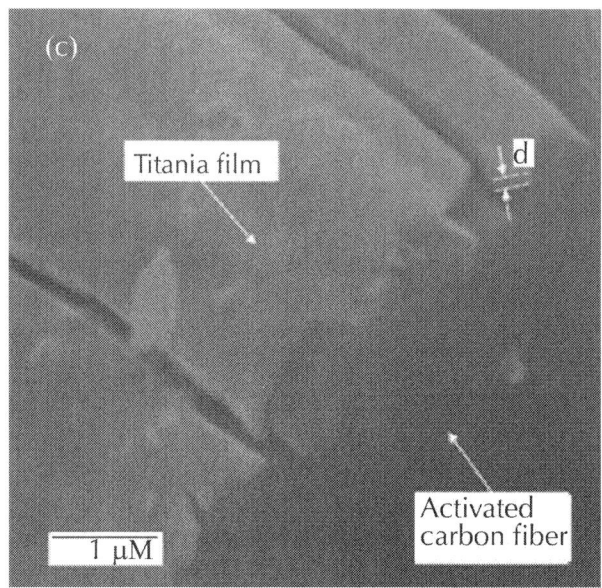

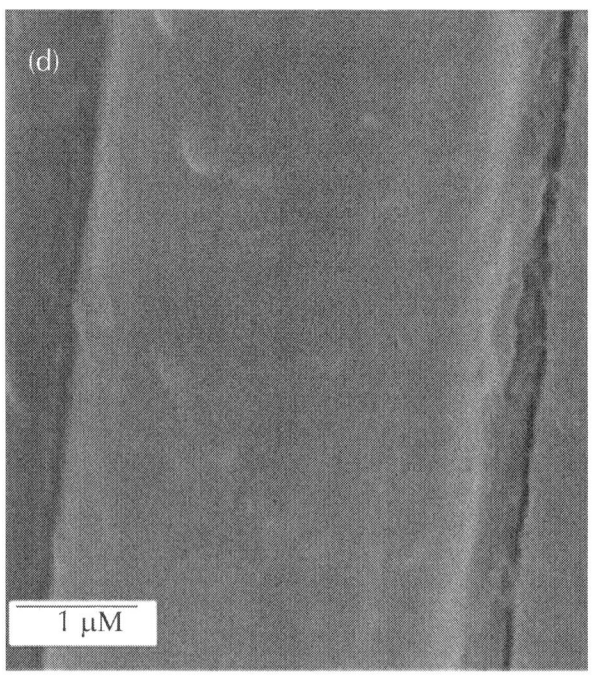

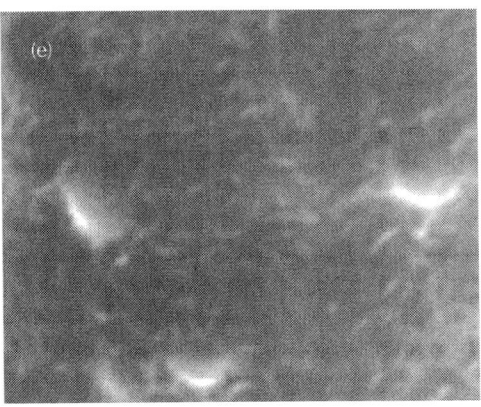

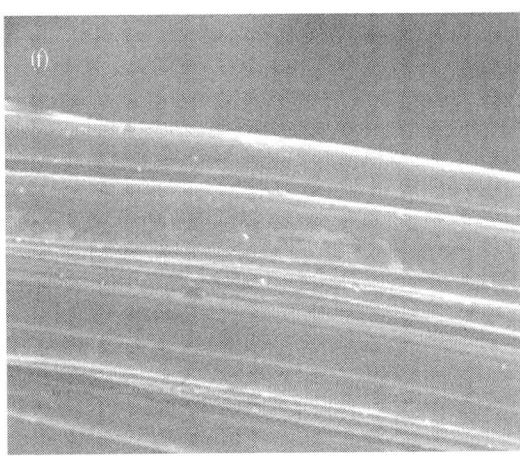

Figure 17: SEM micrographs of the TiO$_2$/ACFs photocatalyst and unmodified ACFs, (a) general view of the photocatalyst, 100x; (b) TiO$_2$ coating on single carbon fiber, 2000x; (c) cross-sectional view of TiO$_2$ coating, 20,000x; (d) surface of TiO$_2$ coating, 20,000x; (e) surface of TiO$_2$ coating, 50,000x; and (f) single unmodified activated carbon fiber 4000x [253].

TiO$_2$: Graphene Photocatalyst System

Meanwhile, the beginning of graphene supply is an idea to resolve the limitation brought by the TiO$_2$:AC photocatalyst. Recently, functionalized grapheme based semiconductor photocatalyst has attracted attention because of its larger definite surface area, higher electron conductivity, and adsorption [248] (Figure 18).

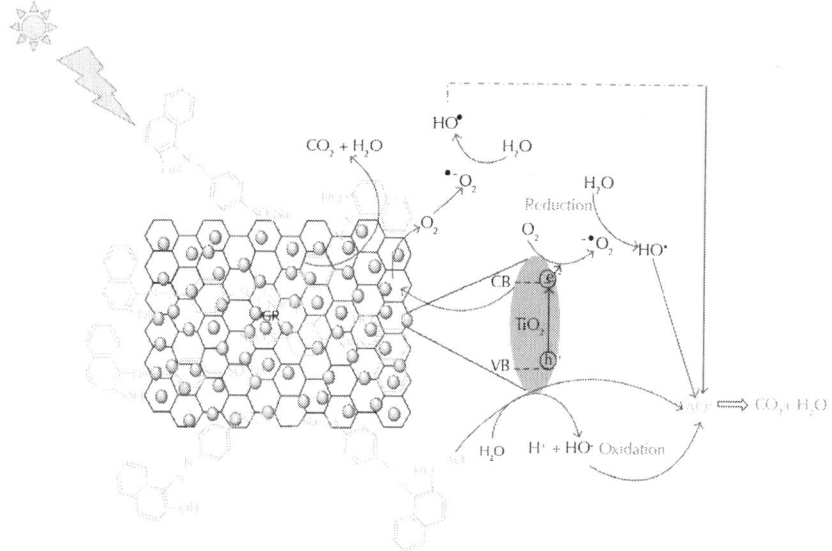

Figure 18: Schematic illustrations on the photodegradation mechanism of the synthetic dye on TiO_2/graphene nanocomposite [248].

A lot of the hard work is used for the combination of TiO_2-graphene photocatalysis hybrid system [295, 296]. Considering advantages due to a higher specified surface region, graphene appeared like a better help to prepare the overloaded nanoparticle metal oxides to attain an identical division not including aggregation. Betterment of the photocatalytic actions of TiO_2-graphene hybrid is associated with huge two-dimensional planar graphene structure supporting the dye adsorption plus squeezing electron hole rejoining because of higher electrical conduction property as has been indicated by the morphological analysis of TiO_2/grapheme photocatalyst [226, 254] (Figure 19).

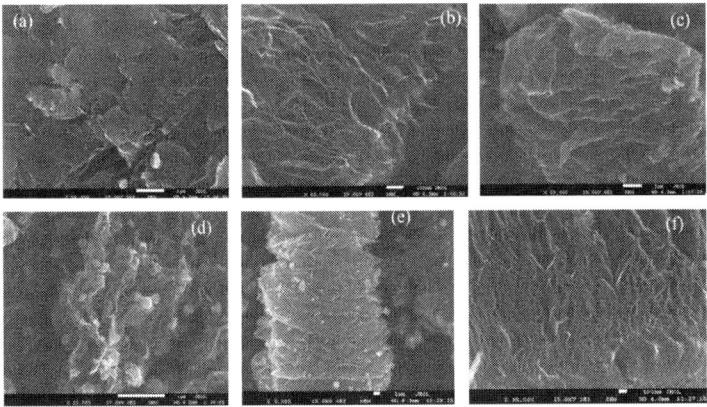

Figure 19: FE-SEM images of TiO_2/graphene nanocomposites at different TiO_2 contents 1 wt% (a), 3 wt% (b), 5 wt% (c), and 10 wt% TiO_2/graphene heat treated at 450°C under reducing (N_2/H_2) (e) and nitrogen (f) [254].

Additionally, absorption of additional catalyst particles into a particularly graphene leaf at individual places can supply better dynamically in achieving the choosy catalytic otherwise logical procedures and adjusts the composition also the morphology of photocatalysts to enhance their photocatalytic results [297].

TiO_2 : CNT Photocatalyst System

Barring graphene, carbon nanotube (CNT) has been regarded as a more attractive catalytic support than activated carbon because of combination of electronic, adsorption, and specific semiconducting characteristics [298]. Studies of TiO_2:CNTs reveal a considerable synergy effect with the metal oxides and carbon phases [298, 299]. Going further, researchers have shown that CNTs can enhance the adsorption and photocatalytic activity of TiO_2 in the presence of ultraviolet [300]. Single walled CNTs execute enhanced and selective photocatalytic oxidation of pH [299]. Heterostructure CNT consists of TiO_2-xNx and C prepared by carbonization of electron-spun polyacrylonitrile nanofibers containing stabilized titanium oxoacetate [301]. So, CNTs can be used as a reliable material for environmental pollution clearing and can be used to improve the photocatalytic efficiency of TiO_2.

TiO$_2$: AC Semiconductor Doped Photocatalyst System

The term "doping" means an additional semiconductor having unusual parallel valence bands and conduction of energy levels with the TiO$_2$ over the outer surface of TiO$_2$ has been authorized the outcome to progress the photocatalytic results. The method of using coupled semiconductors is the enhancement of the photocatalytic with good organization by raising the charge division and increasing the photoresponding domain. Some researches upon doping an additional semiconductor over outer surface of TiO$_2$: AC includes SnO$_2$, V$_2$O$_5$, ZnO, ZrO$_2$, and CsS [239, 260, 302, 303]. The entire of it confirms the superior photodegradation efficiency as compared to the TiO$_2$: AC. Recent researches have reported the modification of TiO$_2$: AC by semi-conductor doping that has led to the improvement of removal of specific containments in the gas resulting in the changes of the physical and chemical properties of the carbon materials. Iron (Fe$_3$O$_4$) dopant TiO$_2$: AC has been found to be a promoter to photocatalysis with a good performance on degradation of Congo red and methyl orange [32]. Some researchers have combined the Fe$_3$O$_4$ compound with TiO$_2$: AC aiming at preparing photocatalyst with magnetic core and photoactive encapsulation [234]. They proved that hybrid systems had magnetic properties and could be separated by magnetic materials. Furthermore, the magnetic hybrid photocatalyst can also be magnetically distributed by a discontinuous magnetic field in an interrupted system [304]. But they found that the photocatalytic activity of the hybrid system declined because the magnetic particles experienced light dissolution [305].

TiO$_2$: AC Nonmetal Doped Photocatalyst System

Doping of nonmetal, such as N, P, B, C, S, F, chlorine, and bromine, has been also widely used to improve the photocatalytic or to extend the photoabsorption into visible light of TiO$_2$: AC system [258]. Among them, the P doped TiO$_2$: AC has recently attracted increasing interest due to its enhanced shows a little band gap. It has an adsorption property in the visible light region [306]. Also, it has been found that phosphorous-doped TiO$_2$: AC prepared by the buffer solution method with NaH$_2$PO$_4$ as precursor showed a greater photocatalytic activity of acetaldehyde organic burning under visible light absence than the pure sample [307]. It synthesizes the phosphorous-doped TiO_2 : AC with high crystallinity and large surface area of hydrothermal process. Methylene blue demoted performance on phosphorous-doped sample was pointedly increased and superior to the commercial Phosphorous 25 [308].

CONCLUSIONS

AC has been employed as an adsorbent for the control of many environmental pollutants due to its high pore density and large uncovered surface area-to-volume ratio. Many AC researches have targeted primarily with concentration in lower to higher concentrations than that associated with indoor aqueous quality. Loading of semiconductors on AC has drawn vast interest while the higher adsorption abilities of AC can assist in the direction of improvement of organic pollutant about the enzyme, enhancing the toxin transmission method thus enhancing the photocatalytic results. In future, photocatalysis reaction in visible light can be addressed as the main challenge, as reflected by recent intensive scientific endeavors. The state-of-the-art accomplishments in visible light will be induced in selection of organic transformations by AC heterogeneous photocatalysis. In addition, the recent strides are to bridge between AC photocatalysis system and other area of the catalyst with the aim of overcoming the existing limitation of photocatalyst by developing more creative approaches.

ACKNOWLEDGMENTS

This work is financially supported by University Malaya Research Grant (UMRG RP022-2012E; UMRG RP022-2012A) and Fundamental Research Grant Scheme (FRGS: FP049-2013B) by University of Malaya and Ministry of High Education (MOE), Malaysia.

REFERENCES

1. C. Baleizão, B. Gigante, H. Garcia, and A. Corma, "Ionic liquids as green solvents for the asymmetric synthesis of cyanohydrins catalysed by VO(salen) complexes," Green Chemistry, vol. 4, no. 3, pp. 272–274, 2002.

2. V. L. Coluim, "The potential environmental impact of engineered nanomaterials," Nature Biotechnology, vol. 21, pp. 1166–1170, 2003.

3. R. H. Hurt, M. Monthioux, and A. Kane, "Toxicology of carbon nanomaterials: status, trends, and perspectives on the special issue," Carbon, vol. 44, no. 6, pp. 1028–1033, 2006.

4. V. Subramanian, H. Zhu, R. Vajtai, P. M. Ajayan, and B. Wei, "Hydrothermal synthesis and pseudocapacitance properties of MnO_2 nanostructures," Journal of Physical Chemistry B, vol. 109, no. 43, pp. 20207–20214, 2005.

5. W. Seiler and P. J. Crutzen, "Estimates of gross and net fluxes of carbon between the biosphere and the atmosphere from biomass burning," Climatic Change, vol. 2, no. 3, pp. 207–247, 1980.

6. M. F. Philip, L. Niwton, and M. F. Fernondo, "Rainforest burning and the global carbon budget: biomass, combustion efficiency, and charcoal formation in the Brazilian Amazon," Journal of Geophysical Research: Atmospheres, vol. 98, pp. 16733–16743, 1980.

7. Y. Kuzyakos, I. Subbotino, H. Chen, I. Bagomolova, and X. Xu, "Black carbon decomposition and incorporation into soil microbial biomass estimated by ^{14}C labeling," Soil Biology and Biochemistry, vol. 41, no. 2, pp. 210–219, 2009.

8. A. Demirbas, "Carbonization ranking of selected biomass for charcoal, liquid and gaseous products,"Energy Conversion and Management, vol. 42, no. 10, pp. 1229–1238, 2001.

9. E. N. Chidumayo, "Woody biomass structure and utilisation for charcoal production in a Zambian Miombo woodland," Bioresource Technology, vol. 37, no. 1, pp. 43–52, 1991.

10. U. Hamer, B. Marschner, S. Brodowski, and W. Amelong, "Interactive priming of black carbon and glucose mineralisation," Organic Geochemistry, vol. 36, no. 7, pp. 823–830, 2004.

11. J. C. Akunna, C. Bizeau, and R. Moletta, "Nitrate and nitrite reductions with anaerobic sludge using various carbon sources: glucose, glycerol, acetic acid, lactic acid and methanol," Water Research, vol. 27, no. 8, pp. 1303–1312, 1993.

12. J. Wang, L. Fang, D. Lopez, and H. Tobias, "Highly selective and sensitive amperometric biosensing of glucose at ruthenium-dispersed carbon paste enzyme electrodes," Analytical Letters, vol. 26, no. 9, pp. 1819–1830, 1993.

13. C. Bourdilon, J. P. Bourgeois, and D. Thomas, "Covalent linkage of glucose oxidase on modified glassy carbon electrodes. Kinetic phenomena," Journal of the American Chemical Society, vol. 102, no. 12, pp. 4231–4235, 1980.

14. Y. Onal, S. Schimpt, and P. Claus, "Structure sensitivity and kinetics of d-glucose oxidation to d-gluconic acid over carbon-supported gold catalysts," Journal of Catalysis, vol. 223, no. 1, pp. 122–133, 2004.

15. J. J. Manya, E. Velo, and L. Puigjaner, "Kinetics of biomass pyrolysis: a reformulated three-parallel-reactions model," Industrial & Engineering Chemistry Research, vol. 42, no. 3, pp. 433–441, 2003.

16. R. D. Cortright, R. R. Davda, and J. A. Dumesic, "Hydrogen from catalytic reforming of biomass-derived hydrocarbons in liquid water," Nature, vol. 418, no. 6901, pp. 964–967, 2002.

17. D. Sutton, B. Kelleher, and J. R. H. Ross, "Review of literature on catalysts for biomass gasification," Fuel Processing Technology, vol. 73, no. 3, pp. 155–173, 2001.

18. T. R. Carlson, T. P. Vispute, and G. W. Huber, "Green gasoline by catalytic fast pyrolysis of solid biomass derived compounds," ChemSusChem, vol. 1, no. 5, pp. 397–400, 2008.

19. H. Mehdi, V. Fabos, R. Taba, A. Bodor, L. T. Mika, and I. T. Horvalth, "Integration of homogeneous and heterogeneous catalytic processes for a multi-step conversion of biomass: from sucrose to Levulinic acid, -valerolactone, 1,4-pentanediol, 2-methyl-tetrahydrofuran, and alkanes," Topics in Catalysis, vol. 48, no. 1–4, pp. 49–54, 2008.

20. T. R. Carlson, G. A. Tompsett, W. C. Conner, and G. W. Huber, "Aromatic production from catalytic fast pyrolysis of biomass-

derived feedstocks," Topics in Catalysis, vol. 52, no. 3, pp. 241–252, 2009.

21. J.-P. Tessonnier, A. Villa, O. Majoulet, D. S. Su, and R. Schlögl, "Defect-mediated functionalization of carbon nanotubes as a route to design single-site basic heterogeneous catalysts for biomass conversion,"Angewandte Chemie, vol. 48, no. 35, pp. 6543–6546, 2009.

22. J. P. T. Dalsgaard, C. Lightfool, and V. Christensen, "Towards quantification of ecological sustainability in farming systems analysis," Ecological Engineering, vol. 4, pp. 181–189, 1995.

23. M. Hoogwijk, A. Faaij, R. van den Broek, G. Berndes, D. Gielen, and W. Turkenburg, "Exploration of the ranges of the global potential of biomass for energy," Biomass and Bioenergy, vol. 25, no. 2, pp. 119–133, 2003.

24. Y. Cong, M. Chen, T. Xu, Y. Zhang, and Q. Wang, "Tantalum and aluminum co-doped iron oxide as a robust photocatalyst for water oxidation," Applied Catalysis B: Environmental, vol. 147, pp. 733–740, 2014.

25. D. Zhang, M. Wen, S. Zhang, et al., "Au nanoparticles enhanced rutile TiO_2 nanorod bundles with high visible-light photocatalytic performance for NO oxidation," Applied Catalysis B, pp. 610–616, 2014.

26. Z. F. Huang, J. J. Zou, L. Pan et al., "Synergetic promotion on photoactivity and stability of $W_{18}O49/TiO_2$ hybrid," Applied Catalysis B: Environmental, vol. 147, pp. 167–174, 2014.

27. B. Weng, S. Liu, N. Zhang, Z.-R. Tang, and Y.-J. Xu, "A simple yet efficient visible-light-driven CdS nanowires-carbon nanotube 1D–1D nanocomposite photocatalyst," Journal of Catalysis, vol. 309, pp. 146–155, 2014.

28. S. N. R. Inturi, T. Boningari, M. Suidan, and P. G. Smirniotis, "Visible-light-induced photodegradation of gas phase acetonitrile using aerosol-made transition metal (V, Cr, Fe, Co, Mn, Mo, Ni, Cu, Y, Ce, and Zr) doped TiO_2," Applied Catalysis B: Environmental, vol. 144, pp. 333–342, 2014.

29. C. F. Chang and C. Y. Man, "Magnetic photocatalysts of copper phthalocyanine-sensitized titania for the photodegradation of dimethyl phthalate under visible light," Colloids and Surfaces A:

Physicochemical and Engineering Aspects, vol. 441, pp. 255–261, 2014.

30.   S. Kumar, B. Kumar, T. Surender, and V. Shanker, "g-C_3N_4/NaTaO$_3$ organic–inorganic hybrid nanocomposite: High-performance and recyclable visible light driven photocatalyst," Materials Research Bulletin, vol. 49, pp. 310–318, 2014.

31.   J. Gamage McEvoy, W. Cui, and Z. Zhang, "Synthesis and characterization of Ag/AgCl-activated carbon composites for enhanced visible light photocatalysis," Applied Catalysis B: Environmental, vol. 144, pp. 702–712, 2014.

32.   D. H. Quinones, A. Rey, P. M. Alvarez, F. J. Beltran, and P. K. Plucinski, "Enhanced activity and reusability of TiO_2 loaded magnetic activated carbon for solar photocatalytic ozonation," Applied Catalysis B: Environmental, vol. 144, pp. 96–106, 2014.

33.   Z. Cherkezova-Zheleva, D. Paneva, M. Tsvetkov et al., "Preparation of improved catalytic materials for water purification," Hyperfine Interactions, vol. 226, no. 1–3, pp. 517–527, 2014.

34.   J. Yun, H. I. Kim, and Y. S. Lee, "A hybrid gas-sensing material based on porous carbon fibers and a TiO_2 photocatalyst," Journal of Materials Science, vol. 48, pp. 8320–8328, 2013.

35.   B. Adeli and F. Taghipour, "A review of synthesis techniques for gallium-zinc oxynitride solar-activated photocatalyst for water splitting," ECS Journal of Solid State Science and Technology, vol. 2, no. 7, pp. Q118–Q126, 2013.

36.   Z. Ou-Yang, H. L. Xu, C. Xiong et al., "Prearation of TiO_2 supported activated carbon and its application in papermaking wastewater," Advanced Materials Research, vol. 791, p. 7, 2013.

37.   P. Muthirulan, M. Meenakshisundararam, and N. Kannan, "Beneficial role of ZnO photocatalyst supported with porous activated carbon for the mineralization of alizarin cyanin green dye in aqueous solution," Journal of Advanced Research, vol. 4, no. 6, pp. 479–484, 2013.

38.   M. Smits, D. Huygh, B. Craeye, and S. Lenaerts, "Effect of process parameters on the photocatalytic soot degradation on self-cleaning cementitious materials," Catalysis Today, vol. 230, pp. 250–255, 2014.

39.   M. J. Nurhidayatullaili, B. Samira, and B. A. H. Sharifah, "Recent

advances in heterogeneous photocatalytic decolorization of synthetic dyes," The Scientific World Journal, vol. 2014, Article ID 692307, 25 pages, 2014.

40. M.-M. Titirice, A. Thomas, and M. Antonietti, "Aminated hydrophilic ordered mesoporous carbons,"Journal of Materials Chemistry, vol. 17, pp. 3412–3418, 2013.

41. V. Budarin, J. H. Clark, J. J. E. Hardy et al., "Starbons: new starch-derived mesoporous carbonaceous materials with tunable properties," Angewandte Chemie, vol. 45, no. 23, pp. 3782–3786, 2006.

42. Q. Wang, H. Li, L. Chen, and X. Huang, "Monodispersed hard carbon spherules with uniform nanopores," Carbon, vol. 39, no. 14, pp. 2211–2214, 2001.

43. C. Poonjarernsilp, N. Sano, and H. Tomon, "Hydrothermally sulfonated single-walled carbon nanohorns for use as solid catalysts in biodiesel production by esterification of palmitic acid," Applied Catalysis B: Environmental, vol. 147, pp. 726–732, 2014.

44. Y. F. Zhang, "Research on low-carbon architectural development based on green life cycle," Applied Mechanics and Materials, vol. 443, pp. 263–267, 2014.

45. N. M. Mubarak, J. N. Sahu, E. C. Abdullah, and N. S. Jayakumar, "Removal of heavy metals from wastewater using carbon nanotubes," Separation and Purification Reviews, vol. 43, no. 4, pp. 311–338, 2014.

46. A. Fraczek-Szczypta, "Carbon nanomaterials for nerve tissue stimulation and regeneration," Materials Science and Engineering C, vol. 34, pp. 35–49, 2014.

47. M. Vikkisk, I. Kruusenberg, U. Joost, E. Shulga, I. Kink, and K. Tammeveski, "Electrocatalytic oxygen reduction on nitrogen-doped graphene in alkaline media," Applied Catalysis B: Environmental, vol. 147, pp. 369–376, 2014.

48. K. Takeba, M. Matsumoto, Y. Shida, and H. Nakazawa, "Determination of phenol in honey by liquid chromatography with amperometric detection," Journal of the Association of Official Analytical Chemists, vol. 73, no. 4, pp. 602–604, 1990.

49. J. Yang, Y. Yu, J. Jan, S.-T. Tua, and E. Dahlquistc, "Effects of SO_2 on

CO_2 capture using a hollow fiber membrane contactor," Applied Energy, vol. 112, pp. 755–764, 2013.

50. R. A. Shawabkeh, M. Al-Harthi, and S. M. Al-Ghamdi, "The synthesis and characterization of microporous, high surface area activated carbon from palm seeds," Energy Sources A: Recovery, Utilization and Environmental Effects, vol. 36, no. 1, pp. 93–103, 2014.

51. K. Labus, S. Gryglewicz, and J. Machnikowski, "Granular KOH-activated carbons from coal-based cokes and their CO_2 adsorption capacity," Fuel, vol. 118, pp. 9–15, 2014.

52. Y.-Z. Xiang, X. Luo, W. Y. Han et al., "Preparation and catalytic activities of hydrolysis catalyst for carbonyl sulfide in light hydrocarbon," Modern Chemistry, vol. 33, p. 48, 2013.

53. T. Li, Y. Ren, and C. Wei, "Study on preparation and properties of PVA-SA-PHB-AC composites carrier for microorganism immobilizatin," Journal of Applied Polymer Science , vol. 131, 2014.

54. B. Zhang, X. P. Zhuang, B. Cheng, N. Wang, and Y. Ni, "Carbonaceous nanofiber-supported sulfonated poly(ether ether ketone) membranes for fuel cell applications," Materials Letters, vol. 115, pp. 248–251, 2014.

55. W. Kangwanwatana, C. Saiwan, and P. Tontiwachwuthikul, "Study of CO_2 adsorption using adsorbent modified with piperazine," Chemical Engineering Transactions, vol. 35, pp. 403–408, 2013.

56. A. R. Milbrandt, D. M. Heimiller, A. D. Perry, and C. B. Field, "Renewable energy potential on marginal lands in the United States," Renewable and Sustainable Energy Reviews, vol. 29, pp. 473–481, 2014.

57. T. Samus, B. Lang, and H. Rohn, "Assessing the natural resource use and the resource efficiency potential of the Desertec concept," Solar Energy, vol. 87, no. 1, pp. 176–183, 2013.

58. A. Tavangar, B. Tan, and K. Venkatakishnan, "Transport properties of two finite armchair graphene nanoribbons," Nanoscale Research Letters, vol. 8, p. 1, 2013.

59. M. Vuk evi , B. Pejic, A. Kalijadis, et al., "Carbon materials from waste short hemp fibers as a sorbent for heavy metal ions—

mathematical modeling of sorbent structure and ions transport," Chemical Engineering Journal, vol. 235, pp. 284–292, 2014.

60. Y. Xiong, Z. Zhang, X. Wang, B. Liu, and J. Lin, "Hydrolysis of cellulose in ionic liquids catalyzed by a magnetically-recoverable solid acid catalyst," Chemical Engineering Journal, vol. 235, pp. 349–355, 2014.

61. Y. Song, J. Zhoa, L. Zhang, and X. Wu, "Homogenous modification of cellulose with acrylamide in NaOH/urea aqueous solutions," Carbohydrate Polymers, vol. 73, no. 1, pp. 18–25, 2008.

62. Z. Zhang, W. Wang, X. Liu et al., "Kinetic study of acid-catalyzed cellulose hydrolysis in 1-butyl-3-methylimidazolium chloride," Bioresource Technology, vol. 112, pp. 151–155, 2012.

63. S. Ali, K. Andrea, and S. Valentin, "Formation and degradation pathways of intermediate products formed during the hydropyrolysis of glucose as a model substance for wet biomass in a tubular reactor,"Engineering in Life Sciences, vol. 3, no. 12, pp. 469–473, 2003.

64. M. Bicker, D. Kaiser, L. Ott, and H. Vogel, "Dehydration of D-fructose to hydroxymethylfurfural in sub- and supercritical fluids," Journal of Supercritical Fluids, vol. 36, no. 2, pp. 118–126, 2005.

65. V. Budarin, R. Luque, D. J. Macquarrie, and J. H. Clark, "Towards a bio-based industry: Benign catalytic esterifications of succinic acid in the presence of water," Chemistry, vol. 13, no. 24, pp. 6914–6919, 2007.

66. R. Demir-Cakan, N. Baccile, M. Antonietti, and M.-M. Titirici, "Carboxylate-rich carbonaceous materials via one-step hydrothermal carbonization of glucose in the presence of acrylic acid," Chemistry of Materials, vol. 21, no. 3, pp. 484–490, 2009.

67. A. Thomas, P. Kuhn, J. Weber, M.-M. Titirici, and M. Antonietti, "Porous polymers: enabling solutions for energy applications," Macromolecular Rapid Communications, vol. 30, no. 4-5, pp. 221–236, 2009.

68. M. Iguchi, T. M. Aida, M. Watanabe, and R. L. Smith Jr., "Dissolution and recovery of cellulose from 1-butyl-3-methylimidazolium chloride in presence of water," Carbohydrate Polymers, vol. 92, no. 1, pp. 651–658, 2013.

69.   G. Fan, C. Liao, T. Fang, M. Wang, and G. Song, "Hydrolysis of cellulose catalyzed by sulfonated poly(styrene-co-divinylbenzene) in the ionic liquid 1-n-butyl-3-methylimidazolium bromide," Fuel Processing Technology, vol. 116, pp. 142–148, 2013.

70.   R. Göbel, Z. L. Xie, M. Neumann et al., "Synthesis of mesoporous carbon/iron carbide hybrids with unusually high surface areas from the ionic liquid precursor [Bmim][FeCl$_4$]," CrystEngComm, vol. 14, no. 15, pp. 4946–4951, 2012.

71.   G. Centi, P. Lanzafame, and S. Perathoner, "Analysis of the alternative routes in the catalytic transformation of lignocellulosic materials," Catalysis Today, vol. 167, no. 1, pp. 14–30, 2011.

72.   F. Tao, H. Song, J. Yang, and L. Chou, "Catalytic hydrolysis of cellulose into furans in MnCl2-ionic liquid system," Carbohydrate Polymers, vol. 85, no. 2, pp. 363–368, 2011.

73.   S. Yu, M. B. Heather, L. Guosheng et al., "Accelerated cellulose depolymerization catalyzed by paired metal chlorides in ionic liquid solvent," Applied Catalysis A: General, vol. 391, no. 1-2, pp. 436–442, 2011.

74.   W.-H. Hsu, Y.-Y. Lee, W.-H. Kelven, and C.-W. Wu, "Cellulosic conversion in ionic liquids (ILs): effects of H$_2$O/cellulose molar ratios, temperatures, times, and different ILs on the production of monosaccharides and 5-hydroxymethylfurfural (HMF)," Catalysis Today, vol. 174, no. 1, pp. 65–69, 2011.

75.   J. Potvin, E. Sorlien, J. Hegner, B. DeBoef, and B. L. Lucht, "Effect of NaCl on the conversion of cellulose to glucose and levulinic acid via solid supported acid catalysis," Tetrahedron Letters, vol. 52, no. 44, pp. 5891–5893, 2011.

76.   D. Liu and E. Y.-X. Chen, "Ubiquitous aluminum alkyls and alkoxides as effective catalysts for glucose to HMF conversion in ionic liquids," Applied Catalysis A: General, vol. 435–436, pp. 78–85, 2012.

77.   H. Abou-Yousef, E. B. Hassan, and P. Steele, "Rapid conversion of cellulose to 5-hydroxymethylfurfural using single and combined metal chloride catalysts in ionic liquid," Journal of Fuel Chemistry and Technology, vol. 41, no. 2, pp. 214–222, 2013.

78.   F. Tao, H. Song, and L. Chou, "Efficient conversion of cellulose into furans catalyzed by metal ions in ionic liquids," Journal of Molecular Catalysis A: Chemical, vol. 357, pp. 11–18, 2012.

79. Z. Wei, Y. Li, D. Thurshara, Y. Liu, and Q. Ren, "Novel dehydration of carbohydrates to 5-hydroxymethylfurfural catalyzed by Ir and Au chlorides in ionic liquids," Journal of the Taiwan Institute of Chemical Engineers, vol. 42, no. 2, pp. 363–370, 2011.

80. A. Amarasekara and O. S. Owereh, "Homogeneous phase synthesis of cellulose carbamate silica hybrid materials using 1-n-butyl-3-methylimidazolium chloride ionic liquid medium," Carbohydrate Polymers, vol. 78, no. 3, pp. 635–638, 2009.

81. H. Nawaz, P. A. R. Pires, and O. A. El Seoud, "Kinetics and mechanism of imidazole-catalyzed acylation of cellulose in LiCl/N,N-dimethylacetamide," Carbohydrate Polymers, vol. 92, no. 2, pp. 997–1005, 2013.

82. Z. Zhang and Z. K. Zhaoa, "Solid acid and microwave-assisted hydrolysis of cellulose in ionic liquid,"Carbohydrate Research, vol. 344, no. 15, pp. 2069–2072, 2009.

83. P. Yang, H. Kobayashi, and A. Fukuoka, "Recent developments in the catalytic conversion of cellulose into valuable chemicals," Chinese Journal of Catalysis, vol. 32, no. 5, pp. 716–722, 2011.

84. L. Peng, L. Lin, J. Zhang, J. Shi, and S. Liu, "Solid acid catalyzed glucose conversion to ethyl levulinate,"Applied Catalysis A: General, vol. 397, no. 1-2, pp. 259–265, 2011.

85. S. Suganuma, K. Nakajima, M. Kitano et al., "Synthesis and acid catalysis of cellulose-derived carbon-based solid acid," Solid State Sciences, vol. 12, no. 6, pp. 1029–1034, 2010.

86. M. Sevilla and A. B. Fuertes, "The production of carbon materials by hydrothermal carbonization of cellulose," Carbon, vol. 47, no. 9, pp. 2281–2289, 2009.

87. M. Kaldstrom, N. Kumar, and D. Y. Murzin, "Valorization of cellulose over metal supported mesoporous materials," Catalysis Today, vol. 167, pp. 91–95, 2011.

88. B. Girisuta, K. Dussan, D. Haverty, J. J. Leahy, and M. H. B. Hayes, "A kinetic study of acid catalysed hydrolysis of sugar cane bagasse to levulinic acid," Chemical Engineering Journal, vol. 217, pp. 61–70, 2013.

89. A. Tanksale, J. N. Beltramini, and G. M. Lu, "A review of catalytic hydrogen production processes from biomass," Renewable and Sustainable Energy Reviews, vol. 14, no. 1, pp. 166–182, 2010.

90. S. G. Wettstein, D. Martin Alonso, E. I. Gürbüz, and J. A. Dumesic, "A roadmap for conversion of lignocellulosic biomass to chemicals and fuels," Current Opinion in Chemical Engineering, vol. 1, no. 3, pp. 218–224, 2012.

91. D. M. Alonso, J. Q. Bond, and J. A. Dumesic, "Catalytic conversion of biomass to biofuels," Green Chemistry, vol. 12, no. 9, pp. 1493–1513, 2010.

92. X. Tong, Y. Ma, and Y. Li, "Biomass into chemicals: conversion of sugars to furan derivatives by catalytic processes," Applied Catalysis A: General, vol. 385, pp. 1–13, 2010.

93. X. Du, Y. Liu, J. Wang, Y. Cao, and K. Fan, "Catalytic conversion of biomass-derived levulinic acid into -valerolactone using iridium nanoparticles supported on carbon nanotubes," Chinese Journal of Catalysis, vol. 34, no. 5, pp. 993–1001, 2013.

94. L. E. Manzer, "Catalytic synthesis of -methylene- -valerolactone: a biomass-derived acrylic monomer,"Applied Catalysis A: General, vol. 272, no. 1-2, pp. 249–256, 2004.

95. F. Chambon, F. Rataboul, C. Pinel, A. Cabiac, E. Guillon, and N. Essayem, "Cellulose hydrothermal conversion promoted by heterogeneous Brønsted and Lewis acids: remarkable efficiency of solid Lewis acids to produce lactic acid," Applied Catalysis B, vol. 105, no. 1-2, pp. 171–181, 2011.

96. C. Aandine, G. Emmanuella, C. Flora, P. Catherine, R. Franck, and E. Nadine, "Cellulose reactivity and glycosidic bond cleavage in aqueous phase by catalytic and non catalytic transformations," Applied Catalysis A: General, vol. 402, pp. 1–10, 2011.

97. H. Jessica, C. P. Kyle, D. Brenston, and L. L. Brett, "Conversion of cellulose to glucose and levulinic acid via solid-supported acid catalysis," Tetrahedron Letters, vol. 51, no. 17, pp. 2356–2358, 2010.

98. H.-S. Qian, S.-H. Yu, L.-B. Luo, J.-Y. Gong, L.-F. Fei, and X.-M. Liu, "Catalytic conversion of biomass-derived levulinic acid into -valerolactone using iridium nanoparticles supported on carbon nanotubes,"Chemistry of Materials, vol. 18, pp. 2012–2018, 2012.

99. V. L. Budarin, J. H. Clark, R. Luque, and D. J. Macquarrie, "Versatile mesoporous carbonaceous materials for acid catalysis," Chemical Communications, no. 6, pp. 634–636, 2007.

100. M.-M. Titirica and M. Antonietti, "Chemistry and materials options of sustainable carbon materials made by hydrothermal carbonization," Chemical Society Reviews, vol. 39, no. 1, pp. 103–116, 2010.

101. H. Cai, C. Li, A. Wang, G. Xu, and T. Zhang, "Zeolite-promoted hydrolysis of cellulose in ionic liquid, insight into the mutual behavior of zeolite, cellulose and ionic liquid," Applied Catalysis B: Environmental, vol. 123–124, pp. 333–338, 2012.

102. I. Jiménez-Morales, J. Santamaría-González, P. Maireles-Torres, and A. Jiménez-López, "Aluminum doped SBA-15 silica as acid catalyst for the methanolysis of sunflower oil," Applied Catalysis B: Environmental, vol. 105, no. 1-2, pp. 199–205, 2011.

103. M. Soorholtz, R. J. White, T. Zimmermann et al., "Direct methane oxidation over Pt-modified nitrogen-doped carbons," Chemical Communications, vol. 49, no. 3, pp. 240–242, 2013.

104. S.-H. Yu, X. Cui, L. Li, et al., "From starch to metal/carbon hybrid nanostructures: hydrothermal metal-catalyzed carbonization," Advanced Materials, vol. 16, no. 18, pp. 1636–1640, 2004.

105. M. Kenichiro, K. Hirokazu, I. Koji, K. Takasuku, F. Atsushi, and T. Seiichi, "Immobilization of pectinase and lipase on macroporous resin coated with chitosan for treatment of whitewater from papermaking,"Bioresource Technology, vol. 123, pp. 616–619, 2011.

106. A. Sinag, T. Yumak, V. Balci, and A. Kruse, "Catalytic hydrothermal conversion of cellulose over $SnO_2$and ZnO nanoparticle catalysts," Journal of Supercritical Fluids, vol. 56, no. 2, pp. 179–185, 2011.

107. H. Zhang, Y.-T. Cheng, T. P. Vispute, R. Xiao, and G. W. Huber, "Catalytic conversion of biomass-derived feedstocks into olefins and aromatics with ZSM-5: the hydrogen to carbon effective ratio,"Energy and Environmental Science, vol. 4, no. 6, pp. 2297–2307, 2011.

108. H.-S. Qian, A. Markus, and S.-H. Yu, "Hybrid "golden fleece": synthesis and catalytic performance of uniform carbon nanofibers and silica nanotubes embedded with a high population of noble-metal nanoparticles," Advanced Functional Materials, vol. 17, pp. 637–643, 2007.

109. W. Deng, M. Liu, Q. Zhang, and Y. Wang, "Direct transformation of cellulose into methyl and ethyl glucosides in methanol and ethanol media catalyzed by heteropolyacids," Catalysis Today, vol. 164, no. 1, pp. 461–466, 2011.

110. A. Kruse, A. Funke, and M.-M. Titirici, "Hydrothermal conversion of biomass to fuels and energetic materials," Current Opinion in Chemical Biology, vol. 17, no. 3, pp. 515–521, 2013.

111. Q. Zhao, L. Wang, S. Zhao, X. Wang, and S. Wang, "High selective production of 5-hydroymethylfurfural from fructose by a solid heteropolyacid catalyst," Fuel, vol. 90, no. 6, pp. 2289–2293, 2011.

112. A. Alahiane, A. Rochdi, M. Taourirte, N. Redwane, S. Sebti, and H. B. Lazrek, "Natural phosphate as Lewis acid catalyst: a simple and convenient method for acyclonucleoside synthesis," Tetrahedron Letters, vol. 42, no. 21, pp. 3579–3581, 2001.

113. M. Zahouily, M. Salah, B. Bahlaouane et al., "Solid catalysts for the production of fine chemicals: the use of natural phosphate alone and doped base catalysts for the synthesis of unsaturated arylsulfones,"Tetrahedron, vol. 60, no. 7, pp. 1631–1635, 2004.

114. M. Mariani, F. Zaccheria, R. Psaro, and N. Ravasio, "Some insight into the role of different copper species as acids in cellulose deconstruction," Catalysis Communications, vol. 44, pp. 19–23, 2014.

115. D. A. Bulushev, S. Beloshapkin, P. E. Plyusnin et al., "Vapour phase formic acid decomposition over PdAu/ -Al$_2$O$_3$ catalysts: effect of composition of metallic particles," Journal of Catalysis, vol. 299, pp. 171–180, 2013.

116. I. G. Baek, S. J. You, and E. Park, "Direct conversion of cellulose into polyols over Ni/W/SiO$_2$–Al2O$_3$,"Bioresource Technology, vol. 114, pp. 684–690, 2012.

117. R. Kourieh, S. Bennici, M. Marzo, A. Gervasini, and A. Auroux, "Investigation of the WO$_3$/ZrO$_2$ surface acidic properties for the aqueous hydrolysis of cellobiose," Catalysis Communications, vol. 19, pp. 119–126, 2012.

118. A. Burcu and G. Gonul, "Isomerizaton of -pinene over H$_3$PW$_{12}$O$_{40}$ catalysts supported on natural zeolite," Chemical Engineering Journal, vol. 168, no. 3, pp. 1311–1318, 2011.

119. R. Ormsby, J. R. Kastner, and J. Miller, "Hemicellulose hydrolysis using solid acid catalysts generated from biochar," Catalysis Today, vol. 190, no. 1, pp. 89–97, 2012.

120. M. Marzo, A. Gervasini, and P. Carniti, "Hydrolysis of disaccharides over solid acid catalysts under green conditions," Carbohydrate Research, vol. 347, no. 1, pp. 23–31, 2012.

121. M. Cheng, T. Shi, H. Guan, S. Wang, X. Wang, and Z. Jiang, "Clean production of glucose from polysaccharides using a micellar heteropolyacid as a heterogeneous catalyst," Applied Catalysis B: Environmental, vol. 107, no. 1-2, pp. 104–109, 2011.

122. X. Tong, Y. Ma, and Y. Li, "Biomass into chemicals: conversion of sugars to furan derivatives by catalytic processes," Applied Catalysis A: General, vol. 385, no. 1-2, pp. 1–13, 2010.

123. R. Weingarten, G. A. Tompsett, W. C. Conner Jr., and G. W. Huber, "Design of solid acid catalysts for aqueous-phase dehydration of carbohydrates: the role of Lewis and Brønsted acid sites," Journal of Catalysis, vol. 279, no. 1, pp. 174–182, 2011.

124. W. Edward, T. K. Yong, A. T. Geoffrey et al., "Conversion of glucose into levulinic acid with solid metal(IV) phosphate catalysts," Journal of Catalysis, vol. 304, pp. 123–134, 2013.

125. J. C. Serrano-Ruiz, A. Pineda, A. M. Balu et al., "Catalytic transformations of biomass-derived acids into advanced biofuels," Catalysis Today, vol. 195, no. 1, pp. 162–168, 2012.

126. L. Hu, Y. Sun, L. Lin, and S. Liu, "12-Tungstophosphoric acid/boric acid as synergetic catalysts for the conversion of glucose into 5-hydroxymethylfurfural in ionic liquid," Biomass and Bioenergy, vol. 47, pp. 289–294, 2012. View at Publisher · View at Google Scholar · View at Scopus

127. Y. Qu, C. Huang, J. Zhang, and B. Chen, "Efficient dehydration of fructose to 5-hydroxymethylfurfural catalyzed by a recyclable sulfonated organic heteropolyacid salt," Bioresource Technology, vol. 106, pp. 170–172, 2012.

128. J. S. Choi, I. K. Song, and W. Y. Lee, "Performance of shell and tube-type membrane reactors equipped with heteropolyacid-polymer composite catalytic membranes," Catalysis Today, vol. 67, no. 1-3, pp. 237–245, 2001.

129. X. Duan, G. Sun, Z. Sun et al., "A heteropolyacid-based ionic

liquid as a thermoregulated and environmentally friendly catalyst in esterification reaction under microwave assistance," Catalysis Communications, vol. 42, pp. 125–128, 2013.

130. M. Cheng, T. Shi, H. Guan, S. Wang, X. Wang, and Z. Jiang, "Clean production of glucose from polysaccharides using a micellar heteropolyacid as a heterogeneous catalyst," Applied Catalysis B, vol. 107, pp. 104–109, 2011.

131. C. Larabi, W. al Maksoud, K. C. Szeto et al., "Thermal decomposition of lignocellulosic biomass in the presence of acid catalysts," Bioresource Technology, vol. 148, pp. 255–260, 2013.

132. M. M. Titrici, A. Markus, and B. Niki, "Hydrothermal carbon from biomass: a comparison of the local structure from poly- to monosaccharides and pentoses/hexoses," Green Chemistry, vol. 10, pp. 1204–1212, 2008.

133. T. Pullawan, A. N. Wilkinson, L. N. Zhang, and S. J. Eichhorn, "Deformation micromechanics of all-cellulose nanocomposites: comparing matrix and reinforcing components," Carbohydrate Polymers, vol. 100, pp. 31–39, 2014.

134. M. Yabushita, H. Kobayashi, and A. Fukuoka, "Catalytic transformation of cellulose into platform chemicals," Applied Catalysis B: Environmental, vol. 145, pp. 1–9, 2014.

135. M. Hara, "Biomass conversion by a solid acid catalyst," Energy and Environmental Science, vol. 3, no. 5, pp. 601–607, 2010.

136. E. Gürbüz, J. Q. Bond, J. A. Dumesic, and Y. Román-Leshkov, "Chapter 8. Role of acid catalysis in the conversion of lignocellulosic biomass to fuels and chemicals," in The Role of Catalysis for the Sustainable Production of Bio-fuels and Bio-chemicals, pp. 261–288, 2013.

137. F. Chambon, F. Rataboul, C. Pinel, A. Cabiac, E. Guillon, and N. Essayem, "Cellulose hydrothermal conversion promoted by heterogeneous Brønsted and Lewis acids: remarkable efficiency of solid Lewis acids to produce lactic acid," Applied Catalysis B: Environmental, vol. 105, no. 1-2, pp. 171–181, 2011.

138. J. Hegner, K. C. Pereira, B. DeBoef, and B. L. Lucht, "Conversion of cellulose to glucose and levulinic acid via solid-supported acid catalysis," Tetrahedron Letters, vol. 51, no. 17, pp. 2356–2358, 2010.

139. P. Rutkowski, "Pyrolytic behavior of cellulose in presence of montmorillonite K10 as catalyst," Journal of Analytical and Applied Pyrolysis, vol. 98, pp. 115–122, 2012. ·

140. P. Lanzafame, D. M. Temi, S. Perathoner, A. N. Spadaro, and G. Centi, "Direct conversion of cellulose to glucose and valuable intermediates in mild reaction conditions over solid acid catalysts," Catalysis Today, vol. 179, no. 1, pp. 178–184, 2012.

141. S. Dora, T. Bhasker, R. Singh, D. V. Naik, and D. K. Adhikari, "Effective catalytic conversion of cellulose into high yields of methyl glucosides over sulfonated carbon based catalyst," Bioresource Technology, vol. 120, pp. 318–321, 2012.

142. B. Guo, Y. Zhang, S.-J. Ha, Y.-S. Jin, and E. Morgenroth, "Combined biomimetic and inorganic acids hydrolysis of hemicellulose in Miscanthus for bioethanol production," Bioresource Technology, vol. 110, pp. 278–287, 2012.

143. A. Bai, X. Zhao, Y. Jin, G. Yang, and Y. Feng, "A novel thermophilic -glucosidase fromCaldicellulosiruptor bescii: characterization and its synergistic catalysis with other cellulases," Journal of Molecular Catalysis B: Enzymatic, vol. 85-86, pp. 248–256, 2013.

144. E. Gürbüz, J. Q. Bond, J. A. Dumesic, and Y. Román-Leshkov, "Chapter 8: role of acid catalysis in the conversion of lignocellulosic biomass to fuels and chemicals," in The Role of Catalysis for the Sustainable Production of Bio-fuels and Bio-chemicals, pp. 261–288, 2013.

145. J. Huang and T. Kunitake, "Nano-precision replication of natural cellulosic substances by metal oxides,"Journal of the American Chemical Society, vol. 125, no. 39, pp. 11834–11835, 2003.

146. J. R. T. Johnson, "Water adsorption and hydrolysis on molecular transition metal oxides and oxyhydroxides," Inorganic Chemistry, vol. 39, no. 15, pp. 3181–3191, 2000.

147. S. Suganuma, K. Nakajima, M. Kitano et al., "Hydrolysis of cellulose by amorphous carbon bearing SO_3H, COOH, and OH groups," Journal of the American Chemical Society, vol. 130, no. 38, pp. 12787–12793, 2008.

148. A. Omegna, J. A. Bokhaven, and R. Prins, "Flexible aluminum coordination in alumino-silicates. Structure of zeolite H-USY and

amorphous silica-alumina," The Journal of Physical Chemistry B, vol. 107, no. 34, pp. 8854–8860, 2003.

149. B. K. Sen and A. V. Saha, "On the nature and structure of "niobic acid" and its pyrolytic products: ^{1}H NMR, I.R., conductivity and Ion exchange studies," Materials Research Bulletin, vol. 16, no. 8, pp. 923–932, 1981.

150. P. Carniti, A. Gervasini, S. Biella, and A. Auroux, "Intrinsic and effective acidity study of niobic acid and niobium phosphate by a multitechnique approach," Chemistry of Materials, vol. 17, no. 24, pp. 6128–6136, 2005.

151. K. Shimomura, L. Dickson, and H. F. Walton, "Separation of amines by ligand exchange, part IV ligand exchange with chelating resins and cellulosic exchangers," Analytica Chimica Acta, vol. 37, pp. 102–111, 1967.

152. C. Buttersack, "Accessibility and catalytic activity of sulfonic acid ion-exchange resins in different solvents," Reactive Polymers, vol. 10, no. 2-3, pp. 143–164, 1989.

153. M. A. Harmer, W. E. Farneth, and Q. Sun, "High surface area nafion resin/silica nanocomposites: a new class of solid acid catalyst," Journal of the American Chemical Society, vol. 118, no. 33, pp. 7708–7715, 1996.

154. M. E. Himmel, S.-Y. Ding, D. K. Johnson et al., "Biomass recalcitrance: engineering plants and enzymes for biofuels production," Science, vol. 315, no. 5813, pp. 804–807, 2007.

155. T. E. Takasuka, A. J. Book, G. R. Lewin, C. R. Currie, and B. G. Fox, "Aerobic deconstruction of cellulosic biomass by an insect-associated Streptomyces," Scientific Reports, vol. 3, article 1030, 2013.·

156. K. Igarashi, "Cellulases: cooperative biomass breakdown," Nature Chemical Biology, vol. 9, no. 6, pp. 350–351, 2013.

157. M. Sun and Y. Zhang, "Study on the preparation of activated carbons with Baijiu Vinasse," Applied Mechanics and Materials, vol. 448–453, pp. 669–673, 2013.

158. T. Heinze and J. Schaller, "New water soluble cellulose esters synthesized by an effective acylation procedure," Macromolecular Chemistry and Physics, vol. 201, pp. 1214–1218, 2000.

159. A. S. Amarasekara and B. Wiredu, "Degradation of cellulose in

dilute aqueous solutions of acidic ionic liquid 1-(1-propylsulfonic)-3-methylimidazolium chloride, and p-toluenesulfonic acid at moderate temperatures and pressures," Industrial and Engineering Chemistry Research, vol. 50, no. 21, pp. 12276–12280, 2011.

160. G. Antova, P. Vasvasova, and M. Zlatanov, "Studies upon the synthesis of cellulose stearate under microwave heating," Carbohydrate Polymers, vol. 57, no. 2, pp. 131–134, 2004.

161. A. Kržan and E. Žagar, "Microwave driven wood liquefaction with glycols," Bioresource Technology, vol. 100, no. 12, pp. 3143–3146, 2009.

162. S. van de Vyver, L. Peng, J. Geboers, et al., "Sulfonated silica/carbon nanocomposites as novel catalysts for hydrolysis of cellulose to glucose," Green Chemistry, vol. 12, no. 9, pp. 1560–1563, 2010.

163. J. Pang, A. Wang, M. Zheng, and T. Zhang, "Hydrolysis of cellulose into glucose over carbons sulfonated at elevated temperatures," Chemical Communications, vol. 46, no. 37, pp. 6935–6937, 2010.

164. D.-M. Lai, L. Deng, J. Li, B. Liao, Q.-X. Guo, and Y. Fu, "Hydrolysis of cellulose into glucose by magnetic solid acid," ChemSusChem, vol. 4, no. 1, pp. 55–58, 2011.

165. L. Shuai and X. Pan, "Hydrolysis of cellulose by cellulase-mimetic solid catalyst," Energy & Environmental Science, no. 5, pp. 6889–6894, 2012.

166. J. A. Geboers, S. van de Vyver, R. Ooms, B. Op de Beeck, P. A. Jacobs, and B. F. Sels, "Chemocatalytic conversion of cellulose: opportunities, advances and pitfalls," Catalysis Science and Technology, vol. 1, no. 5, pp. 714–726, 2011.

167. W. Daengprasert, P. Boonnoun, N. Laosiripojana, M. Gato, and A. Shortipurk, "Application of sulfonated carbon-based catalyst for solvothermal conversion of cassava waste to hydroxymethylfurfural and furfural," Industrial & Engineering Chemistry Research, vol. 50, no. 3, pp. 7903–7910, 2011.

168. J. W. Han and H. Lee, "Direct conversion of cellulose into sorbitol using dual-functionalized catalysts in neutral aqueous solution," Catalysis Communications, vol. 19, pp. 115–118, 2012.

169. P. L. Dhepe and A. Fukroku, "Cracking of cellulose over supported

metal catalysts," Catalysis Surveys from Asia, vol. 11, no. 4, pp. 186–191, 2007.

170. M. Benoit, A. Rodrigues, Q. Zhang et al., "Depolymerization of cellulose assisted by a nonthermal atmospheric plasma," Angewandte Chemie, vol. 38, pp. 9126–9129, 2011.

171. G. Akiyama, R. Matsuda, H. Sato, M. Takata, and S. Kitagawa, "Cellulose hydrolysis by a new porous coordination polymer decorated with sulfonic acid functional groups," Advanced Materials, vol. 23, no. 29, pp. 3294–3297, 2011.

172. P. Boonoun, N. Laosiripojana, C. Muangnapoh et al., "Application of sulfonated carbon-based catalyst for reactive extraction of 1,3-propanediol from model fermentation mixture," Industrial and Engineering Chemistry Research, vol. 49, no. 24, pp. 12352–12357, 2010.

173. W. Namchot, N. Panyacharay, W. Jonglertjunya, and C. Sakdaronnarong, "Hydrolysis of delignified sugarcane bagasse using hydrothermal technique catalyzed by carbonaceous acid catalysts," Fuel, vol. 116, pp. 608–616, 2014.

174. T. Klamrassamee, V. Champreda, V. Reunglek, and N. Laosiripojana, "Comparison of homogeneous and heterogeneous acid promoters in single-step aqueous-organosolv fractionation of eucalyptus wood chips," Bioresource Technology, vol. 147, pp. 276–284, 2013.

175. A. R. Jule and M. W. Schoonover, "Early discoveries in zeolite chemistry and catalysis at Union Carbide, and follow-up in industrial catalysis," Applied Catalysis A, vol. 222, no. 1-2, pp. 261–275, 2001.

176. M. Guisnet, "'Ideal' bifunctional catalysis over Pt-acid zeolites," Catalysis Today, vol. 218-219, pp. 123–134, 2013.

177. M. Stöcker, "Gas phase catalysis by zeolites," Microporous and Mesoporous Materials, vol. 82, no. 3, pp. 257–292, 2005.

178. L. Dixit and T. S. R. P. Rao, "New approach to acid catalysis and hydrocarbon—zeolite interactions,"Studies in Surface Science and Catalysis, vol. 113, pp. 313–319, 1998.

179. L. D. Rollmann, L. A. Green, R. A. Bradway, and H. K. C. Timken, "Adamantanes from petroleum with zeolites," Catalysis Today, vol. 31, no. 1-2, pp. 163–169, 1996.

180. J. Weitkamp, "New directions in zeolite catalysis," Studies in Surface Science and Catalysis, vol. 65, pp. 21–46, 1991.

181. Y. Chen, G. Li, F. Yang, and S.-M. Zhang, "Mn/ZSM-5 participation in the degradation of cellulose under phosphoric acid media," Polymer Degradation and Stability, vol. 96, no. 5, pp. 863–869, 2011.·

182. L. R. Ferreira, S. Lima, P. Neves et al., "Hydrolysis of delignified sugarcane bagasse using hydrothermal technique catalyzed by carbonaceous acid catalysts," Chemical Engineering Journal, vol. 215, p. 772, 2013.

183. M. Hernandez, E. Lima, A. Guzman, M. Vera, O. Novelo, and V. Lara, "A small change in the surface polarity of cellulose causes a significant improvement in its conversion to glucose and subsequent catalytic oxidation," Applied Catalysis B: Environmental, vol. 144, pp. 528–537, 2014.

184. G. Gliozzi, A. Innorta, A. Mancini et al., "Zr/P/O catalyst for the direct acid chemo-hydrolysis of non-pretreated microcrystalline cellulose and softwood sawdust," Applied Catalysis B: Environmental, vol. 145, pp. 24–33, 2014.

185. J. Cejka, G. Centi, J. Perez-Pariente, and M. Horacek, "Preface," Catalysis Today, vol. 179, no. 1, p. 1, 2012.

186. F. Ocampo, J. A. Cunha, M. R. L. Santo et al., "Synthesis of zeolite crystals with unusual morphology: application in acid catalysis," Applied Catalysis A: General, vol. 390, pp. 102–109, 2010.

187. F. F. Brites-Nobrega, A. N. B. Polo, A. M. Benedetti, M. M. D. Leão, V. Slusarski-Santana, and N. R. C. Fernandes-Machadoa, "Evaluation of photocatalytic activities of supported catalysts on NaX zeolite or activated charcoal," Journal of Hazardous Materials, vol. 263, pp. 61–66, 2013.

188. A. M. Azeez, D. Meier, J. Odermatt, and T. Willner, "Effects of zeolites on volatile products of beech wood using analytical pyrolysis," Journal of Analytical and Applied Pyrolysis, vol. 91, no. 2, pp. 296–302, 2011.

189. O. D. Mante, F. A. Agblevor, and R. McClung, "A study on catalytic pyrolysis of biomass with Y-zeolite based FCC catalyst using response surface methodology," Fuel, vol. 108, pp. 451–464, 2013.

190. D. A. Zapata, Y. Huang, M. A. Gonzalez-Borja, and D. E. Resosca, Journal of Catalysis, vol. 308, p. 62, 2013.

191. R. Otomo, T. Yokoi, J. N. Kondo, and T. Tatsumi, "Dealuminated Beta zeolite as effective bifunctional catalyst for direct transformation of glucose to 5-hydroxymethylfurfural," Applied Catalysis A, vol. 470, pp. 318–326, 2014.

192. Y. Yu, X. Li, L. Su, Y. Zhang, Y. Wang, and H. Zhang, "The role of shape selectivity in catalytic fast pyrolysis of lignin with zeolite catalysts," Applied Catalysis A, vol. 447-448, pp. 115–123, 2012.

193. F. Tao, H. Song, and L. Chou, "Efficient conversion of cellulose into furans catalyzed by metal ions in ionic liquids," Journal of Molecular Catalysis A, vol. 357, pp. 11–18, 2012.

194. A. Nzihou, B. Stanmore, and P. Sharrock, "A review of catalysts for the gasification of biomass char, with some reference to coal," Energy, vol. 58, pp. 305–317, 2013. ·

195. A. Saddawi, J. M. Jones, and A. Williams, "Influence of alkali metals on the kinetics of the thermal decomposition of biomass," Fuel Processing Technology, vol. 104, pp. 189–197, 2012.

196. K. R. Vuyyuru and P. Strasser, "Oxidation of biomass derived 5-hydroxymethylfurfural using heterogeneous and electrochemical catalysis," Catalysis Today, vol. 195, no. 1, pp. 144–154, 2012.

197. Q. Gu, J. Long, L. Fan, et al., "Single-site Sn-grafted Ru/TiO$_2$ photocatalysts for biomass reforming: Synergistic effect of dual co-catalysts and molecular mechanism," Journal of Catalysis, vol. 303, pp. 141–155, 2013.

198. M. Dreher, B. Johnson, A. A. Peterson, M. Nachtegaal, J. Wambach, and F. Vogel, "Catalysis in supercritical water: pathway of the methanation reaction and sulfur poisoning over a Ru/C catalyst during the reforming of biomolecules," Journal of Catalysis, vol. 301, pp. 38–45, 2013.

199. H. Wang, L. Zhu, S. Peng, F. Peng, H. Yu, and J. Yang, "High efficient conversion of cellulose to polyols with Ru/CNTs as catalyst," Renewable Energy, vol. 37, no. 1, pp. 192–196, 2012.

200. C. M. Osmundsen, K. Egeblad, and E. Toarning, "Silyted hydrophobic zeolites with enchanced tolerance to hot liquid

water," in New & Future Developments in Catalysis, pp. 73–89, 2011.

201. J. C. Serrano-Ruiz, R. Luque, and J. H. Clark, "Chapter 17—The role of heterogeneous catalysis in the biorefinery of the future," in The Role of Catalysis for the Sustainable Production of Bio-fuels and Bio-chemicals, pp. 557–576, 2013.

202. C. M. Andrew and S. Liu, "Combining bio- and chemo-catalysis: from enzymes to cells, from petroleum to biomass," Trends in Biotechnology, vol. 29, no. 5, pp. 199–204, 2011.

203. M. N. Uddin, W. M. A. WanDaud, and H. F. Abbas, "Potential hydrogen and non-condensable gases production from biomass pyrolysis: insights into the process variables," Renewable and Sustainable Energy Reviews, vol. 27, pp. 204–224, 2013.

204. Y. Zu, P. Yang, J. Wang et al., "Efficient production of the liquid fuel 2,5-dimethylfuran from 5-hydroxymethylfurfural over Ru/Co3O4 catalyst," Applied Catalysis B: Environmental, vol. 146, pp. 244–248, 2014.

205. V. Skoulou and A. Zabaniotou, "Fe catalysis for lignocellulosic biomass conversion to fuels and materials via thermochemical processes," Catalysis Today, vol. 196, no. 1, pp. 56–66, 2012.

206. G. Guan, G. Chen, Y. Kasai et al., "Catalytic steam reforming of biomass tar over iron- or nickel-based catalyst supported on calcined scallop shell," Applied Catalysis B: Environmental, vol. 115-116, pp. 159–168, 2012.

207. T. Yoshikawa, S. Shinohara, T. Yagi et al., "Production of phenols from lignin-derived slurry liquid using iron oxide catalyst," Applied Catalysis B, vol. 146, pp. 289–297, 2014.

208. A. Aho, N. Kumar, A. V. Lashkul, et al., "Catalytic upgrading of woody biomass derived pyrolysis vapours over iron modified zeolites in a dual-fluidized bed reactor," Fuel, vol. 89, no. 8, pp. 1992–2000, 2010.

209. P. V. Aravid and W. Jong, "Evaluation of high temperature gas cleaning options for biomass gasification product gas for solid oxide fuel cells," Progress in Energy and Combustion Science, vol. 38, no. 6, pp. 737–764, 2012. ·

210. L. Pino, A. Vita, M. Laganà, and V. Recupero, "Hydrogen from

biogas: catalytic tri-reforming process with Ni/La—Ce—O mixed oxides," Applied Catalysis B, vol. 148-149, pp. 91–105, 2014.

211. A. Rahmatpour and S. Mohammadian, "Polystyrene-supported TiCl4 as a novel, efficient and reusable polymeric Lewis acid catalyst for the chemoselective synthesis and deprotection of 1,1-diacetates under eco-friendly conditions," Comptes Rendus Chimie, vol. 16, no. 10, pp. 912–919, 2013.

212. H. J. Park, S. H. Park, J. M. Sohn et al., "Steam reforming of biomass gasification tar using benzene as a model compound over various Ni supported metal oxide catalysts," Bioresource Technology, vol. 101, no. 1, pp. S101–S103, 2010.

213. L. Faba, E. Diaz, and S. Ordanez, "Improvement of the stability of basic mixed oxides used as catalysts for aldol condensation of bio-derived compounds by palladium addition," Biomass & Bioenergy, vol. 56, pp. 592–599, 2013.

214. D. Li, C. Ishikawa, M. Koike, L. Wang, Y. Nakagawa, and K. Tomishige, "Production of renewable hydrogen by steam reforming of tar from biomass pyrolysis over supported Co catalysts," International Journal of Hydrogen Energy, vol. 38, no. 9, pp. 3572–3581, 2013.

215. L. V. Antisarri, S. Carbone, A. Gatti, G. Vianello, and P. Nannipieri, "Toxicity of metal oxide (CeO_2, Fe_3O_4, SnO_2) engineered nanoparticles on soilmicrobial biomass and their distribution in soil," Soil Biology and Biochemistry, vol. 60, no. 5, pp. 87–94, 2013.

216. L.-D. Felica, C. Courson, P. U. Foscoto, and A. Kiennemann, "Iron and nickel doped alkaline-earth catalysts for biomass gasification with simultaneous tar reformation and CO_2 capture," An International Journal of Hydrogen Energy, vol. 36, no. 9, pp. 5296–5310, 2011.

217. M. Khan and W. Cao, "Development of photocatalyst by combined nitrogen and yttrium doping,"Materials Research Bulletin, vol. 49, pp. 21–27, 2014.

218. X. Zhang, J. Zhoa, and L. Guo, "Band gap-tunable (CuAg)$_x$In$_{2x}$Zn$_{2(1-2x)}$S$_2$ solid solutions synthesized by hydrothermal method with ultrasonic assistance and their photocatalytic H_2 production performance," Journal of Alloys and Compounds, vol. 582, pp. 617–622, 2014.

219. J. G. Mcevoy, D. A. Bilodeau, W. Cui, and Z. Zhang, "Visible-light-driven inactivation of Escherichia coliK-12 using an Ag/AgCl–activated carbon composite photocatalyst," Journal of Photochemistry and Photobiology A, vol. 267, pp. 25–34, 2013.

220. I. Velo-Gala, J. J. López-Peñalver, M. Sánchez-Polo, and J. Rivera-Utrilla, "Activated carbon as photocatalyst of reactions in aqueous phase," Applied Catalysis B: Environmental, vol. 142-143, pp. 694–704, 2013.

221. C. Andriantsiferana, E. F. Mohamed, and H. Delmas, "Photocatalytic degradation of an azo-dye on TiO_2/activated carbon composite material," Environmental Technologies, vol. 35, no. 1–4, pp. 355–363, 2014.

222. J. Han and C. J. Li, "Steam reforming of biomass gasification tar using benzene as a model compound over various Ni supported metal oxide cataysts," Advanced Materials Research, vol. 750, p. 1864, 2013.

223. R. H. Jie, G.-B. Guo, W.-G. Zhao, and S.-L. An, "Preparation and photocatalytic degradation of methyl orange of nano-powder TiO_2 by hydrothermal method supported on activated carbon," Journal of Synthetic Crystals, vol. 42, pp. 2144–2149, 2013.

224. K. Nagai and T. Abe, "Full-spectrum-visible-light photocatalyst based on the active layer of organic solar cell—towards water splitting and volatile molecule degradation-," Kobunshi Ronbunshu, vol. 70, no. 9, pp. 459–475, 2013.

225. J. Xing, H. B. Jiang, J. F. Chen, et al., "Active sites on hydrogen evolution photocatalyst," Journal of Materials Chemistry A, vol. 1, pp. 15258–15264, 2013.

226. H. Huang, Z. Yue, G. Li et al., "Ultraviolet-assisted preparation of mesoporous WO_3/reduced graphene oxide composites: Superior interfacial contacts and enhanced photocatalysis," Journal of Materials Chemistry A, vol. 1, no. 47, pp. 15110–15116, 2013.

227. H. Gulyas, A. S. O. Argaez, F. Kong, C. L. Jorge, S. Eggers, and R. Otterphol, "Combining activated carbon adsorption with heterogeneous photocatalytic oxidation: lack of synergy for biologically treated greywater and tetraethylene glycol dimethyl ether," Environmental Technology, vol. 34, no. 11, pp. 1393–1403, 2013.

228. N. Sobana, B. Krishnakumar, and M. Swaminathan, "Synergism and effect of operational parameters on solar photocatalytic degradation of an azo dye (Direct Yellow 4) using activated carbon-loaded zinc oxide," Materials Science in Semiconductor Processing, vol. 16, no. 3, pp. 1046–1051, 2013.

229. F. Chekin, S. Bagheri, and S. B. Abd Hamid, "Synthesis of Pt doped TiO_2 nanoparticles: Characterization and application for electrocatalytic oxidation of l-methionine," Sensors and Actuators B: Chemical, vol. 177, pp. 898–903, 2013.

230. A. E. Eliyas, L. Ljutzkanov, I. D. Stambolova, et al., "Visible light photocatalytic activity of TiO_2 deposited on activated carbon," Central European Journal of Chemistry, vol. 11, no. 3, pp. 464–470, 2013.

231. K. Wantala, S. Neramittagapong, A. Neramittagapong, K. Kasipar, S. Khaownetr, and S. Chuichulcherm, "Photocatalytic degradation of alachlor on Fe-TiO_2-immobilized on GAC under black light irradiation using Box-Behnken design," Materials Science Forum, vol. 734, pp. 306–316, 2013.

232. J.-W. Yoon, M.-H. Baek, J.-S. Hong, C.-Y. Lee, and J.-K. Suh, "Photocatalytic degradation of azo dye using TiO_2 supported on spherical activated carbon," Korean Journal of Chemical Engineering, vol. 29, no. 12, pp. 1722–1729, 2012.

233. Y.-F. Zhang, L.-G. Qiu, Y.-P. Yuan, Y.-J. Zhu, X. Jiang, and J.-D. Xiao, "Magnetic Fe_3O_4@C/Cu and Fe_3O_4@CuO core–shell composites constructed from MOF-based materials and their photocatalytic properties under visible light," Applied Catalysis B: Environmental, vol. 144, pp. 863–869, 2014.

234. S. Vivekanandhan, M. Schreiber, C. Mason, A. K. Mohanty, and M. Misra, "Maple leaf (Acer sp.) extract mediated green process for the functionalization of ZnO powders with silver nanoparticles," Colloids and Surfaces B: Biointerfaces, vol. 113, pp. 169–175, 2014.

235. Y. Wang, P. Ren, C. Feng, X. Zheng, Z. Wang, and D. Li, "Photocatalytic behavior and photo-corrosion of visible-light-active silver carbonate/titanium dioxide," Materials Letters, vol. 115, pp. 85–88, 2014.

236. B. Yang and M. D. Bai, App Mech Mater, vol. 448, p. 2946, 2014.

237. D. Zhang, M. Wen, S. Zhang et al., "Au nanoparticles enhanced rutile TiO_2 nanorod bundles with high visible-light photocatalytic performance for NO oxidation," Applied Catalysis B: Environmental, vol. 147, pp. 610–616, 2014.

238. S.-Y. Ye, Y.-C. Fang, X.-L. Song, S.-C. Luo, and L. M. Ye, "Decomposition of ethylene in cold storage by plasma-assisted photocatalyst process with TiO_2/ACF-based photocatalyst prepared by gamma irradiation," Chemical Engineering Journal, vol. 225, pp. 499–508, 2013.

239. C. Liu, Z. Lei, Y. Yang, H. Wang, and Z. Zhang, "Improvement in settleability and dewaterability of waste activated sludge by solar photocatalytic treatment in Ag/TiO_2-coated glass tubular reactor,"Bioresource Technology, vol. 137, pp. 57–62, 2013.

240. Y. Yang, S. Zhan, X. Goa, et al., "Degradation of toluene using modified TiO_2 as photocatalysts,"Advanced Materials Research, vol. 669, pp. 7–18, 2013.

241. I. W. Mwangi, J. C. Ngila, P. Ndungu, T. A. M. Msagati, and J. N. Kamau, "Immobilized Fe (III)-doped titanium dioxide for photodegradation of dissolved organic compounds in water," Environmental Science and Pollution Research, vol. 20, no. 9, pp. 6028–6038, 2013.

242. Q. L. Yu and H. J. H. Brouwers, "Design of a novel photocatalytic gypsum plaster: with the indoor air purification property," Advanced Materials Research, vol. 651, pp. 751–756, 2013.

243. W.-D. Zhang, B. Xu, and L.-C. Jiang, "Functional hybrid materials based on carbon nanotubes and metal oxides," Journal of Materials Chemistry, vol. 20, no. 31, pp. 6383–6391, 2010. ·

244. R. C. Wang and C. W. Yu, "Phenol degradation under visible light irradiation in the continuous system of photocatalysis and sonolysis," Ultrasonics Sonochemistry, vol. 20, no. 1, pp. 553–564, 2013.

245. S.-S. Dong, J.-B. Zhang, L.-L. Gao, Y.-L. Wang, and D.-D. Zhou, "Preparation of spherical activated carbon-supported and Er^{3+}:$YAlO_3$-doped TiO_2 photocatalyst for methyl orange degradation under visible light," Transactions of Nonferrous Metals Society of China, vol. 22, no. 10, pp. 2477–2483, 2012.·

246. H.-J. Shang, Y.-L. Zhou, Y. Zhao et al., Modern Chem Indust, vol. 32, p. 59, 2012.

247. S. Bagheri, K. Shameli, and S. Bee Abd Hamid, "Synthesis and characterization of anatase titanium dioxide nanoparticles using egg white solution via sol-gel method," Journal of Chemistry, vol. 2013, Article ID 848205, 5 pages, 2013.

248. P. Muthirulan, C. N. Devi, and M. M. Sundaram, "TiO_2 wrapped graphene as a high performance photocatalyst for acid orange 7 dye degradation under solar/UV light irradiations," Ceramics International, vol. 40, no. 4, pp. 5945–5957, 2014. · ·

249. S. M. Sun, "Studies on photodegradation kinetics of rhodamine B by titanium dioxide—carbon composite materials," Advanced Materials Research, vol. 531, pp. 59–62, 2012.

250. Z. Zhang, Y. Xu, X. Ma et al., "Microwave degradation of methyl orange dye in aqueous solution in the presence of nano-TiO_2-supported activated carbon (supported-TiO_2/AC/MW)," Journal of Hazardous Materials, vol. 209-210, pp. 271–277, 2012.

251. W. Zhoa, J. Zhang, X. Zhu et al., "Enhanced nitrogen photofixation on Fe-doped TiO_2 with highly exposed (1 0 1) facets in the presence of ethanol as scavenger," Applied Catalysis B: Environmental, vol. 144, pp. 468–477, 2014.

252. M.-H. Baek, J.-W. Yoon, J.-S. Hong, and J.-K. Suh, "Application of TiO_2-containing mesoporous spherical activated carbon in a fluidized bed photoreactor—adsorption and photocatalytic activity,"Applied Catalysis A: General, vol. 450, pp. 222–229, 2013.

253. P. Fu, Y. Luan, and X. Dai, "Preparation of activated carbon fibers supported TiO_2 photocatalyst and evaluation of its photocatalytic reactivity," Journal of Molecular Catalysis A, vol. 221, no. 1, pp. 81–86, 2004.

254. A. I. Adel, R. A. Geioushya, B. Houcine, A. A. Saleh, A.-H. Ali, and W. B. Detlef, "TiO_2 decoration of graphene layers for highly efficient photocatalyst: Impact of calcination at different gas atmosphere on photocatalytic efficiency," Applied Catalysis B: Environmental, vol. 129, pp. 62–70, 2013.

255. T. S. Jamil, M. Y. Ghaly, N. A. Fathy, T. A. Abd El-Halim, and L. Osterund, "Enhancement of TiO_2 behavior on photocatalytic oxidation of MO dye using TiO_2/AC under visible irradiation and sunlight radiation," Separation and Purification Technology, vol. 98, pp. 270–279, 2012.

256. M. He and H. Xia, "TiO$_2$/activated carbon fibers photocatalyst: effects of coating precedures on the microstructure, adhesion property and photocatalytic ability," Advanced Materials Research, vol. 518, p. 764, 2012.

257. Q. Yang, Y. Liao, and L. Mao, "Kinetics of photocatalytic degradation of gaseous organic compounds on modified TiO$_2$/AC composite photocatalyst," Chinese Journal of Chemical Engineering, vol. 20, no. 3, pp. 572–576, 2012.

258. L.-Y. Hao, X.-X. Lin, D.-G. Fu, and X. Ji, "Photocatalytic properties of magnetic activated carbon supported F-doped TiO$_2$," Journal of Inorganic Materials, vol. 28, pp. 997–1002, 2013.

259. C.-Y. Kuo, C.-H. Wu, and S.-T. Chen, "Desalination and water treatment," In Press.

260. M.-H. Baek, J.-S. Hong, J.-W. Yoon, and J.-K. Suh, "Photocatalytic degradation of humic acid by Fe-TiO$_2$ supported on spherical activated carbon with enhanced activity," International Journal of Photoenergy, vol. 2013, Article ID 296821, 5 pages, 2013.

261. H.-J. Jung, J.-S. Hong, and J.-K. Suh, "A comparison of fenton oxidation and photocatalyst reaction efficiency for humic acid degradation," Journal of Industrial and Engineering Chemistry, vol. 19, no. 4, pp. 1325–1330, 2013.

262. S. Li and G. Ye, "Photocatalytic degradation of formaldehyde by TiO$_2$ nanoparticles immobilized in activated carbon fibers," Advanced Materials Research, vol. 482–484, pp. 2539–2542, 2012.

263. S.-Y. Dong, F. Tian, X.-F. Pan, A.-X. Tian, C.-L. Ma, and T.-L. Huang, "Preparation of oxygen-vacantTiO2-x and activated carbon fiber composite using a single step thermal plasma method for low-concentration elemental mercury removal," Modern Chemical Industry, vol. 32, p. 55, 2012.

264. S. M. Sun, "Enhancement of TiO$_2$ behavior on photocatalytic oxidation of MO dye using TiO$_2$/AC under visible irradiation and sunlight radiation," Applied Mechanics and Materials, vol. 485, p. 161, 2012.

265. N. Riaz, F. K. Chong, B. K. Dutta, Z. B. Man, M. S. Khan, and E. Nurlaela, "Photodegradation of orange II under visible light using Cu-Ni/TiO$_2$: effect of calcination temperature," Chemical Engineering Journal, vol. 185-186, pp. 108–119, 2012. · ·

266. M. A. Gondal, C. Li, X. Chang et al., "Facile preparation of magnetic C/TiO$_2$/Ni composites and their photocatalytic performance for removal of a dye from water under UV light irradiation," Journal of Environmental Science and Health A, vol. 47, no. 4, pp. 570–576, 2012.

267. W. Den and C.-C. Wang, "Enhancement of adsorptive chemical filters via titania photocatalysts to remove vapor-phase toluene and isopropanol," Separation and Purification Technology, vol. 85, pp. 101–111, 2012.

268. W. Zhou, P. Zhang, and W. Liu, "Decolarization of C.I Reactive Red 2 by UV/TiO$_2$/PAC and visible light/TiO$_2$/PAC system," International Journal of Photoenergy, Article ID 325902, 2012.

269. S.-Y. Lu, Y.-L. Huang, Q.-L. Wang, X.-D. Li, and J.-H. Yan, "Photocatalytic degradation of gaseous 1,2-dichlorobenzene over TiO$_2$/AC photocatalysts," Acta Physica, vol. 27, no. 9, p. 2191, 2011.

270. X.-Y. Xing, D.-F. Zhao, G.-F. Zhu, and Y.-M. Liu, "Preparation of Ti-FAC composite photocatalyst by hydrothermal-sol impregnation method and its photocatalytic degradation of methylene blue in aqueous solution," Journal of Synthetic Crystals, vol. 40, p. 963, 2011.

271. B. Tryba, A. W. Morawski, and M. Inagaki, "Application of TiO$_2$-mounted activated carbon to the removal of phenol from water," Applied Catalysis B: Environmental, vol. 41, no. 4, pp. 427–433, 2003.

272. Y. Gao and H. Liu, "Preparation and catalytic property study of a novel kind of suspended photocatalyst of TiO$_2$-activated carbon immobilized on silicone rubber film," Materials Chemistry and Physics, vol. 92, no. 2-3, pp. 604–608, 2005.

273. W. K. Jo, S. H. Shin, and E. S. Hwang, "Removal of dimethyl sulfide utilizing activated carbon fiber-supported photocatalyst in continuous-flow system," Journal of Hazardous Materials, vol. 191, no. 1–3, pp. 234–239, 2011.

274. Y. Wang, Z. Hu, Y. Chen, G. Zhao, Y. Liu, and Z. Wen, "A novel approach towards high-performance composite photocatalyst of TiO$_2$ deposited on activated carbon," Applied Surface Science, vol. 255, no. 7, pp. 3953–3958, 2009.

275. H. Slimen, A. Houas, and J. P. Nogier, "Elaboration of stable anatase TiO_2 through activated carbon addition with high photocatalytic activity under visible light," Journal of Photochemistry and Photobiology A: Chemistry, vol. 221, no. 1, pp. 13–21, 2011. ··

276. W. Li and S. Liu, "Bifunctional activated carbon with dual photocatalysis and adsorption capabilities for efficient phenol removal," Adsorption, vol. 18, no. 2, pp. 67–74, 2012.

277. M. Janus, E. Kusiak, and A. W. Morawski, "Carbon modified TiO_2 photocatalyst with enhanced adsorptivity for dyes from water," Catalysis Letters, vol. 131, no. 3-4, pp. 506–511, 2009.

278. M. Janus, B. Tryba, M. Inagaki, and A. W. Morawski, "New preparation of a carbon-TiO_2 photocatalyst by carbonization of n-hexane deposited on TiO_2," Applied Catalysis B: Environmental, vol. 52, no. 1, pp. 61–67, 2004.

279. D. D. Cui, J.-C. Jiang, K. Sun, and X.-C. Lu, "Photocatalytic degradation of gaseous 1,2-dicholorobenzene over TiO_2/AC photocatalyst," Functional Materials, vol. 42, p. 438, 2011.

280. M. Toyoda, T. Yano, T. Tomoki, Y. Amao, and M. Inagaki, "Effects of carbon coating on Ti_nO_{2n-i} for decomposition of iminoctadine triacetate in aqueous solution under visible light," Journal of Advanced Oxidation Technologies, vol. 9, no. 1, pp. 49–52, 2006.

281. M. Toyoda, Y. Nanbu, T. Kito, M. Hirano, and M. Inagaki, "Preparation and performance of anatase-loaded porous carbons for water purification," Desalination, vol. 159, no. 3, pp. 273–282, 2003.

282. D. Huang, Y. Miyamoto, T. Matsumoto et al., "Preparation and characterization of high-surface-area TiO_2/activated carbon by low-temperature impregnation," Separation and Purification Technology, vol. 78, no. 1, pp. 9–15, 2011. ··

283. A. Bolvin, S. Amellal, M. Schiavon, and M. T. Genuchten, "2,4-Dichlorophenoxyacetic acid (2,4-D) sorption and degradation dynamics in three agricultural soils," Environmental Pollution, vol. 138, no. 1, pp. 92–99, 2005.

284. U. Aruldoss, L. J. Kennedy, J. Judith Vijaya, and G. Sekaran, "Photocatalytic degradation of phenolic syntan using TiO_2 impregnated activated carbon," Journal of Colloid and Interface Science, vol. 355, no. 1, pp. 204–209, 2011.

285. M. V. Shankar, S. Anandan, N. Venkatachalam, B. Arabindoo, and V. Murugesan, "Fine route for an efficient removal of 2,4-dichlorophenoxyacetic acid (2,4-D) by zeolite-supported TiO_2," Chemosphere, vol. 63, no. 6, pp. 1014–1021, 2006.

286. M. Cristina Yeber, J. Rodríguez, J. Freer, N. Durán, and H. D. Mansilla, "Photocatalytic degradation of cellulose bleaching effluent by supported TiO_2 and ZnO," Chemosphere, vol. 41, no. 8, pp. 1193–1197, 2000.

287. K. Byrappa, A. K. Subramani, S. Ananda et al., "Impregnation of ZnO onto activated carbon under hydrothermal conditions and its photocatalytic properties," Journal of Materials Science, vol. 41, no. 5, pp. 1355–1362, 2006.

288. J. Matos, E. Garcia-Lopez, L. Palmisano, A. Garcia, and G. Marci, "Influence of activated carbon in TiO_2 and ZnO mediated photo-assisted degradation of 2-propanol in gas–solid regime," Applied Catalysis B: Environmental, vol. 99, no. 1-2, pp. 170–180, 2010.

289. M. I. Litter and N. Quici, "Photochemical advanced oxidation processes for water and wastewater treatment," Recent Patents on Engineering, vol. 4, no. 3, pp. 217–241, 2010.

290. S. H. Yao, Y. F. Jia, and S. L. Zhao, "Photocatalytic oxidation and removal of arsenite by titanium dioxide supported on granular activated carbon," Environmental Technology, vol. 33, pp. 983–988, 2012.

291. R. Jiang, H.-Y. Zhu, G.-M. Zeng, L. Xiao, and Y.-J. Guan, "Synergy of adsorption and visible light photocatalysis to decolor methyl orange by activated carbon/nanosized CdS/chitosan composite," Journal of Central South University of Technology, vol. 17, no. 6, pp. 1223–1229, 2010.

292. S.-Y. Ye, Q.-M. Tian, X.-I. Song, and S.-C. Luo, "Photoelectrocatalytic degradation of ethylene by a combination of TiO_2 and activated carbon felts," Journal of Photochemistry and Photobiology A, vol. 208, no. 1, pp. 27–35, 2009.

293. S.-M. Lam, J.-C. Sin, and A. R. Mohamed, "Parameter effect on photocatalytic degradation of phenol using TiO_2-P25/activated carbon (AC)," Korean Journal of Chemical Engineering, vol. 27, no. 4, pp. 1109–1116, 2010.

294. K. M. Joshi and V. S. Shrivasta, "Removal of hazardious textile dyes

from aqueous solution by using commercial activated carbon with TiO_2 and ZNO as photocatalyst," International Journal of ChemTech Research, vol. 2, no. 1, pp. 427–435, 2010.

295. J.-C. Sin, S.-M. Lam, and A. R. Mohamed, "Optimizing photocatalytic degradation of phenol by TiO_2/GAC using response surface methodology," Korean Journal of Chemical Engineering, vol. 28, no. 1, pp. 84–92, 2011.

296. Y. Zhang, Z.-R. Tang, X. Fu, and Y.-J. Xu, "TiO_2-graphene nanocomposites for gas-phase photocatalytic degradation of volatile aromatic pollutant: is TiO_2-graphene truly different from other TiO_2-carbon composite materials?" ACS Nano, vol. 4, no. 12, pp. 7303–7314, 2010.

297. X.-Y. Zhang, H.-P. Li, X.-L. Cui, and Y. Lin, "Graphene/TiO_2 nanocomposites: synthesis, characterization and application in hydrogen evolution from water photocatalytic splitting," Journal of Materials Chemistry, vol. 20, pp. 2801–2806, 2010.

298. K. Woan, G. Pyrgiotakis, and W. Sigmund, "Photocatalytic carbon-nanotube-TiO_2 composites," Advanced Materials, vol. 21, no. 21, pp. 2233–2239, 2009.

299. Y.-J. Xu, Y. Zhuang, and X. Fu, "New insight for enhanced photocatalytic activity of TiO_2 by doping carbon nanotubes: a case study on degradation of benzene and methyl orange," The Journal of Physical Chemistry C, vol. 114, pp. 2669–2676, 2010.

300. C.-Y. Yen, Y.-F. Lin, C.-H. Hung et al., "The effects of synthesis procedures on the morphology and photocatalytic activity of multi-walled carbon nanotubes/TiO_2 nanocomposites," Nanotechnology, vol. 19, no. 4, Article ID 045604, 2008.

301. Y. Yu, J. C. Yu, J.-G. Yu et al., "Enhancement of photocatalytic activity of mesoporous TiO_2 by using carbon nanotubes," Applied Catalysis A: General, vol. 289, no. 2, pp. 186–196, 2005.

302. J.-W. Shi, H.-J. Cui, J.-W. Chen et al., "TiO_2/activated carbon fibers photocatalyst: Effects of coating procedures on the microstructure, adhesion property, and photocatalytic ability," Journal of Colloid and Interface Science, vol. 388, pp. 201–208, 2012.

303. Z.-H. Liao, J.-J. Chen, K. F. Yao, F. H. Zhao, and R.-X. Li, "Progress of nanometer-TiO_2 photocatalyst immobilization," Journal of Inorganic Materials, vol. 19, no. 1, 2004.

304. Y. Li, S. Sun, M. Ma, Y. Ouyang, and W. Yan, "Kinetic study and model of the photocatalytic degradation of rhodamine B (RhB) by a TiO_2-coated activated carbon catalyst: effects of initial RhB content, light intensity and TiO_2 content in the catalyst," Chemical Engineering Journal, vol. 142, no. 2, pp. 147–155, 2008.

305. Y. Ao, J. Xu, D. Fu, L. Ba, and C. Yuan, "Deposition of anatase titania onto carbon encapsulated magnetite nanoparticles," Nanotechnology, vol. 19, no. 40, Article ID 405604, 2008.

306. X. Hong, Z. Wang, W. Cai et al., "Visible-light-activated nanoparticle photocatalyst of iodine-doped titanium dioxide," Chemistry of Materials, vol. 17, no. 6, pp. 1548–1552, 2005.

307. X.-X. Lin, X. Ji, D.-G. Fu, and L.-Y. Hao, "Photocatalytic properties of magnetic activated carbon supported F-doped TiO_2," International Journal of Inorganic Materials, vol. 28, no. 9, pp. 997–1002, 2013.

308. D. Huang, Y. Miyamoto, J. Ding et al., "A new method to prepare high-surface-area N-TiO_2/activated carbon," Materials Letters, vol. 65, no. 2, pp. 326–328, 2011.

Carbon Nanotube-Enzyme Biohybrids in A Green Hydrogen Economy

Anne De Poulpiquet[1], Alexandre Ciaccafava[1],
Saïda Benomar[1], Marie-Thérèse Giudici-Orticoni[1], and
Elisabeth Lojou[1]

[1]Bioénergétique et Ingénierie des Protéines, CNRS - AMU - Institut de Microbiologie de la Méditerranée, France

INTRODUCTION

Alternative energy pathways to replace depleting oil reserves and to limit the effects of global warming by reducing the atmospheric emissions of carbon dioxide are nowadays required. Dihydrogen appears as an attractive candidate because it represents the highest energy output relative to the molecular weight (120 MJ kg^{-1} against

50 MJ kg^{-1} for natural gas), and because its combustion delivers only water and heat. Whereas the main renewable sources of energy available in nature (solar, wind, geothermal…) need to be transformed, dihydrogen is able to transport and store energy. Dihydrogen can be produced from renewable energies, indirectly from photosynthesis *via* biomass transformation, or directly by bacteria. It can be converted into electricity using fuel cell technology. From all these properties and because it does not compete with food and water resources, dihydrogen has been defined as third generation biofuel. It thus emerges as a new fully friendly environmental energy vector. The use of dihydrogen as an energy carrier is not a new idea. Let us simply remember that Jules Verne, a famous French visionary novelist, wrote early in 1874: "I believe that O$_2$ and H$_2$ will be in the future our energy and heat sources" [1]. His prediction simply relied on the discovery a few years before of the fuel cell concept by C. Schönbein, then W. Groove, who demonstrated that when stopping water electrolysis, a current flow occurred in the reverse way [2]. However in order to implement the dihydrogen economy and replace fossil fuels, there are significant technical challenges that need to be overcome in each of the following domains:

- dihydrogen production and generation,
- dihydrogen storage and transportation,
- dihydrogen conversion to electrical energy.

As opposed to widespread opinions, natural dihydrogen sources exist alone on the earth's surface. Local and continuous emanations of dihydrogen can be observed in cratonic zones, ophiolitic rocks or oceanic ridges [3].

Dihydrogen is effectively produced in the upper mantle of the earth through natural oxidation of iron (II)-rich minerals, like ferromagnesians, by water of the hydrosphere. The ferrous iron is oxidized in ferric iron and water is concurrently reduced in dihydrogen,

as given by following equation: $2Fe^{2+}$ (mineral) $+ 2H^+$ (water) à $2Fe^{3+}$ (mineral) $+ H_2$.

The same reaction can occur with other ions like Mn^{2+}. Exploitation of these sources remains however difficult so far as dihydrogen does not accumulate on the earth subsurface, especially for two reasons. First because as a powerful energy source dihydrogen is quickly consumed (biologically or abiotically), and second because as the lightest and most mobile gas it is not much retained by Earth's attraction and escapes in the atmosphere.

Combined with water and hydrocarbons dihydrogen is nevertheless the most abundant element on earth. Green means to ecologically convert H containers into dihydrogen still remain however challenging. The energetic volume density of dihydrogen is low (10.8 MJ m^{-3} against 40 MJ m^{-3} for natural gas) so that storage and transportation appear as bottlenecks for large scale development in transportation

For example. Conversion of dihydrogen to electricity in fuel cells presents high electrical efficiency (more than 50% against less than 30% for gas engines), but requires the use of catalysts both for H_2 oxidation and O_2 reduction. These are mainly based on platinum catalysts, which are highly expensive, weakly available on earth, and non-biodegradable. Extensive researches thus aim to decrease the amount of platinum catalysts in fuel cells. Following the discovery of carbon nanotubes (CNTs) [4, 5], their large scale availability opened a new avenue in these three domains. Due to their intrinsic properties, such as high stability, high electrical and thermal conductivities [6] and high developed surface areas, carbon nanotubes constitute attractive materials, able to enhance the credibility of a hydrogen economy.

Besides, platinum catalysts are inhibited by very low amount of CO and S (0.1% of CO is sufficient to decrease one hundred fold the catalytic activity of Pt in ten minutes!), thus requiring strong steps of H_2purification [7]. They are not specific to either O_2 or H_2 catalysis, thus requiring the use of a membrane to separate the anodic and cathodic compartments. Nafion® perfluoronated membrane is currently the only really performing polymer [8], increasing its cost. Replacement of platinum-based catalysts is thus highly needed. In that way, a new concept appeared less than five years ago, when looking at the pathways microorganisms use for the production of ATP, their own energetic source [9,10]. As an example, the hyperthermophilic, microaerophilic bacterium *Aquifex aeolicus*, couples H_2oxidation to O_2 reduction *via* a membrane quinone pool (Figure 1).

The redox coupling generates a proton gradient through the cell membrane for ATP synthesis. Clearly, this pathway can be considered as an *"in vivo* biofuel cell". The question rises if we could take benefit of bacterial energetic pathways for our own energetic needs. The idea thus emerged that microorganisms or enzymes could be used instead of chemical catalysts for the development of efficient electricity producing devices. These innovative batteries called biofuel cells rely on enzymes highly specific for various fuels and oxidants [11]. A mandatory condition is that these enzymes have to be immobilized onto electrodes. One of the most common biofuel cell uses glucose oxidase and laccase, two enzymes specific for glucose oxidation and oxygen reduction, respectively. A few years ago, a new concept of biofuel cells appeared based on enzymes specific of dihydrogen oxidation. This biohydrogen economy relies on the opportunity to use low-cost materials for efficient conversion of solar energy to dihydrogen and of dihydrogen to electricity.

Many microorganisms biosynthesize hydrogenase, the metalloenzyme that catalyzes the dihydrogen conversion. At least two modes of application of dihydrogen-metabolizing protein catalysts are nowadays considered within dihydrogen as a future energy carrier. Hydrogenases may be used as catalysts in dihydrogen production by coupling oxygenic photosynthesis to biological dihydrogen production [12]. Hydrogenases can also be used directly as anode catalysts in biofuel cells instead of chemical catalysts [13]. The improved knowledge of hydrogenase structure and of catalytic mechanisms allows nowadays to design the development of biofuel cells functioning as Proton Exchange Membrane (PEM) fuel cells.

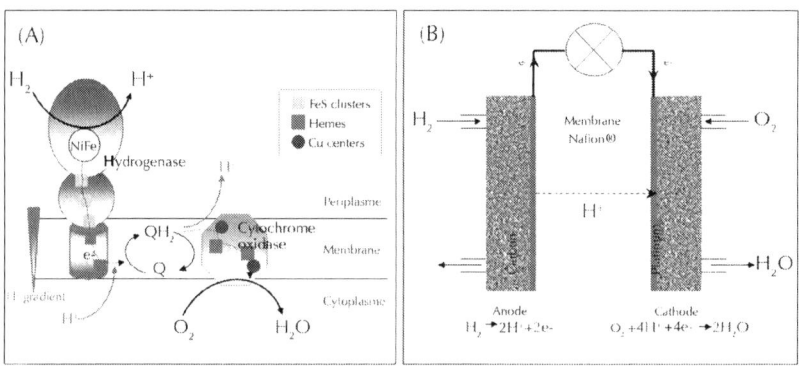

Figure 1: A) Energetic metabolism of the bacterium *Aquifex aeolicus*: H_2 oxidation in the periplasm is coupled to O_2 reduction in the cytoplasm *via* a membrane quinone pool to generate a trans-membrane proton gradient for ATP synthesis; (B) General view of a chemical PEM fuel cell.

For all these innovative concepts, one of the key points is the increase in power density, thus in the current density furnished by a redox couple displaying a large as possible potential difference. Apart from the improvement in enzyme stability, the increase in the current densities supposes an optimization of both the interfacial electron transfer rate and the amount of connected enzymes at the electrode. Carbon nanotubes which develop large surface areas and can be functionalized constitute an attractive platform for such enzyme immobilization. CNTs are described as graphene sheets rolled into tubes. They exist under various structural configurations (single-walled (SWCNTs), multi-walled (MWCNTs)) differing in electrical properties, thus tuning the platform properties for enzyme immobilization. The end of the tubes is capped by a fullerene-type hemisphere that yields selective functionalization of the CNTs [14].

With the objective of dihydrogen as a future green energy vector, this review focuses on the last developments in the fuel -and more especially biofuel- cell field thanks to the advantageous use of carbon nanotubes. In a first part, carbon nanotubes for H_2 storage enhancement are discussed. Then fuel cells in which carbon nanotubes help to decrease the amount of high cost noble metal catalysts are described. Green H_2 economy is then emphasized considering the key role of hydrogenase, the enzyme responsible for dihydrogen conversion. This requires the functional immobilization of the biocatalysts onto electrodes. The use of carbon nanotubes in this immobilization step is underlined, including the modes of carbon nanotube functionalization and enzyme or microbes grafting. Then the advantages of developing biofuel cells in which chemical catalysts are replaced by enzymes or microbes are described. A short review of the sugar/O_2 biofuel cells, the most widely investigated biofuel cell, is given with a particular attention on the devices based on carbon nanotube-modified bioelectrodes. The last developments based on carbon nanotube networks for hydrogenase immobilization, or mimicking synthetic complex immobilization, in view of efficient dihydrogen catalytic oxidation are finally described in order to allow the design of a future H_2/O_2 biofuel cell.

CARBON NANOTUBES: AN ATTRACTIVE CARBON MATERIAL

The discovery of carbon nanotubes (CNTs) has induced breakthroughs in many scientific domains, including H_2 economy, biosensors, bio electrochemistry. This is due to their remarkable properties, such as good electronic, mechanical and thermal properties. Their nanometric size compares with that of proteins and enzymes, offering the possibility of electrical connection. Their large developed surface area allows the development of devices in smaller volumes. SWCNTs are sp2 hybridized carbon in a hexagonal honeycomb structure that is rolled into hollow tube morphology [15]. MWCNTs are multiple concentric tubes encircling each other [5].

Depending on the chirality, CNTs can be metallic or semiconducting. The distinction between metallic and semiconducting is very important for application, but the physical separation of allotropes is one of the most difficult challenges to overcome. In MWCNTs, a single metallic layer results in the entire nanotubes metallic behavior. Most often mixtures of these two forms are present in CNTs preparation. More information on the physical and electronic structures can be found in many published reviews [16]. CNTs are produced by various methods such as arc discharge, laser ablation, and chemical vapor deposition (CVD). Commercially CNTs are generally produced by CVD during the pyrolysis of hydrocarbon gases at high temperature. The control of synthesis parameters (reagent gas, T°, metal catalysts) allows for the control of CNT properties. Metal impurities may remain in the CNTs sample, thus requiring purification steps. CNTs may be treated to functionalize the surface.

CARBON NANOTUBES FOR SAFE AND EFFICIENT H_2 STORAGE

The use of H_2 in fuel cells to generate electricity has been proved early in the middle of the nineteenth century. Surprisingly this discovery by C. Schönbein in 1839 of current generation by use of H_2 and O_2 in sulphuric acid was applied by NASA only late in 1960. Despite intensive studies over the last two decades, fuel cells still suffer from high cost and low durability. The first difficulty responsible for this slow large scale development lies on dihydrogen storage and transportation, both regarded as bottlenecks considering dihydrogen specific volumic density as a gas. For convenience the gas must be intensely pressurized to several hundred atmospheres and stored in a pressure vessel. The ways to store dihydrogen with minimum hazard are under liquid state under cryogenic temperatures (at a temperature of -253 °C), or more efficiently in a solid state. Storage of dihydrogen in hydride form uses an alloy that can absorb and hold large amounts of dihydrogen by bonding with hydrogen and forming hydrides. A dihydrogen storage alloy is capable of absorbing and releasing dihydrogen without compromising its own structure, according to the reaction: $M + H_2 \leftrightarrow MH_2$, where M represents the metal and H, hydrogen. Qualities that make these alloys useful include their ability to absorb and release large amounts of dihydrogen gas many times without deteriorating, and their selectivity toward dihydrogen only. In addition, their absorption and release rates can be controlled by adjusting temperature or pressure. The dihydrogen storage alloys in common use occurs in four different forms: AB_5 (e.g., $LaNi_5$), AB (e.g., $FeTi$), A_2B (e.g., Mg_2Ni) and AB_2 (e.g., ZrV_2). Metal hydrides, such as MgH_2, Mg_2NiH_4 or $LiBH_4$, constitute secure reserves of dihydrogen [17-19]. Dihydrogen is released from MH_2 upon increase in temperature and/or decrease in pressure.

Table 1: H density as a function of storage method

Material	H_2gas, 200 bar	H_2liquid, -253 C	Mg_H2	Mg_2Ni_H4	$FeTi_H2$	$LaNi_{5H}6$
H-atom per $cm^3(x10^{22})$	0.99	4.2	6.5	5.9	6.0	5.5

Much progress has been made during the last years in that domain, including the highlight of the advantages offered by using CNTs. An efficient approach appears to be the formulation of new carbon/ transition metal catalyst composites of specific composition and molecular structure, which can greatly stimulate and improve the chemical reactions involving dihydrogen relocation in alkali-metal aluminium materials. Absorption kinetics and dihydrogen storage capacity were shown to be enhanced by mixing MH_2 with SWCNTs as a result of an increase in interfacial area, decrease in MH_2 particle agglomeration and nano-platform for efficient H_2 diffusion [20, 21]. The hydriding and dehydriding kinetics of SWCNT/catalyzed sodium aluminium composite were found to be much better than those of the material ground without carbon additives. Temperature of H_2 desorption was lowered [22]. The presence of carbon creates new dihydrogen transition sites and the high dihydrogen diffusivity of the nanotubes facilitates hydrogen atom transition. Faster thermal energy transfer through the nanotubes may also help reduce hydriding and dehydriding times.

Dihydrogen can be stored through physisorption on CNTs, based on Van der Waals interaction. Based on the surface area of a single graphene sheet, the maximum value for the storage of dihydrogen capacity is around 3 wt%. Dihydrogen can also be stored through chemisorption in CNTs matrix. If the ϖ-bonding between carbon atoms were fully utilized, every carbon atom could be a site for chemisorption of one hydrogen atom. Dillon et al. first reported in 1997 dihydrogen storage in SWCNT networks [23]. Both SWCNTs and MWCNTs store dihydrogen in microscopic pores on the tubes [24, 25]. Similar to metal hydrides in their mechanism for storing and releasing dihydrogen, the carbon nanotubes hold the potential to store a significant volume of dihydrogen. The storage capacity is dependent on many parameters of the CNTs, including their structure, structure defects, pretreatment, purification, geometry (surface area, tube diameter, length), arrangement of tubes in bundles, storage pressure, temperature, Dihydrogen uptake varies linearly with tube diameter, because the uptake is proportional to the surface area, i.e. the number of carbon atoms. The adsorption sites exist inside and outside the tube, between tubes in bundles, between the shells in MWCNTs. For dihydrogen storage into the tube dihydrogen must pass through the CNT wall or the tube must be opened. Hydrogen forms stable C-H bonds on SWCNT surface at room temperature that can dissociate above 200°C. According to SWCNT diameter 100% hydrogenation can be obtained, thus more than 7 wt % dihydrogen storage capacity, which is above the target fixed by the US Department of Energy's Office of Energy Efficiency and Renewable Energy [26].

CARBON NANOTUBES FOR A DECREASE IN THE AMOUNT OF NOBLE METAL CATALYSTS IN FUEL CELLS

Among the different types of fuel cells, PEM fuel cell operates at low temperatures around 100°C. For small portable application requiring less than 10 kW, they are more suitable than higher powering solid oxide fuel cells (functioning at 700°C) due to the possible use of usual materials for electronic connectors (mainly based on carbon) and membrane. However the necessary use of platinum-based catalysts on electronic connectors to accelerate the rate of dihydrogen oxidation and oxygen reduction is a real brake towards the fuel cell development. Platinum is scarce enough on earth to be a limiting factor in case of large scale development of fuel cells. Consequently platinum currently accounts for 25% in the total cost of a fuel cell. Over the past five years, the price of platinum has ranged from just below $800 to more than $2,200 an ounce. Carbon black particles offer a high surface area support, able to decrease the amount of platinum particles. But they suffer from mass transfer limitations and strong carbon corrosion.

Among the low-cost alternatives to platinum, carbon appears to be the most promising. Due to their nano-structure and unique chemical and physical properties, CNTs have appeared as ideal supporting materials to improve both catalytic activity and electrode stability. The enhancement of fuel cell performances by using CNT/Pt or Pt-alloy catalysts may arise from:

- higher dispersion of Pt nanoparticles,
- increased electron transfer rates,
- porous structure of CNT layers.

Various CNT-Pt composites were used to reduce the platinum amount while preserving high catalytic activity in PEM fuel cells. Platinum nanodots sputter-deposited on a CNT-grown carbon paper [27], or deposited on functionalized MWCNTs [28] exhibited great improvement in cell performance compared to platinum on carbon black. This was primarily attributed to high porosity and high surface area developed by the CNT layer. Compared to a commercial Pt/carbon black catalyst, Pt/SWCNT films cast on a rotating disk electrode was shown to exhibit a lower onset potential and a higher electron-transfer rate constant for oxygen reduction. Improved stability of the SWCNT support was also confirmed from the minimal change in the oxygen reduction current during repeated cycling over a period of 36 h [29]. Platinum particles deposited on MWCNT encapsulated in micellar surfactant were also explored as efficient catalysts for fuel cells [30, 31]. An in situ synthetic method was reported for preparing and decorating metal nanoparticles at sidewalls of sodium dodecyl sulfate micelle functionalized SWCNTs/MWCNTs. Accelerated durability evaluation was carried out by conducting 1500 potential cycles between 0.1 and 1.2 V at 80°C. These Nano composites were demonstrated to yield a high fuel cell performance with enhanced durability. The membrane electrode assembly with Pt/MWCNTs showed superior performance stability with a power density degradation of only 30% compared to commercial Pt/C (70%) after potential cycles. Identically electro catalytically active platinum nanoparticles on CNTs with enhanced nucleation and stability have been demonstrated through introduction of electron-conducting polyaniline (PANI) [32]. A bridge between the Pt nanoparticles and MWCNTs walls was demonstrated with the presence of platinum nitride bonding and ϖ-ϖ bonding. The synthesized PANI was found to wrap around the CNT as a result of ϖ-ϖ bonding, and highly dispersed Pt nanoparticles were loaded onto the CNT with narrowly distributed particle sizes ranging from 2.0 to 4.0 nm. The Pt-PANI/CNT catalysts were electro active and exhibited excellent electrochemical stability, therefore constitute promising potential applications in proton exchange membrane fuel cells. Strong evidence thus emerges that CNTs/Pt composites are efficient as catalysts for fuel cells. Although platinum content has been dramatically decreased, industrials consider that further optimization is mandatory for a large scale fuel cell production. In addition Nafion® membrane between the cathodic and anodic compartment delays the large scale application of fuel cells, due to cost and problem of mass transfer. Breakthrough research towards these two bottlenecks could surely enforce a hydrogen economy.

TOWARDS A GREEN H₂ ECONOMY: CARBON NANOTUBES FOR ENZYME AND MICROBE IMMOBILIZATION

Replacement of chemical catalysts is thus nowadays highly needed in view of the development of a green energy economy. Microorganisms contain many biocatalysts, namely enzymes, which are highly efficient and specific towards various substrate conversions. Given they are produced in large enough quantities, these enzymes could be used as catalysts in biotechnological devices. A mandatory condition to develop heterogeneous catalysis is to succeed in the functional immobilization and in the stabilization of the enzymes on solid supports. The redox active site of enzymes is indeed buried inside the protein moiety so that the enzymatic property can be maintained under environmental stresses. Specific channels are often involved to allow the substrate to reach the active site. Complex but highly organized electron transfer chains occur for energetic metabolism. Electron transfer between two physiological partners associated with transformation of the substrate involves specific recognition site. The game for a bio electrochemist that aims to get the highest electron transfer rate for heterogeneous catalysis is to reproduce at the electrode interface the physiological electron transfer recognition process. Given the usual size of an enzyme (5-10 nm), electron transfer cannot occur *via* electron tunneling from the active site to the surface of the enzyme. In some enzymes, electron relays, one being located at the protein surface, act as a conductive line for electron shuttling. If the electrode interface is built so that it fits the surface electron relay environment, one can expect to favor a direct electrical connection of the enzyme onto the electrode. In case of direct electron transfer failure, an artificial redox mediator that acts as a fast redox system and shuttles electrons between the enzyme and the electrode can be used (Figure 2) [13, 33].

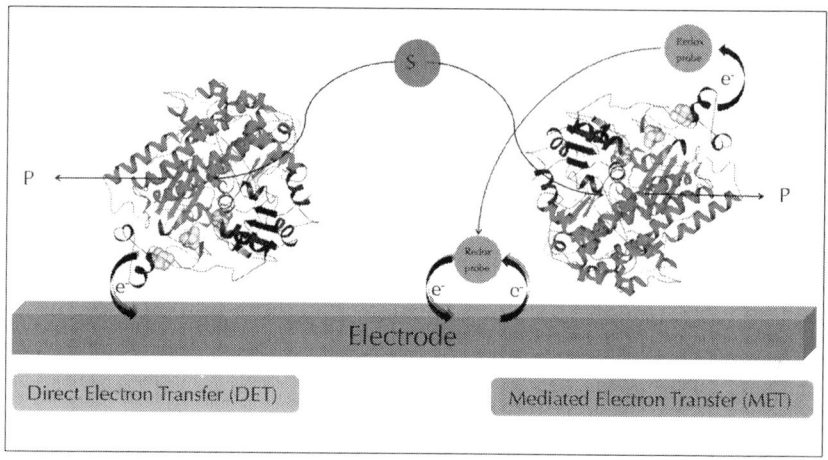

Figure 2: Interfacial electron transfer between an enzyme and an electrode can be achieved by direct (left) or mediated (right) electron transfer process.

Direct electron transfer process is preferred to mediated one, because it is not limited by the affinity between the enzyme and the redox mediator, and because it avoids the co-immobilization of enzyme and mediator. It is furthermore expected to yield the highest power density because enzymes, as biocatalysts, transform their substrate into products with very low over voltages. However it requires the knowledge of the protein structure and the construction of a tuned electrochemical interface that fits the electron transfer site.

There are many strategies for efficient enzyme immobilization onto electrochemical interfaces, including simple physical adsorption, covalent attachment, cross-linking or entrapment in polymers. The objectives are to optimize the immobilization procedure so that the efficiency of the enzyme and its stability are preserved. Moreover, due to the size of enzymes compared to chemical catalysts, large surface area interfaces baring many anchorage sites are required to obtain high catalytic currents.

To reach these goals, 3D structures are preferred, and CNT-based electrodes are very popular, both SWCNTs and MWCNTs. CNTs can be directly grown onto electrode surface, or adsorbed on it, or imbedded in polymer coating. In most cases, higher activity was reported for enzymes physically adsorbed onto CNTs [34]. Hydrophobic interactions between the enzyme and the CNT walls and ϖ-ϖ interactions between side walls of CNTs and aromatic rings of the enzyme are thought to be the driving force for direct adsorption of enzymes on CNTs [35]. Electrostatic interaction between the defect sites of CNTs and protonated amino residues of the enzyme plays also a role in the adsorption process [35]. CNTs are quite easily functionalized, allowing covalent, thus stable specific attachment of enzymes. The oxidation in strong acidic solutions at high temperature was demonstrated to remove the end caps and shorten the lengths of the CNTs. The length of the CNTs was shown to be a function of the oxidation duration [36]. Acid treatment also adds oxide groups, primarily carboxylic acids, to the tube ends and defect sites [37]. The control of reactants and/or oxidation conditions may control the locations and density of the functional groups on the CNTs, which can be used to control the location and density of the attached enzymes [37]. Covalent immobilization is induced by carbodiimide reaction between the free amine groups on the enzyme surface and carboxylic groups generated by side wall oxidation of CNTs.

H_2SO_4 / HNO_3 → OH → EDC / NHS → Protein / $-NH_2$ → NH

Further chemical reactions can be performed at the oxide groups generated on the oxidized CNTs to functionalize with groups such as amides, thiols, etc…From an electrochemical point of view, the side walls of CNTs were suggested to behave as basal plane of pyrolytic graphite, while their open ends resemble the edge planes [38, 39]. But recent work has demonstrated that the side wall may be responsible for electrochemical activity [40]. It has been furthermore suggested that the uncovered surface of CNTs promotes the accessibility of the substrate to the enzyme [41]. It is also interesting to note that the open spaces between CNTs are accessible to large species such as entire bacteria [42], opening the way for the development of fuel cells using whole microorganisms instead of purified enzymes. The cost and complexity of CNT manufacturing seem to be still clogging issues in that field.

Abundant literature exists on the ways CNTs are architectured for efficient enzyme immobilization, including those specific for development of enzymatic fuel cells. Enzymes and proteins as various as glucose oxidase and dehydrogenase, tyrosinase, laccase and bilirubin oxidase, peroxidase, haemoglobin and myoglobin, *i.e.* flavin, copper or heme containing active sites, have been studied. Whereas direct electron transfer between protein or enzyme and an electrochemical interface has been for long time supposed to be restricted to small proteins (<15kDa) possessing active sites exposed to the surface (it is the case for many cytochromes as example [43]), the use of CNT-modified electrodes has greatly enhanced the number and kinds of enzymes able to be directly connected to an electrode. Enzymes as large as one hundred kDa, with many cofactors are now considered for direct electron transfer. Consequently, recent works during the last years focus and report on direct communication between enzymes and electrode interface through CNT network. The induced porosity of the film depends on the type of CNTs used. But generally the nanometric size of the CNTs compared to the size of enzymes favors a direct electronic connection of the enzyme whatever its orientation [44]. The physical properties of CNTs, including high electrical conductivity, explain why CNT layers can be built up on electrodes most often yielding high rate direct electron transfer for enzymatic product transformation. Many researchers report on the increase in electro active surface area by use of CNT coatings that contribute to an increase in the direct electron transfer process [45-52]. CNTs are usually deposited on electrodes as thick films. Alternatively, layer-by-layer (LBL) process induces a quite stable protein film with nice electrocatalytic properties [53-55]. LBL is based on electrostatic interaction between oppositely charged monolayers in an alternating assembling. Although CNTs greatly amplify the current response, layer-by-layer architecture suffers from weak stability of the build-up and decrease in electron transfer for the upper layers. Besides vertically aligned CNTs were suggested to act as molecular wires that ensure the electrical communication between enzyme and electrode [56-58]. The carboxylic functions induced by acidic treatment of CNTs can be used for further chemical modifications. Amine- [59-61], thionin- [62, 63], diazonium salts [64, 65], pyrene [66, 67] (Figure 3) or other ϖ-ϖ stacking interactions [68] were used to functionalize CNTs. These modifications were demonstrated to be efficient platforms for enzyme immobilization. Mixing CNTs with surfactant [69-71] was claimed to assist in the dispersion of CNTs while avoiding oxidative functionalization which may disrupt their ϖ-network. Polymer modified

CNTs [72, 73] and sol-gel-CNT nanocomposite films [74] were proved to behave as friendly platforms for enzyme encapsulation.

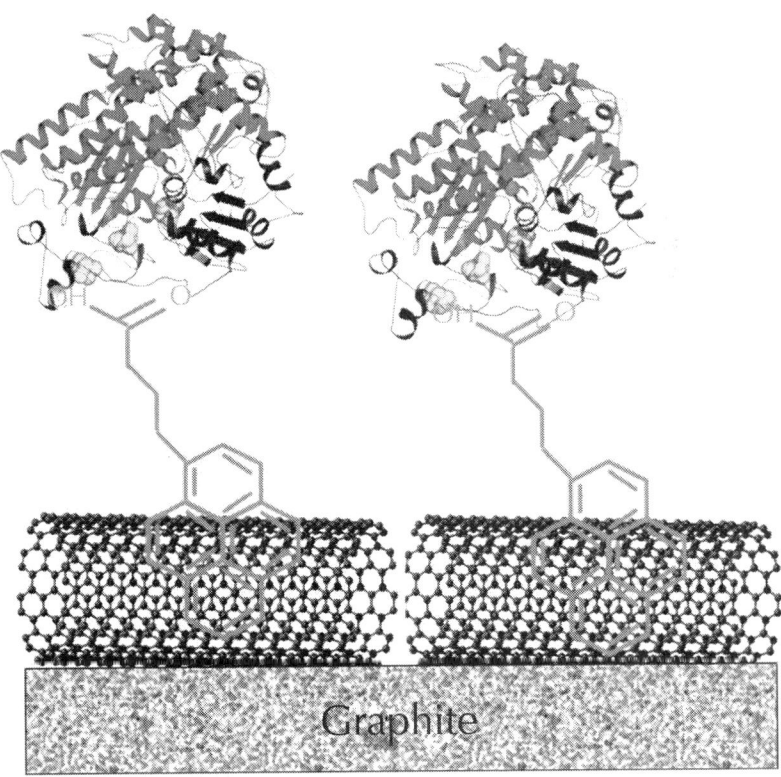

Figure 3: Schematic drawing of the build-up of enzyme on SWCNTs *via* ϖ-ϖ interactions.

Many enzymes however cannot be electrically connected to the electrode interface and require redox mediator to electrochemically follow substrate conversion. In that case, electrode kinetics is mainly dependant on mediator kinetics, so that the choice of the redox mediator mainly impedes the power density. Another issue is that the mediator can be co-immobilized with the enzyme at the electrode, while still being capable of efficient interaction with the enzyme. CNTs have also been used for building networks enabling co-immobilization of enzymes and redox mediators. In that way, one of the most popular redox entities is osmium polymer which forms hydrogels with enzymes allowing both charge transfer reaction between enzyme and mediators and diffusion of substrate and product [75].

Composite CNT/osmium films were used to immobilize bacteria [76], or enzymes [77]. By optimizing the CNT and polymer amounts, enhanced current responses were obtained linked to a promotion of the electron transfer within the composite. Various phenothiazine derivatives were also used to form Nano hybrids with CNTs acting as efficient redox mediator platforms [78-80]. Phenothiazine derivatives strongly adsorb onto CNTs leading to great enhancement of redox dye loading onto the electrode, but also to improved electrochemical sensing devices. Another strategy involves the use of a redox polymer as redox mediator platform. Electro polymerization of the redox conducting polymer onto CNTs enhances the amount of redox units and the electrical conductivity of the coating [81]. An interesting construction has also been obtained by immobilization of physiological cofactor onto CNT layers *via* ϖ-ϖ interactions, then immobilization of the enzyme [82]. The covalent coupling between the enzyme and its natural cofactor which was immobilized onto CNTs was proved to be efficient towards mediated substrate catalysis. This overview of multiple architectures involving enzymes and CNTs highlights the deep efforts engaged in the last years for efficient biocatalyst immobilization that open avenues towards biotechnological devices.

CARBON NANOTUBES FOR BIOLOGICAL PRODUCTION OF DIHYDROGEN

Apart from replacement of noble metal catalysts in fuel cells, a new green technology for production of dihydrogen is required. It currently relies on steam reforming of hydrocarbons under high temperature and pressure conditions, which starts from fossil fuels, thus producing greenhouse gases. Dihydrogen production *via* water electrolysis appears as a renewable solution given that the energy input comes from a renewable source, ideally solar energy. Many bacteria gain energy by the oxidation of dihydrogen assisted by a number of complex mechanisms. Various species evolve H_2 under anaerobic conditions. This is also a human being process since bacteria in our digestive tract produce H$_2$, though not detectable because immediately recycled by other bacteria. Photosynthetic organisms such as microalgae and cyanobacteria are very efficient in water splitting [83]. They possess photosensitizers for photon capture and charge separation, and enzymes for water oxidation to oxygen and water reduction to dihydrogen. This chemical activity relies on the expression of very efficient enzymes, called hydrogenases [84], which catalyze with high turn-over (one molecule of hydrogenase produces up to 9000 molecules of H$_2$ per second at neutral pH and 37°C) and low overvoltage the conversion of protons into dihydrogen and the oxidation of dihydrogen. The sequences of 450 hydrogenases are now available. Hydrogenases differ in size, structure, electrons donors. They also differ by their position in the cell (soluble in the periplasm, membrane-bound), and by their activity preferentially towards H$_2$ oxidation or protons reduction. Hydrogenase active site is composed of non-noble metals such as iron and nickel, unlike platinum catalyst necessary for the chemical electrolysis of water. Three distinct classes can be split which differ from the type of metal content in the active site: [NiFe], [FeFe] and [Fe] hydrogenases. [NiFe] and [FeFe] hydrogenases possess dinuclear active centers which are connected through thiolate bridges. [NiFe] hydrogenase (Figure 4) is the most usual hydrogenase in microorganisms. It is composed of two subunits. The larger subunit harbors the [NiFe] active site. The small subunit contains FeS clusters. Electrons are transferred to the active site along these FeS clusters distant less than 10 Å that act as a conductive line. [FeFe] hydrogenases are monomeric. In addition to the active site they contain additional domains which accommodate FeS clusters.

In order to use these biocatalysts for green dihydrogen production, two main research domains are currently concerned: the understanding of the catalytic mechanisms of H$_2$ production, and the optimization of enzyme immobilization. Adsorption onto graphite electrodes [85, 86] was largely used to study the mechanisms by which hydrogenases produce H$_2$. Grafting of hydrogenase onto gold electrode modified by thiolated Self-Assembled-Monolayer [87] allowed efficient proton reduction into dihydrogen in aqueous buffer solutions. Hydrogenase is also considered as a promising biocatalyst for photobiological production of dihydrogen when coupled to a photocatalyst [88]. Hybrid complexes of hydrogenases with TiO$_2$ nanoparticles [89, 90] were studied for H$_2$ production. The optimized system was shown to produce H_2 at a turnover frequency of approximately 50 (mol H_2) s^{-1} (mol total hydrogenase)$^{-1}$ at pH 7 and 25 °C, even under the typical solar irradiation of a northern European sky. Cd-based nanorods [91, 92] were recently studied. The CdS nanorod/hydrogenase complexes photocatalyzed reduction of protons to H_2 at a hydrogenase turnover frequency of 380-900 s^{-1} and photon conversion efficiencies of up to 20% under illumination at 405 nm. Cd-based complexes allowed photoproduction of dihydrogen for a couple of hours, but still suffer from quick inhibition of hydrogenase.

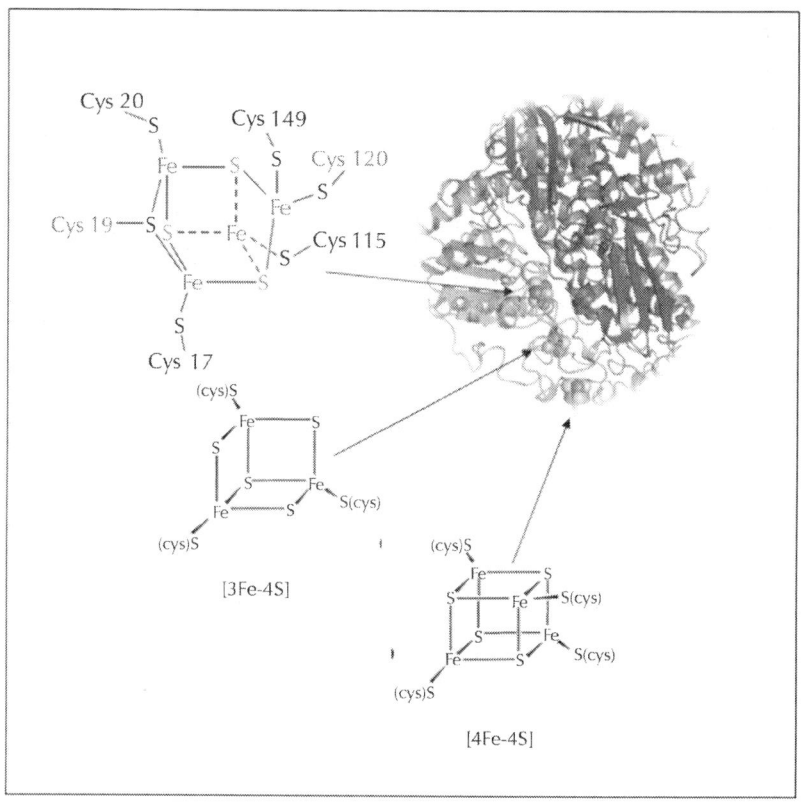

Figure 4: Structure of an oxygen-tolerant [NiFe] hydrogenase.

Although a very attractive way, little work has been done towards enhancement of green hydrogen production using CNTs. Three studies from the same group reported however catalytically active hydrogenase-SWCNT biohybrids [93, 94]. Surfactant-suspended SWCNTs were shown to spontaneously self-assemble with hydrogenase. Photoluminescence excitation and Raman spectroscopy showed that SWCNTs act as molecular wires to make electrical contact with at least one of the FeS electron relay. Hydrogenase was demonstrated to be strongly attached to the SWCNTs and to mediate electron injection into nanotubes. The displacement of the surfactant by hydrogenase to gain access to the SWCNTs was strongly suggested by photoluminescence studies. Furthermore, Raman studies of charge transfer complexes between hydrogenase and either metallic (m) or semiconducting (s) SWCNTs revealed a difference in oxygen deactivation of hydrogenase according to the SWCNT species. m-SWCNTs most probably interact with hydrogenase to produce a more oxygen-tolerant species. The study further suggested that purified m-SWCNTs or s-SWCNTs, rather than mixed preparation, would be more suitable for hydrogenase-SWCNTs biohybrids. The formation of these catalytically active biohybrids in addition with the intrinsic properties developed by CNT networks on electrodes certainly accounts for the improved dihydrogen production observed in the following studies. Kihara *et al.* immobilized hydrogenase on a SWCNT-forest with a unique dense structure of vertically aligned millimetre-scale height SWCNTs [95]. Hydrogenase was demonstrated to spontaneously assemble between adjacent nanotubes. The maximum rate of dihydrogen production was reported to be 720 nmol/min/(mg hydrogenase) and the electron transfer efficiency was estimated to be 32%. It is two thousand fold higher than reported before using the same hydrogenase on Langmuir-Blodgett film [96]. Nevertheless, one key point in the development of biotechnological devices is the long term stability of enzymes. If these biological catalysts are very efficient *in vivo*, they often suffer from weak stability when extracted from their physiological environment. Enzyme encapsulation in silica-derived sol-gel materials has been demonstrated to stabilize many enzymes. This procedure was applied to hydrogenase [97]. The majority of hydrogenase was shown to be entrapped in the gel and protected against proteolysis. Hydrogenase/sol-gel pellets retained 60% of the specific mediated activity for H_2 production displayed by hydrogenase in solution. The gel-encapsulated enzyme retained its activity for long periods, *i.e.* 80% of the activity after four weeks at room temperature. Notably, by doping the hydrogenase-containing sol-gel materials with MWCNTs Zadvorny *et al.* demonstrated a 50% increase in dihydrogen production [98]. Furthermore stabilization of hydrogenase was proved through encapsulation process.

One alternative for green hydrogen production is to synthesize metal complexes that mimic the active site of enzymes. Huge work has been done in that field in order to obtain bioinspired models that could produce H_2 as efficiently as hydrogenase, while being much more stable [99]. The most performing complex involves mononuclear nickel diphosphine complex. This complex is inspired from the active sites of both [NiFe] and [FeFe] hydrogenases and displays remarkable catalytic proton reduction in organic solvent [100]. Le Goff *et al.* took benefice from this complex and from the results obtained by immobilization of hydrogenase on CNT networks [44]. The authors successfully immobilized the nickel complex onto carbon nanotube networks by covalent coupling [101]. Such construction was demonstrated to be very efficient for dihydrogen production in aqueous solution, evolving dihydrogen with overvoltage less than 20 mV and exceptional stability.

CARBON NANOTUBES FOR BIOFUEL CELLS: AN ATTRACTIVE GREEN ALTERNATIVE

Beside researches towards decrease in chemical catalyst amount and discovery of less expensive catalysts (as alloys for example), a new concept emerged early in 1964 by Yahiro *et al.* [102]. A fuel cell was constructed using usual O_2 reduction at platinum modified electrode in the cell cathodic compartment, but using glucose as a fuel in the anodic compartment. The innovative idea was the use of an enzyme specific for fuel oxidation instead of platinum. For glucose oxidation, glucose oxidase was tested as the anodic catalyst. The fuel cell delivered 30 nA cm^{-2} at 330 mV. a very low power density indeed but the proof of concept of biofuel cell was born. Generally speaking these biofuel cells function as fuel cells but used enzymes instead of noble metals as catalysts (Figure 5). They are referred as enzymatic biofuel cells. Microorganisms can also be used as catalysts, defining microbial fuel cells. Microbial biofuel cells use the metabolism of microorganisms under anaerobic conditions to oxidize fuel [103-104]. Although a promising concept, little is known yet about the mechanisms by which fuel is oxidized at the anode. The involvement of nanowires, electron transfer mediators, either membrane-bound or excreted, is supposed to be responsible for the cell current. Enzymatic biofuel cells are however more efficient because no mass transfer limitations across the cell membrane exist.

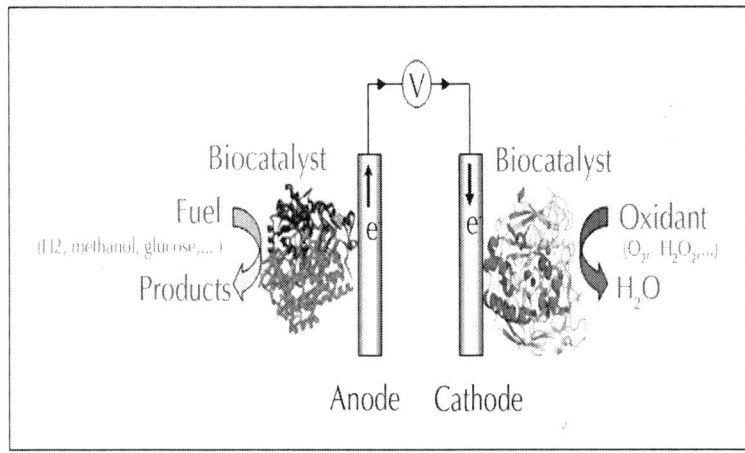

Figure 5: Schematic representation of an enzymatic biofuel cell.

The advantages of enzymatic biofuel cells over fuel cells are multiple. Biocatalysts are widespread, then *a priori* inexpensive, and biodegradable. Enzymes are highly efficient and specific to their substrates. The substrate specificity decreases reactant cross-over, and might theoretically allow to design fuel cells with no membrane between the anodic and cathodic compartments. Both costs are reduced and the design is simplified. A large variety of fuels and oxidants can be used to feed the biofuel cells, as opposed to the poor available fuels and oxidant in classical fuel cells (dihydrogen, methanol, oxygen). Indeed, many enzymes are nowadays characterized which differ by their natural abundant substrates. Dihydrogen, but also various inexpensive sugars can thus be used as efficient fuels at the anode. Furthermore, the involvement of cascades of enzymes can enhance the cell performance because of the summation of the electrons from each enzymatic reaction [105]. Finally, biofuel cells can deliver power under soft working conditions, as enzymes usually perform their enzymatic reactions at mild pH and temperature. Nevertheless, some extremophilic enzymes operate in extreme acidic or basic pH, as well as at high temperatures (around 90°C) or high pressure, offering the possibility to develop biofuel cell devices for special applications requiring extreme working conditions [106]. The applications of biofuel cells are still in their infancy. They are mainly thought to power small portable devices. Remarkable progress has been reported for implantable biofuel cells during the last year to power drug pumps, glucose sensors, vision devices [107-109].

The most common redox couple that has been used in biofuel cells is sugar/O_2, essentially because of sugar and O_2 abundance in nature and their essential role in living metabolism. In particular, glucose is an important metabolite and a source of energy for many living organisms. In that field, CNTs have been widely used, both at the anode and cathode. Glucose/O_2 biofuel cell is thus a very pertinent investigation field to investigate the role of CNTs. A view of some typical results is presented in Table 2.

Table 2: Performances of glucose/O_2 fuel cells

Enzymes Anode / Cathode	Mediators Anode / Cathode	Power density $\mu W\ cm^{-2}$	Ref
Gox / Laccase	Ferrocene / -	15	[111]
GDH / BOD	PQQ / -	23	[82]
Gox / Pt	Ferrocenecarboxalde-hyde / -	51	[112]
GDH/ BOD	Poly(brilliant cresyl blue) / -	54	[113]
GDH / laccase	Azine dies / -	58	[114]
Gox / Pt	Benzoquinone / -	77	[52]
Gox / Laccase	Ferrocene / ABTS	100	[115]
Gox / BOD	Ferrocene methanol / ABTS	120	[116]
GDH / Laccase	- / -	131	[117]
CDH / Pt	Os complex / -	157	[118]
Gox / Laccase	- / -	1300	[119]

Gox: Glucose oxidase; GDH: Glucose dehydrogenase; BOD: Bilirubin oxidase; ABTS: 2, 2′-azino-bis (3-ethylbenzothiazoline-6-sulfonate) diammonium; CDH: cellobiose dehydrogenase.

Data highlight that kinetics of bio electrochemical reactions, thus power density, largely depends on the experimental conditions, *i.e.* enzyme and mediators, T°, pH, concentration of substrate, electrolyte and type of electrode construction. Highest values are obtained with mediator less fuel cells, reaching power densities upper than 1 mW cm^{-2} which is sufficient to power small electrical devices. It appears that direct connection of copper enzymes, namely laccase or BOD, for oxygen reduction at the cathode can be quite easily obtained with the help of CNT network. Direct connection of enzymes for glucose oxidation is conversely hardly observed, even on CNT coatings. From literature examination direct connection of Gox at electrode interfaces is still controversial. Due to the peculiar structure of Gox, a dimer with flavin adenine dinucleotide active site buried within a thick and isolated protein shell, it is understandable that electrical connection of Gox could be unexpected. A recent work concluded that CNTs were capable to electrically connect Gox, but this connection was unfruitful for glucose catalytic oxidation [110].

CARBON NANOTUBES FOR BIOELECTROOXIDATION OF H$_2$: TOWARDS H$_2$/O$_2$ BIOFUEL CELLS

We already described above hydrogenases, the enzymes that convert with high specificity and efficiency protons into dihydrogen. Most of these biocatalysts are also efficient in the oxidation of dihydrogen into protons. Consequently this allows to imagine biofuel cells in which the fuel would be dihydrogen, exactly as in PEM fuel cells. As hydrogenases are able to oxidize dihydrogen with very low overvoltage, the open circuit voltage for the biofuel cell using oxygen at the cathode, is expected to be not far from the thermodynamic one, *i.e.* 1.23 V. Hence, high power densities are expected, provided that a strong and efficient electrical connection between hydrogenase and electrode can be achieved. Simple adsorption of hydrogenase was performed in a first step, because it allowed a direct oxidation of dihydrogen without any redox mediators [120]. Catalytic mechanisms associated with dihydrogen oxidation at the active site were largely studied. The effect of strong hydrogenase inhibitors such as oxygen and CO were explored by this mean, leading to nice developments in engineering of more tolerant hydrogenases [121] or use of naturally resistant hydrogenases [122, 123]. However, this immobilization procedure relies on a monolayer of enzyme, which furthermore suffers from quick desorption. Otherwise, multilayer enzymatic films require a redox mediator so that even the last layer far from the electrode could be connected. Other immobilization processes are thus needed, that can favor an enhancement in both the amount of connected hydrogenases as well as their stability, while preserving their functionality.

Carbon nanotube networks constituted technological breakthroughs in that way. All the recent developments using immobilization of hydrogenases onto carbon nanotubes point out improved catalytic currents essentially related to an increase in the active area of the electrode. The respective role of metallic-SWCNTs against semiconducting one was explored for dihydrogen oxidation by immobilized hydrogenase [124]. A higher oxidation process was revealed when the nanotube mixture was enriched in metallic SWCNT. The study furthermore suggested no need of oxygenated SWCNTs for efficient anchoring of hydrogenases. The catalytic current enhancement was claimed to be due to an increase in active electrode surface area and an improved electronic coupling between hydrogenase redox active sites and the electrode surface. In most cases, however, CNTs are used as a mixture of metallic and semi-conducting tubes. Oxidation of the mixture yields the defects and functionalities described above in this review. Advantage is gained due to these chemical functions quite easily generated on the surface of the carbon nanotubes. Electrodes modified by carbon nanotubes are thus expected to offer numerous anchoring sites for stable hydrogenase immobilization. The literature provides a few examples of efficient immobilization of hydrogenase on carbon nanotubes coatings bearing various functionalities. Both SWCNTs and MWCNTs are used. Notably, more and more articles are devoted nowadays to this domain in hydrogenase research. A bionanocomposite made of the hydrogenase, MWCNTs and a thiopyridine derivative was proved to form stable monolayers when transferred by Langmuir-Blodgett method on indium tin oxide electrode surfaces [125]. A greater amount of electroactive hydrogenase towards dihydrogen oxidation was demonstrated to be adsorbed on the Langmuir-Blodgett films. De Lacey and co-workers grew MWCNTs on electrode by chemical vapor deposition of acetylene [65]. A high density of vertically aligned carbon nanotubes was obtained, which were functionalized by electroreduction of a diazonium salt for covalent binding of hydrogenase. High coverage of electroactive enzyme was measured, suggesting that almost all the functionalized CNT surface was accessible to hydrogenase. Great stabilization of the catalytic current for H_2 oxidation was obtained, with no decrease in current density after one month. Another work by Heering and co-workers studied a gold electrode pre-treated by polymyxin then a multilayer of carbon nanotubes [126]. Polymyxin was shown to help in the stable attachment of hydrogenase on the gold electrode. Using adsorption of hydrogenase on a nanotube layer pretreated with polymyxin the current density for H_2 oxidation was an order of magnitude higher than at the gold electrode only modified by polymixin. This result was supposed to origin from greater surface area even though only the top of the nanotube layer was supposed to be accessible to the enzyme. The catalytic current was stable with time, at least for two hours under continuous cycling, and several days upon storage under ambient conditions. AFM visualization of hydrogenase immobilized onto polymyxin-treated SWCNT layer on SiO_2 revealed that hydrogenase was structurally intact and preferentially adsorbed on the sidewalls of the CNTs rather than on SiO_2 [126].

In our laboratory, we immobilized the [NiFe] hydrogenase from a mesophilic anaerobic bacterium (the sulfate reducing bacterium *Desulfovibrio fructosovorans* Df) by adsorption onto SWCNT films [44]. The current for direct H_2 oxidation was shown to increase with the amount of SWCNTs in the coating (Figure 6).

Because non-turnover signals were not detected for hydrogenase in these conditions, the increase in surface area was evaluated using a redox protein as a probe. It was shown that SWCNTs induced one order larger surface area. The same hydrogenase was entrapped in methylviologen functionalized polypyrrole films coated onto SWCNTs and MWCNTs [127]. Although no direct electrical hydrogenase connection was observed, an efficient dihydrogen oxidation through a mediated process occurred. It was concluded that the entrapment of hydrogenase into the redox polymer coated onto CNTs combined the electron carrier properties of redox probes, the flexibility of polypyrroles, and the high electroactive area developed by CNTs.

The reason why no direct connection could be observed is however not clearly understood yet. In our group we handled immobilization of hydrogenase on a film obtained by electro polymerization of a phenothiazine dye on a SWCNT coating [81]. The phenothiazine dye was shown to be able to mediate dihydrogen oxidation but also to serve as an anchor for the enzyme when adsorbed or when electro polymerized. Higher current density than in the absence of SWCNT was observed.

In addition, a wider potential window for dihydrogen oxidation was reached as well as very stable electrochemical signals with time. We postulated that the conductive polymer which was electro polymerized onto CNTs could play a multiple role: enhancement of the electro active surface area, enhancement of redox mediator units due to phenothiazine monomers entrapped in the polymer matrix, enhancement of hydrogenase anchorage sites. We have already mentioned in this review the advantages of a direct electron transfer over a mediated one for H_2 oxidation, including gain in over-potential values, less interferences due to enzyme specificity, absence of redox mediators that could be difficult to co-immobilize with the enzyme

Functionalized carbon nanotube films were evaluated in our group as platforms for various hydrogenases that present a very different environment of FeS cluster electron relay. Dihydrogen oxidation was studied at gold electrodes modified with functionalized self-assembled-monolayers [128]. As expected, dihydrogen oxidation process was demonstrated to be driven by electrostatic or hydrophobic interactions according to the specific environment of the surface electron relay. Interestingly, at CNT coatings, although CNTs were negatively charged, direct electrical connection of hydrogenases that present a negatively charged patch around the FeS surface electron relay was observed [44, 123]. In other words, despite unfavourable electrostatic interactions, direct electron transfer process for dihydrogen oxidation was achieved. One important conclusion was that on such CNT films, the nanometric size of the CNTs allows a population of hydrogenases to be directly connected to a neighbouring nanotube, hence allowing direct electron transfer for H_2 oxidation, whatever the orientation of the enzyme.

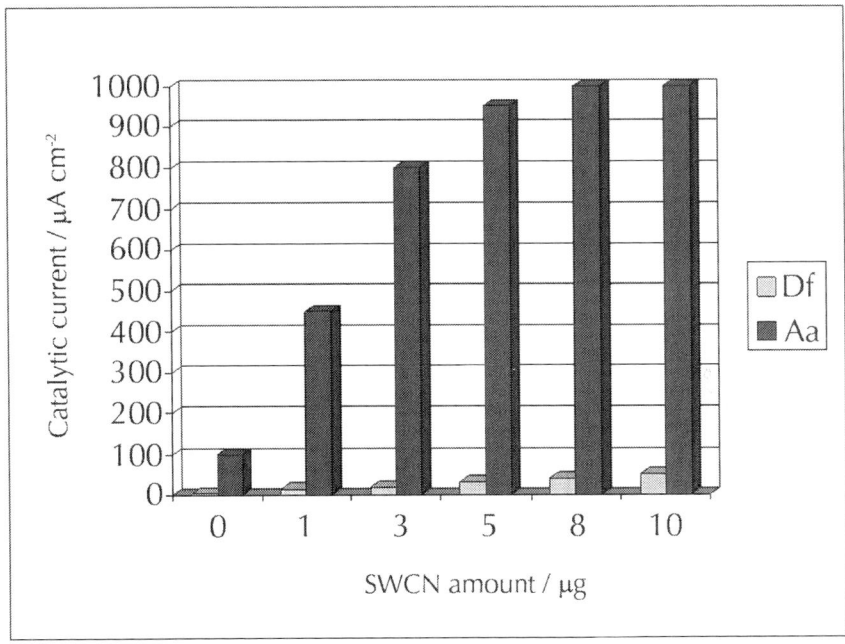

Figure 6: Comparative evolution of the catalytic current for dihydrogen oxidation with the amount of SWCNTs deposited at a graphite electrode in the case of hydrogenases from *Aquifex aeolicus* (Aa) or *Desulfovibrio fructosovorans* (Df). Catalytic currents are measured using voltammetry under H_2 at 60 and 25°C for Aa and Df respectively.

However, the extreme oxygen sensitivity of hydrogenases used in the former studies yielded an intensive research towards more resistant enzymes. During the last years, four [NiFe] membrane-bound hydrogenases have been discovered from aerobic or extremophilic organisms [128, 129-132]. They have been demonstrated to oxidize H_2 in the presence of oxygen and CO. The crystallographic structure of three of them has been resolved, showing that an uncommon [4Fe-3S] cluster proximal to the active site prevents deleterious oxygen attack. Of course, the sensitivity to oxygen, and also to CO, of most hydrogenases known before was a strong limitation for their potential use in biotechnological devices. Therefore these resistant biocatalysts open new avenues towards a biohydrogen economy. No doubt that these researches will increase in the next future. To date, two main studies report the immobilization of resistant hydrogenase on CNT-modified electrodes. Krishnan *et al.* very recently modified MWCNTs by pyrenebutyric acid, and demonstrated it was an efficient platform for stable O_2-resistant hydrogenase linkage [133]. In our group, original use of a hyperthermophilic O_2- and CO-resistant hydrogenase allowed the increase in the catalytic current for direct H_2 oxidation on a large range of temperature up to 70°C. Attempts to enhance the number of electrically connected hydrogenase succeeded by use of coatings of chemically oxidized SWCNTs [123]. Values as high as 1 mA cm^{-2} were reached depending on the amount of SWCNTs used in the coating (Figure 6). For the lowest amounts of SWCNTs, the increase in the catalytic current was demonstrated to be essentially due to the increase in surface area. However the catalytic current rapidly reached a plateau, although the peak current for the redox probe still increased, suggesting rapid saturation of the surface.

DESIGN OF A H$_2$/O$_2$ BIOFUEL CELL BASED ON CARBON NANOTUBES-MODIFIED ELECTRODES

H_2/O_2 biofuel cells did not get much attention before O_2 and CO resistant hydrogenases were proved to be efficient for H_2 oxidation when immobilized onto electrode surfaces. Even though more and more efficient hydrogenase immobilization procedures are nowadays reported, few H_2/O_2 biofuel cells are described. An early study by Armstrong's group in 2006 [134] demonstrated that simple adsorption on graphite electrode of hydrogenase at the anode and laccase (a copper protein for O_2 reduction) at the cathode, allowed a wristwatch to run for 24h. Power density of around 5 µW cm^{-2} at 500 mV was delivered with no membrane between the two compartments providing hydrogenase was extracted from *Ralstonia metallireducens*. As this is an aerobic bacterium, the result underlined that the H_2/O_2 biofuel cell could operate only with O_2 resistant hydrogenase. In 2010, the same group improved the device by using another O_2 resistant hydrogenase from *Escherichia coli* and bilirubin oxidase (BOD), another copper protein more efficient than laccase towards oxygen reduction because being able to function at neutral pH [135]. The oxygen reductase was covalently linked to the graphite electrode which had been modified by diazonium salt reduction. The power density was enhanced compared to the former study reaching 63 μW cm^{-2}. But most of all, this work provided a nice understanding of the operating conditions of such H_2/O_2 fuel cells involving hydrogenase as anode catalyst.

Due to the understanding of how hydrogenases could be efficiently connected at CNT-coated electrodes, a huge step jumped over very recently. First, using covalent attachment of both O_2 resistant hydrogenase and BOD on pyrene derivative functionalized MWCNTs, a membrane-less biofuel cell was designed fed with a non-explosive 80/20 dihydrogen/air mixture [133]. This biofuel cell displayed quite a good stability with time and a much higher power density than reported before. Indeed, an average power density of 119 μW cm^{-2} was measured. Low solubility of oxygen and weak affinity of BOD for oxygen was shown to limit the cathodic current.

Secondly in our group, a more performant H_2/O_2 mediatorless biofuel cell was constructed based on one step covalent attachment directly on SWCNTs of an hyperthermophilic O_2 resistant hydrogenase at the anode and BOD at the cathode [136] (Figure 7).

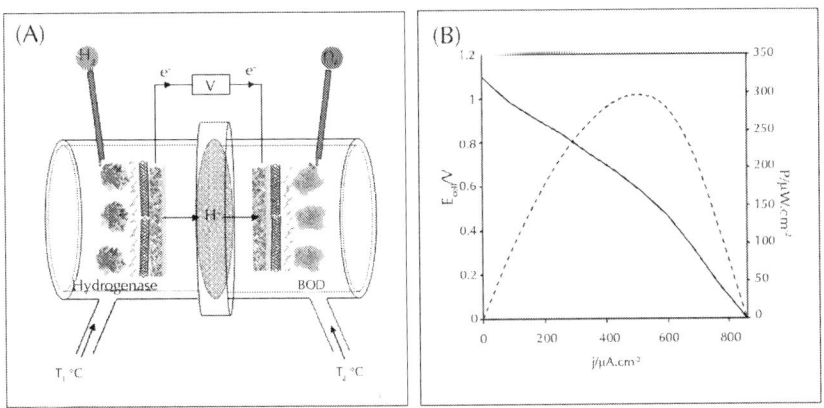

Figure 7: (A) Schematic representation of H_2/O_2 biofuel cell with O_2 resistant hydrogenase at the anode and bilirubin oxidase (BOD) at the cathode. Each half cell, separated by a Nafion® membrane, is independently thermoregulated with waterbaths. (B) Performance of the H_2/O_2 biofuel cell.

Taking advantage of temperature, the biofuel cell delivered power densities up to 300 μW cm^{-2} at 0.6V with an OCV of 1.1V, which is the highest performance ever reported. Furthermore, promising stability of the biofuel cell during 24h of continuous use lets us consider this device as an alternative power supply for small portable applications. The analysis of the fuel cell parameters during polarization, allows us to define the potential window in which the fuel cell fully operates. Interestingly, in Armstrong's group [135] and in our group, different approaches on the settings of biofuel cell working conditions, led to similar observations of an unexpected increasing anodic potential.

This high oxidizing potential generates an inactive state of hydrogenase active site. It is worth noticing that this hydrogenase inactivation occurred under anaerobic conditions in our group while it was under aerobic conditions in Armstrong's group. Consequently, dramatic loss in power densities was observed. By applying negative potential to the anode, and thus providing electrons to the active site, we were unable to reactivate hydrogenase. Another protocol used by Armstrong, consisted to add a second hydrogenase coated anode, unconnected to the system but present in the anodic half-cell which was consequently unaffected by the oxidizing potential but still in presence of O_2. This second anode, under H_2 oxidation was used as an electron supplier and connected to the first electrode. This procedure reactivated hydrogenase and allowed full recovery of OCV. It is of relevant interest to overcome hydrogenase inactivation in H_2/O_2 biofuel cell.

FUTURE DIRECTIONS

As reviewed in this chapter, many of the CNTs based technologies are promising for the development of a green hydrogen economy. Not only abiotic dihydrogen storage, but also microbial dihydrogen production and use of this green dihydrogen in biofuel cells can take advantages of the outstanding properties of CNTs. In all these applications, CNTs appear to play multiple roles including increase in surface area, increase in electron transfer rate, and increase in directly connected enzymes. Possible protection against oxygen damage of enzymes has even been strongly suggested. Use of CNTs thus allows to architecture three-dimensional nanostructured interfaces which can be an alternative to strictly orientated proteins or enzymes for high direct electron transfer interfacial processes. The ease in obtaining tuned surface functionalizations is one of the very attracting points in view of the development of efficient bioelectrodes.

This is in particular the case for biofuel cells using dihydrogen as a fuel. During the last years, tremendous research on hydrogenase, the key enzyme for dihydrogen conversion, has led to the discovery, then control of some hydrogenases presenting properties that allow their use in biotechnological devices. During this year, based on these new resistant enzymes and on improved knowledge of how CNTs can enhance direct current densities, two H_2/O_2 biofuel cells have been reported. Although these biofuel cells constitute the first device using hydrogenases, they already deliver sufficient power density for small portable applications. No doubt that this research field will gain more and more interest in a next future.

However, various directions might be followed to further improve the biological system in such a way it could be commercially available. One is the enhancement of long-term stability of the device, which is obviously the critical point shared by (bio) fuel cells, yet. Search for more stable enzymes in the biodiversity or enzyme engineering has to be explored. Protection of enzymes by various encapsulation procedures could be another solution given efficient interfacial electron transfer can be reached. The use of whole microorganisms with controlled and driven metabolism, or at least immobilization of naturally encapsulated enzymes will be a next step. As an example, reconstitution of proteoliposomes with a membrane-bound hydrogenase was proved to enhance the stability of the enzyme [137]. This could be a novel route for preserving enzymes in their physiological environment, hence enhancing their stability. New enzymes, with outstanding properties (T°, pH, inhibitors, substrate affinity...) have to be discovered and studied. Notably, two very recent publications report on a new thermostable bilirubin oxidase and a tyrosinase which present outstanding resistances to serum constituents [138,139]. These two new enzymes appear to be able to efficiently replace the currently used BOD for implantable applications of biofuel cells.

More sophisticated materials interfaces, constituted of mixtures of CNTs with other conducting materials could bring a hierarchical porosity necessary for both enzyme immobilization and substrate diffusion. Carbon fibers, mesoporous carbon templates could be used to build very interesting new electrochemical interfaces. This diversity in potential carbon materials for efficient enzyme immobilization would be a key step to go through the difficulties linked to CNTs, *i.e.* effective cost for separation and purification as well as possible toxicity. Finally, to avoid the membrane between the cathodic and anodic compartments, and build a miniaturized biofuel cell, unusual cell designs, such as microfluidic or flow-through systems, are likely to open new avenues. All these future developments will certainly require a multidisciplinary approach, coupling electrochemists with biochemists and physicists, and coupling methods such as electrochemistry and spectrometry, electrochemistry and molecular genetics or electrochemistry and materials chemistry. This multidisciplinary willingness will help in the elucidation of the interactions between enzymes and nanostructured materials at the nanoscale and yield innovative Nano biotechnological approaches and applications.

ACKNOWLEDGEMENTS

We gratefully acknowledge the contribution of Marielle Bauzan (Fermentation Plant Unit, IMM, CNRS, Marseille, France) for growing the bacteria, Dr Marianne Guiral, Dr Marianne Ilbert and Pascale Infossi for fruitful discussions. This work was supported by research grants from CNRS, Région PACA and ANR.

REFERENCES

1.   J. Verne, 1874 L'île mystérieuse

2.   W. Grove, 1838 on a new voltaic combination. Philosophical Magazine and Journal of Science13 430

3.   J. L. Charlou, J. P. Donval, C. Konn, H. Ondréas, Y. Fouquet, P. Jean-Baptiste, E. Fourré, 2010 High production and fluxes of H2 and CH4 and evidence of abiotic hydrocarbon synthesis by serpentinization on ultramafic-hosted hydrothermal systems on Mid-Atlantic Ridge. Rona P., Devey C., Dyment J. Murton B. Editors, "Diversity of hydrothermal systems on slow spreading

ocean ridges" Edited by AGU Geophysical monograph series 188 265 296

4. A. Oberlin, M. Endo, T. Koyama, 1976 Filamentous growth of carbon through benzene decomposition. Journal of Crystal Growth 32 3 335 349

5. S. Iijima, 1991 Helical microtubules of graphitic carbon. Nature 354 56

6. R. Saito, G. Dresselhaus, M. Dresselhaus, 1998 Physical Properties of Carbon Nanotubes Imperial College Press, London.

7. Othmer. Kirk, 1996 Encyclopedia of Chemical Technology, (4th ed.) Wiley and Sons, New York.

8. C. Heitner-Wirguin, 1996 Recent advances in perfluorinated ionomer membranes: structure, properties and applications. Journal of Membrane Science 120 1 1 33

9. J. Cracknell, K. Vincent, F. Armstrong, 2008 Enzymes as working or inspirational electrocatalysts for fuel cells and electrolysis. Chemical Review 7 108 2439 2461

10. M. Guiral, L. Prunetti, C. Aussignargues, A. Ciaccafava, P. Infossi, M. Ilbert, E. Lojou, M. T. Giudici-Orticoni, 2012 The hyperthermophilic bacterium *Aquifex aeolicus*: from respiratory pathways to extremely resistant enzymes and biotechnological applications. Advances in Microbiological Physiology; to be edited.

11. I. Ivanov, T. Vidakovic-Koch, K. Sundmacher, 2010 Recent Advances in Enzymatic Fuel Cells: Experiments and Modeling Energies 3 4 803 846

12. P. Tran, V. Artero, M. Fontecave, 2010 Water electrolysis and photoelectrolysis on electrodes engineered using biological and bio-inspired molecular systems. Energy & Environmental Science 3 6 727 747

13. E. Lojou, 2011 Hydrogenases as catalysts for fuel cells: Strategies for efficient immobilization at electrode interfaces. Electrochimica Acta 56 28 10385 10397

14. J. M. Schnorr, T. M. Swager, 2011 Emerging Applications of Carbon Nanotubes. Chemistry of Matererials 23 3 646 657

15. S. Iijima, T. Ichihashi, 1993 Single-shell carbon nanotubes of 1-nm diameter. Nature 363 603 605

16. M. Dresselhaus, G. Dresselhaus, A. Jorio, 2004 Unusual properties and structure of carbon nanotubes. Annual Review of Material Research 34 247

17. M. Botzung, S. Chaudourne, O. Gillia, C. Perret, M. Latroche, A. Percheron-Guegan, P. Marty, 2008 Marty Simulation and experimental validation of a hydrogen storage tank with metal hydrides. International Journal of Hydrogen Energy 33 1 98 104

18. J. Vajo, W. Li, P. Liu, 2010 Thermodynamic and kinetic destabilization in LiBH4/Mg2NiH4: promise for borohydride-based hydrogen storage. Chemical Communications 46 36 6687 6689

19. C. Li, P. Peng, D. Zhou, L. Wan, 2011 Research progress in LiBH4 for hydrogen storage: A review. International Journal of Hydrogen Energy 36 22 14512 14526

20. X. Yao, C. Wu, A. Du, G. Lu, H. Cheng, S. Smith, J. Zou, Y. He, 2006 Mg-based nanocomposites with high capacity and fast kinetics for hydrogen storage. Journal of Physical Chemistry B 110 24 11697 11703

21. Y. Luo, P. Wang, L. P Ma , M. Cheng, 2007 Enhanced hydrogen storage properties of MgH2 co-catalyzed with NbF5 and single-walled carbon nanotubes. Scripta Materialia 56 9 765 768

22. C. Wu, P. Wang, X. Yao, C. Liu, D. Chen, G. Lu, H. Cheng, 2006 Effect of carbon/noncarbon addition on hydrogen storage behaviour of magnesium hydride. Journal of Alloys Compounds 414 259

23. A. Dillon, K. Jones, T. Bekkedahl, C. Kiang, D. Bethune, M. Heben, 1997 Storage of hydrogen in single-walled carbon nanotubes. Nature 386 6623 377 379

24. F. Ding, B. Yakobson, 2011 Challenges in hydrogen adsorptions: from physisorption to chemisorption. Frontiers of Physics 6 2 142 150

25. A. Nikitin, X. Li, Z. Zhang, H. Ogasawara, H. Dai, A. Nilsson, 2008 Hydrogen storage in carbon nanotubes through the formation of stable C-H bonds. Nano Letters 8 1 162 167

26. U. Sahaym, M. Norton, 2008 Advances in the application of nanotechnology in enabling a hydrogen economy. Journal of Material Sciences 43 16 5395 549

27. Z. Tang, C. Poh, K. K. Lee, Z. Tian, D. Chua, J. Lin, 2010 Enhanced catalytic properties from platinum nanodots covered carbon nanotubes for proton-exchange membrane fuel cells. Journal of Power Sources 195 1 155 159

28. J. Lin, V. Kamavaram, A. Kannan, 2010 Synthesis and characterization of carbon nanotubes supported platinum nanocatalyst for proton exchange membrane fuel cells. Journal of Power Sources 195 2 466 470

29. A Kongkanand, S Kuwabata, G Girishkumar, P Kamat, 2006 Single-wall carbon nanotubes supported platinum nanoparticles with improved electrocatalytic activity for oxygen reduction reaction. Langmuir 22 5 2392 2396

30. J. Lin, C. Mason, A. Adame, X. Liu, X. Peng, A. Kannan, 2010 Synthesis of Pt nanocatalyst with micelle-encapsulated multi-walled carbon nanotubes as support for proton exchange membrane fuel cells Electrochimica Acta 55 22 6496 6500

31. C. Lee, Y. C. Ju, P. T. Chou, Y. C. Huang, L. C. Kuo, J. C. Oung, 2005 Preparation of Pt nanoparticles on carbon nanotubes and graphite nanofibers via self-regulated reduction of surfactants and their application as electrochemical catalyst. Electrochemistry Communications 7 4 453 458

32. D. He, C. Zeng, C. Xu, N. Cheng, H. Li, S. Mu, 2011 Polyaniline-Functionalized Carbon Nanotube Supported Platinum Catalysts. Langmuir 27 9 5582 5588

33. R. Ludwig, W. Harreither, F. Tasca, L. Gorton, 2011 Cellobiose Dehydrogenase: A Versatile Catalyst for Electrochemical Applications. ChemPhysChem. 11 13 2674 2697

34. J. Cang-Rong, G. Pastorin, 2009 The influence of carbon nanotubes on enzyme activity and structure: investigation of different immobilization procedures through enzyme kinetics and circular dichroism studies. Nanotechnology 20 25 255102

35. K. Matsuura, T. Saito, T. Okasaki, S. Oshima, M. Yumura, S. Iijima, 2006 Selectivity of water-soluble proteins in single-walled carbon nanotube dispersions. Chemical Physics Letters 4-6 429 497 502

36. F. Patolsky, Y Weizmann, I. Willner, 2004 Long-range electrical contacting of redox enzymes by SWCNT connectors. Angewandte Chemistry International Edition 16 43 2113 2117

37. J. Liu, A. Chou, W. Rahmat, M. Paddon-Row, J. Gooding, 2005 Achieving direct electrical connection to glucose oxidase using aligned single walled carbon nanotube arrays. Electroanalysis 17 1 38 46

38. J. Wang, 2005 Carbon-nanotube based electrochemical biosensors: A review. Electroanalysis 17 1 7 14

39. G. Wildgoose, C. Banks, H. Leventis, R. Compton, 2006 Chemically modified carbon nanotubes for use in electroanalysis. Microchimica Acta 152 3-4 187 214

40. I. Dumitrescu, P. Unwin, J. Macpherson, 2009 Electrochemistry at carbon nanotubes: perspective and issues. Chemical Communications 45 6886 4901

41. P. Ji, H. Tan, X. Xu, W. Feng, 2010 Lipase Covalently Attached to Multiwalled Carbon Nanotubes as an Efficient Catalyst in Organic Solvent. AIChE Journal 56 11 3005 3011

42. V. Upadhyayula, V. Gadhamshetty, 2010 Appreciating the role of carbon nanotube composites in preventing biofouling and promoting biofilms on material surfaces in environmental engineering: A review. Biotechnological Advances 28 6 802 816

43. E. Lojou, P. Luciano, S. Nitsche, P. Bianco, 1999 Poly(ester-sulfonic acid):modified carbon electrodes for the electrochemical study of c-type cytochromes. Electrochimica Acta 44 19 3341 3352

44. E. Lojou, X. Luo, M. Brugna, N. Candoni, S. Dementin, M. T. Giudici-Orticoni, 2008 Biocatalysts for fuel cells: efficient hydrogenase orientation for H2 oxidation at electrodes modified with carbon nanotubes. Journal of Biological Inorganic Chemistry 13 7 1157 1167

45. S. Minteer, P. Atanassov, H. Luckarift, G. Johnson, 2012 New materials for biological fuel cells Material Today 15 4 166 173

46. M. Weigel, E. Tritscher, F. Lisdat, Direct electrochemical conversion of bilirubin oxidase at carbon nanotube-modified glassy carbon electrodes Electrochemistry Communications 2007 9 4 689 693

47. M. Pumera, B. Smid, 2007 Redox protein noncovalent functionalization of double-wall carbon nanotubes: Electrochemical binder-less glucose biosensor. Journal of Nanosciences and Nanotechnology 7 10 3590 3595

48. I. Willner, Y. M. Yan, B. Willner, R. Tel-Vered, 2009 Integrated Enzyme-Based Biofuel Cells-A Review. Fuel Cells 09 1 7 24

49. A. Ueda, D. Kato, R. Kurita, T. Kamata, H. Inokuchi, S. Umemura, S. Hirono, O. Niwa, 2011 Efficient Direct Electron Transfer with Enzyme on a Nanostructured Carbon Film Fabricated with a Maskless Top-Down UV/Ozone Process Journal of the American Chemical Society 133 13 4840 4846

50. F. Tasca, L. Gorton, W. Harreither, D. Haltrich, R. Ludwig, G. Nöll, 2088 Direct electron transfer at cellobiose dehydrogenase modified anodes for biofuel cells Journal of Physical Chemistry C 112 26 9956 9961

51. W. Zheng, H. Zhao, H. Zhou, X. Xu, M. Ding, Y. Zheng, 2010 Electrochemistry of bilirubin oxidase at carbon nanotubes Journal of Solid State Electrochemistry 14 2 249 254

52. W. Zheng, H. Zhou, Y. Zheng, N. Wang, 2008 A comparative study on electrochemistry of laccase at two kinds of carbon nanotubes and its application for biofuel cell Chemical Physics Letters 381 385

53. L. Zhao, H. Liu, N. Hu, 2006 Assembly of layer-by-layer films of heme proteins and single-walled carbon nanotubes: electrochemistry and electrocatalysis Analytical Bioanalytical Chemistry 384 2 414 422

54. G. Liu, Y. Lin, 2006 Amperometric glucose biosensor based on self-assembling glucose oxidase on carbon nanotubes Electrochemistry Communications 8 2 251 256

55. R. Iost, F. Crespilho, 2012 Layer-by-layer self-assembly and electrochemistry: Applications in biosensing and bioelectronics Biosensors Bioelectronics 31 1 1 10

56. J. Gooding, R. Wibowo, J. Liu, W. Yang, D. Losic, S. Orbons, F. Mearns, J. Shapter, D. Hibbert, 2003 Protein electrochemistry using aligned carbon nanotube arrays. Journal of the American Chemical Society 125 30 9006 9007

57. X. Yu, D. Chattopadhyay, I. Galeska, F. Papadimitrakopoulos, J. Rusling, 2003 Peroxidase activity of enzymes bound to the ends of single-wall carbon nanotube forest electrodes Electrochemistry Communications 5 5 408 411

58.   M. Esplandiu, M. Pacios, L. Cyganek, J. Bartroli, M. Del Valle, 2009 Enhancing the electrochemical response of myoglobin with carbon nanotube electrodes. Nanotechnology

59.   P. Santhosh, A. Gopalan, K. Lee, 2006 Gold nanoparticles dispersed polyaniline grafted multiwall carbon nanotubes as newer electrocatalysts: Preparation and performances for methanol oxidation Journal of Catalysis 238 1 177 185

60.   E. Nazaruk, M. Karaskiewicz, K. Zelechowska, J. Biernat, J. Rogalski, R. Bilewicz, 2012 Powerful connection of laccase and carbon nanotubes Material for mediator-free electron transport on the enzymatic cathode of the biobattery. Electrochemistry Communications 14 1 67 70

61.   K. Sadowska, K. Stolarczyk, J. Biernat, K. Roberts, J. Rogalski, R. Bilewicz, 2010 Derivatization of single-walled carbon nanotubes with redox mediator for biocatalytic oxygen electrodes Bioelectrochemistry 80 1 73 80

62.   D. Jeykumari, S. Narayanan, 2008 Fabrication of bienzyme nanobiocomposite electrode using functionalized carbon nanotubes for biosensing applications Biosensors Bioelectronics 23 11 1686 1693

63.   Z. Wang, M. Li, P. Su, Y. Zhang, Y. Shen, D. Han, A. Ivaska, L. Niu, 2008 Direct electron transfer of horseradish peroxidase and its electrocatalysis based on carbon nanotube/thionine/gold composites Electrochemistry Communications 10 2 306 310

64.   F. Le Floch, A. Thuaire, G. Bidan, J. P. Simonato, 2009 The electrochemical signature of functionalized single-walled carbon nanotubes bearing electroactive groups. Nanotechnology 20 14 45705 45705

65.   M. Alonso-Lomillo, O. Rüdiger, A. Maroto-Valiente, M. Velez, I. Rodriguez-Ramos, F. Munoz, V. Fernandez, A. De Lacey, 2007 Hydrogenase-coated carbon nanotubes for efficient H2 oxidation. Nano Letters 7 6 1603 1608

66.   M. Jönsson-Niedziolka, A. Kaminska, M. Opallo, 2010 Pyrene-functionalised single-walled carbon nanotubes for mediatorless dioxygen bioelectrocatalysis Electrochimica Acta 55 28 8744 8750

67.   V. Karachevtsev, S. Stepanian, A. Glamazda, M. Karachevtsev,

V. Eremenko, O. Lytvyn, L. Adamowicz, 2011 Noncovalent Interaction of Single-Walled Carbon Nanotubes with 1-Pyrenebutanoic Acid Succinimide Ester and Glucose oxidase. Journal of Physical Chemistry C 115 43 21072 21082

68.   C Lau, E Adkins, R Ramasamy, H Lackarift, G Johnson, P Atanassov, 2012 Design of Carbon Nanotube-Based Gas-Diffusion Cathode for O2 Reduction by Multicopper Oxidases. Advanced Energy Materials 2 1 162 168

69.   H. Xu, H. Xiong-Y, Q. Zeng-X, L. Jia, Y. Wang, S. Wang-F, 2009 Direct electrochemistry and electrocatalysis of heme proteins immobilized in single-wall carbon nanotubes-surfactant films in room temperature ionic liquids. Electrochemistry Communications 11 2 286 289

70.   Y. Yan, W. Zheng, M. Zhang, L. Wang, L. Su, L. Mao, 2005 Bioelectrochemically functional nanohybrids through co-assembling of proteins and surfactants onto carbon nanotubes: Facilitated electron transfer of assembled proteins with enhanced faradic response. Langmuir 21 14 65606566

71.   S. Cosnier, R. Ionescu, M. Holzinger, 2008 Aqueous dispersions of SWCNTs using pyrrolic surfactants for the electro-generation of homogeneous nanotube composites. Application to the design of an amperometric biosensor. Journal of Materials Chemistry 18 42 5129 5133

72.   M. Gao, L. Dai, G. Wallace, 2003 Biosensors based on aligned carbon nanotubes coated with inherently conducting polymers. Electroanalysis 15 13 1089 1094

73.   C. Y. Tsai, C. S. Li, W. S. Liao, 2006 Electrodeposition of polypyrrole-multiwalled carbon nanotube-glucose oxidase nanobiocomposite film for the detection of glucose. Biosensors Bioelectronics 22 4 495 500

74.   H Chen, S Dong, 2007 Direct electrochemistry and electrocatalysis of horseradish peroxidase immobilized in sol-gel-derived ceramic-carbon nanotube nanocomposite film. Biosensors Bioelectronics 8 22 1811 1815

75.   A. Heller, 2006 Electron-conducting redox hydrogels: design, characteristics and synthesis. Current Opinion in Chemical Biology 10 6 664 672

76. S Timur, U Anik, D Odaci, L Gorton, 2007 Development of a microbial biosensor based on carbon nanotube (CNT) modified electrodes. Electrochemistry Communications 7 9 1810 1815

77. J. Song, H. Shin, C. Kang, 2011 A Carbon Nanotube Layered Electrode for the Construction of the Wired Bilirubin Oxidase Oxygen Cathode. Electroanalysis 23 12 2941 2948

78. I. Tiwari, M. Singh, 2011 Preparation and characterization of methylene blue-SDS- multiwalled carbon nanotubes nanocomposite for the detection of hydrogen peroxide. Microchimica Acta 174 3-4 223 230

79. S. Pakapongpan, R. Palangsuntikul, W. Surareungchai, 2011 Electrochemical sensors for hemoglobin and myoglobin detection based on methylene blue-multiwalled carbon nanotubes nanohybrid-modified glassy carbon electrode. Electrochimica Acta 56 19 6831 6836

80. T. Hoshino, S. Sekiguchi, H. Muguruma, 2012 Amperometric biosensor based on multilayer containing carbon nanotube, plasma-polymerized film, electron transfer mediator phenothiazine, and glucose dehydrogenase. Bioelectrochemistry 84 1

81. A. Ciaccafava, P. Infossi, M. T. Giudici-Orticoni, E. Lojou, 2010 Stabilization role of a phenothiazine derivative on the electrocatalytic oxidation of hydrogen via *Aquifex aeolicus* hydrogenase at graphite membrane electrodes. Langmuir 26 23 18534 18541

82. C. Tanne, G. Göbel, F. Lisdat, 2010 Development of a (PQQ)-GDH-anode based on MWCNT-modified gold and its application in a glucose/O2-biofuel cell. Biosensors Bioelectronics 26 2 530 535

83. E. Reisner, 2011 Solar Hydrogen Evolution with Hydrogenases: From Natural to Hybrid Systems European Journal of Inorganic Chemistry 7 1005 1016

84. P. Vignais, B. Billoud, 2007 Occurrence, classification, and biological function of hydrogenases: An overview. Chemical Reviews 107 10 4206 4272

85. A. Parkin, C. Cavazza, J. Fontecilla-Camp, F. Armstrong, 2006 Electrochemical investigations of the interconversions between

catalytic and inhibited states of the [FeFe]-hydrogenase from *Desulfovibrio desulfuricans*. Journal of the American Chemical Society 128 51 16808 16815

86. S. Dementin, V. Belle, P. Bertrand, B. Guigliarelli, G. Adryanczyk-Perrier, A. De Lacey, V. Fernandez, M. Rousset, C. Léger, 2006 Changing the ligation of the distal [4Fe4S] cluster in NiFe hydrogenase impairs inter- and intramolecular electron transfers. Journal of the American Chemical Society 128 15 5209 5218

87. H. Krassen, S. Stripp, Abendroth. G. von, K. Ataka, T. Happe, J. Heberle, 2009 Immobilization of the [FeFe]-hydrogenase CrHydA1 on a gold electrode: Design of a catalytic surface for the production of molecular hydrogen. Journal of Biotechnology 142 1 3 9

88. O. Zadvornyy, J. Lucon, R. Gerlach, N. Zorin, T. Douglas, T. Elgren, J. Peters, 2012 Photo-induced H2 production by [NiFe]-hydrogenase from *T. roseopersicina* covalently linked to a Ru(II) photosensitizer. Journal of Inorganic Biochemistry 106 1 151 155

89. E. Reisner, D. Powell, C. Cavazza, J. Fontecilla-Camps, F. Armstrong, 2009 Visible Light-Driven H2 Production by Hydrogenases Attached to Dye-Sensitized TiO2 Nanoparticles. J. Am. Chem. Soc. 131 51 18457 18466

90. S. Morra, F. Valetti, S. Sadeghi, P. King, T. Meyer, G. Gilardi, 2011 Direct electrochemistry of an [FeFe]-hydrogenase on a TiO2 Electrode. Chemical Communications 47 38 10566 10568

91. K. Brown, M. Wilker, M. Boehm, G. Dukovic, P. King, 2012 Characterization of Photochemical Processes for H2 Production by CdS Nanorod-[FeFe] Hydrogenase Complexes. Journal of the American Chemical Society 143 12 5627 5636

92. K. Brown, S. Dayal, X. Ai, G. Rumbles, P. King, 2010 Controlled Assembly of Hydrogenase-CdTe Nanocrystal Hybrids for Solar Hydrogen Production. Journal of the American Chemical Society 132 28 9672 9680

93. T. Mc Donald, D. Svedruzic, Y. Kim-H, J. Blackburn, S. Zhang, P. King, M. Heben, 2007 Wiring-up hydrogenase with single-walled carbon nanotubes. Nano Letters 7 11 3528 3534

94. J. Blackburn, D. Svedruzic, T. Mc Donald, Y. Kim-H, P. King, M. Heben, 2008 Raman spectroscoipy of charge transfer interaction

between single wall carbon nanotubes and [FeFe] hydrogenase. Dalton Transactions 5454 5461

95. T. Kihara, X. Liu-Y, C. Nakamura, K. Park-M, S. Yasuda-W, D. Qian-J, K. Kawasaki, N. Zorin, S. Yasuda, K. Hata, T. Wakayama, J. Miyake, 2001 Direct electron transfer to hydrogenase for catalytic hydrogen production using a single-walled carbon nanotubes forest. International Journal of Hydrogen Energy 36 13 7523 7529

96. K. Noda, N. Zorin, C. Nakamura, M. Miyake, I. Gogotov, Y. Asada, H. Akutsu, J. Miyake, 1998 Langmuir-Blodgett film of hydrogenase for electrochemical hydrogen production. Thin Solid Films 327 329 639 642

97. T Elgren, O Zadvorny, E Brecht, T Douglas, N Zorin, M Maroney, J Peters, 2005 Immobilization of active hydrogenases by encapsulation in polymeric porous gels. Nano Letters 10 5 2085 2087

98. O. Zadvorny, A. Barrows, N. Zorin, J. Peters, T. Elgren, 2010 High level oh hydrogen production activity achieved for hydrogenase encapsulated in sol-gel material doped with carbon nanotubes. Journal of Materials Chemistry 20 1065

99. M. Fontecave, V. Artero, 2011 Bioinspired catalysis at the crossroads between biology and chemistry: A remarkable example of an electrocatalytic material mimicking hydrogenases. Compte Rendu Chimie 14 4 362 371

100. M. Dubois, D. Dubois, 2009 The roles of the first and second coordination spheres in the design of molecular catalysts for H2 production and oxidation. Chemical Society Reviews 38 1 62 72

101. A. Le Goff, V. Artero, B. Jousselme, P. D. Tran, N. Guillet, R. Metaye, A. Fihri, S. Palacin, M. Fontecave, 2009 From Hydrogenases to Noble Metal-Free Catalytic Nanomaterials for H2 Production and Uptake. Science 326 5958 1384 1387

102. A. Yahiro, S. Lee, D. Kimble, 1964 Bioelectrochemistry I. Enzyme utilizing Bio-fuel cell studies. Biochimica Biophysica Acta 88 375 383

103. Ashley. E. Franks, Kelly. P. Nevin, 2010 Microbial Fuel Cells A Current Review. Energies 3 5 899 919

104. M. Zhou, M. Chi, J. Luo, H. He, T. Jin, 2011 An overview of electrode materials in microbial fuel cells. Journal of Power Sources 196 10 4427 4435

105. D. Sokic-Lazic, S. Minteer, 2008 Citric acid cycle biomimic on a carbon electrode. Biosensors Bioelectronics 24 4 939 944

106. K. Egorova, G. Antranikian, 2005 Industrial relevance of thermophilic Archaea. Current Opinion in Microbiology 8 6 649 655

107. L. Halamkova, J. Halamek, V. Bocharova, A. Szczupak, L. Alfonta, E. Katz, 2012 Implanted Biofuel Cell Operating in a Living Snail. Journal of the American Chemical Society 134 11 5040 5043

108. P. Cinquin, C. Gondran, F. Giroud, S. Mazabrard, A. Pellissier, F. Boucher, J. P. Alcaraz, K. Gorgy, F. Lenouvel, S. Mathé, P. Porcu, S. Cosnier, 2010 A Glucose BioFuel Cell Implanted in Rats. PLoS ONE 5 e10476

109. M. Falk, V. Andoralov, Z. Blum, J. Sotres, D. Suyatin, T. Ruzgas, T. Arnebrant, S. Shleev, 2012 Biofuel cell as a power source for electronic contact lenses. Biosensors Bioelectronics 37 1 38 45

110. A. Zebda, C. Gondran, A. Le Goff, M. Holzinger, P. Cinquin, S. Cosnier, 2011 Mediatorless high-power glucose biofuel cells based on compressed carbon nanotube-enzyme electrodes. Nature Com. 2 370

111. H. Zhao, H. Zhou, J. Zhang, W. Zheng, Y. Zheng, 2009 Carbon nanotube-hydroxyapatite nanocomposite: A novel platform for glucose/O2 biofuel cell. Biosensors Bioelectronics 25 2 463 468

112. J. Liu, X. Zhang, H. Pang, B. Liu, Q. Zou, J. Chen, 2012 High-performance bioanode based on the composite of CNTs-immobilized mediator and silk film-immobilized glucose oxidase for glucose/O2 biofuel cells. Biosensors Bioelectronics 31 1 170 175

113. F. Gao, Y. Yan, L. Su, L. Wang, L. Mao, 2007 An enzymatic glucose/O2 biofuel cell: Preparation, characterization and performance in serum. Electrochemistry Communications 9 5 989 996

114. X. Li, H. Zhou, P. Yu, L. Su, T. Ohsaka, L. Mao, 2008 A miniature glucose/O2 biofuel cell with single-walled carbon nanotubes-modified carbon fiber microelectrodes as the substrate. Electrochemistry Communications 10 6 851 854

115. E. Nazaruk, K. Sadowska, J. Biernat, J. Rogalski, G. Ginalska, R. Bilewicz, 2010 Enzymatic electrodes nanostructured with functionalized carbon nanotubes for biofuel cell applications. Analytical and Bioanalytical Chemistry 398 4 1651 1660

116. J. Lim, P. Malati, F. Bonet, B. Dunn, 2007 Nanostructured sol-gel electrodes for biofuel cells. Journal of the Electrochemical Society 154 2 A140 A145

117. M. Karaskiewicz, E. Nazaruk, K. Zelzchowska, J. Biernat, J. Rogalski, R. Bilewicz, 2012 Fully enzymatic mediatorless fuel cell with efficient naphthylated carbon nanotube-laccase composite cathodes. Electrochemistry Communications 20 124

118. F. Tasca, L. Gorton, W. Harreither, D. Haltrich, R. Ludwig, G. Nöll, 2008 Highly efficient and versatile anodes for biofuel cells based on cellobiose dehydrogenase from *Myriococcum thermophilum*. Journal of Physical Chemistry C 112 35 13668 13673

119. Y. Wang, Y. Yao, 2012 Direct electron transfer of glucose oxidase promoted by carbon nanotubes is without value in certain mediator-free applications. Microchimica Acta 176 3-4 271 277

120. K. Vincent, A. Parkin, F. Armstrong, 2007 Investigating and exploiting the electrocatalytic properties of hydrogenases. Chemical Reviews 107 10 4366 4413

121. P. P. Liebgott, A. de Lacey, B. Burlat, L. Cournac, P. Richaud, M. Brugna, V. Fernandez, B. Guigliarelli, M. Rousset, C. Léger, S. Dementin, 2011 Original Design of an Oxygen-Tolerant [NiFe] Hydrogenase: Major Effect of a Valine-to-Cysteine Mutation near the Active Site. Journal of the American Chemical Society 133 4 986 997

122. J. A. Cracknell, K. A. Vincent, M. Ludwig, O. Lenz, B. Friedrich, F. A. Armstrong, 2008 Enzymatic Oxidation of H2 in Atmospheric O2: The Electrochemistry of Energy Generation from Trace H2 by Aerobic Microorganisms. Journal of the American Chemical Society 130 2 424 425

123. X. J. Luo, M. Brugna, P. Infossi, M. T. Giudici-Orticoni, E. Lojou, 2009 Immobilization of the hyperthermophilic hydrogenase from *Aquifex aeolicus* bacterium onto gold and carbon nanotube electrodes for efficient H2 oxidation. Journal of Biological Inorganic Chemistry 14 8 1275 1288

124. D. Svedruzic, J. Blackburn, R. Tenent, D. J. Rocha, T. Vinzant, M. Heben, P. King, 2011 High-performance hydrogen production and oxidation electrodes with hydrogenase supported on metallic single-wall carbon nanotubes networks. Journal of the American Chemical Society 133 12 4299 4306

125. Q. Sun, N. A. Zorin, D. Chen, M. Chen, X. T. Liu, J. Miyake, J. D. Qian, 2010 Langmuir-Blodgett films of pyridyldithio-modified multiwalled carbon nanotubes as a support to immobilize hydrogenase. Langmuir 26 12 10259 10265

126. F. J. M. Hoeben, I. Heller, S. P. J. Albracht, C. Dekker, S. G. Lemay, H. A. Heering, 2008 Polymyxin-coated Au and Carbon nanotubes electrodes for stable [NiFe]-hydrogenase film voltammetry. Langmuir 24 11 5925 5931

127. J. Baur, A. Le Goff, S. Dementin, M. Holzinger, M. Rousset, S. Cosnier, 2011 Three-dimensional carbon nanotube-polypyrrole-[NiFe] hydrogenase electrodes for the efficient electrocatalytic oxidation of H2. International Journal of Hydrogen Energy 36 19 12096 12101

128. A. Ciaccafava, P. Infossi, M. Ilbert, M. Guiral, S. Lecomte, M. T. Giudici-Orticoni, E. Lojou, 2012 Electrochemistry, AFM, and PM-IRRA Spectroscopy of Immobilized Hydrogenase: Role of a Hydrophobic Helix in Enzyme Orientation for Efficient I I2 Oxidation. Angewandte Chemistry International Edition 51 4 953 956

129. A. Volbeda, P. Amara, C. Darnault, J. M. Mouesca, A. Parkin, M. M. Roessler, F. A. Armstrong, J. C. Fontecilla-Camps, 2012 X-ray crystallographic and computational studies of the O2-tolerant [NiFe]-hydrogenase 1 from *Escherichia coli*. Proceeding of the National Academic Sciences 10 14 5305 5310

130. Y. Shomura, S. K. Yoon, H. Nishihara, Y. Higuchi, 2011 Structural basis for a [4Fe-3S] cluster in the oxygen-tolerant membrane-bound [NiFe]-hydrogenase. Nature 479 7372 253NIL_143.

131. M. Pandelia, V. Fourmond, P. Tron, E. Lojou, P. Bertrand, C. Léger, M. T. Giudici-Orticoni, W. Lubitz, 2010 Membrane-Bound Hydrogenase I from the Hyperthermophilic Bacterium *Aquifex aeolicus:* Enzyme Activation, Redox Intermediates and Oxygen Tolerance. Journal of the American Chemical Society 132 20 6991 7004

132. J. Fritsch, P. Scheerer, S. Frielingsdorf, S. Kroschinsky, B. Friedrich, O. Lenz, C. M. T. Spahn, 2011 The crystal structure of an oxygen-tolerant hydrogenase uncovers a novel iron-sulphur centre. Nature 479 7372 249NIL_134.

133. S. Krishnan, F. A. Armstrong, 2012 Order-of-magnitude enhancement of an enzymatic hydrogen-air fuel cell based on pyrenyl carbon nanostructures. Chemical Science 3 4 1015 1023

134. K. Vincent, J. Cracknell, J. Clark, M. Ludwig, O. Lenz, B. Friedrich, F. Armstrong, 2006 Electricity from low-level H2 in still air- an ultimate test for an oxygen tolerant hydrogenase. Chemistry Communications 5033 5035

135. A. Wait, A. Parkin, G. Morley, Santos. L. dos, F. Armstrong, 2010 Characteristics of enzyme-based hydrogen fuel cells using an oxygen-tolerant hydrogenase as the anodic catalyst. Journal of Physical Chemistry C 114 27 12003 12009

136. A. Ciaccafava, A. de Poulpiquet, V. Techer, M. T. Giudici-Orticoni, S. Tingry, C. Innocent, E. Lojou, 2012 An innovative powerful and mediatorless $H_2/O2$ biofuel cell based on an outstanding bioanode. Electrochemistry Communications 23 25 28

137. P. Infossi, E. Lojou, J. P. Chauvin, G. Herbette, M. Brugna, M. T. Giudici-Orticoni, 2010 *Aquifex aeolicus* membrane hydrogenase for hydrogen bioxidation: role of lipids and physiological partners in enzyme stability and activity. International Journal of Hydrogen Energy 35 19 10778 10789

138. B. Reuillard, A. Le Goff, A. Agnès, A. Zebda, M. Holzinger, S. Cosnier, 2012 Direct electron transfer between tyrosinase and multi-walled carbon nanotubes for bio electrocatalytic oxygen reduction. Electrochemistry Communications 20 19 22

139. F. Durand, C. Kjaergaard, E. Suraniti, S. Gounel, R. Hadt, E. Solomon, 2012 Mano N Bilirubin oxidase from *Bacillus pumilus*: A promising enzyme for the elaboration of efficient cathodes in biofuel cells. Biosensors Bioelectronics 35 1 140 146

Citations

CHAPTER 1

H. V. Lee, S. B. A. Hamid, and S. K. Zain, "Conversion of Lignocellulosic Biomass to Nanocellulose: Structure and Chemical Process," The Scientific World Journal, vol. 2014, Article ID 631013, 20 pages, 2014. doi:10.1155/2014/631013.

CHAPTER 2

Rick Arneil D. Arancon, Carol Sze Ki Lin, King Ming Chan, Tsz Him Kwan, and Rafael Luque, Advances on waste valorization: new horizons for a more sustainable society, DOI: 10.1002/ese3.9.

CHAPTER 3

Stephen R. Hughes, William R. Gibbons, Bryan R. Moser, and Joseph O. Rich (2013). Sustainable Multipurpose Biorefineries for Third-Generation Biofuels and Value-Added Co-Products, Biofuels - Economy, Environment and Sustainability, Prof. Zhen Fang (Ed.), ISBN: 978-953-51-0950-1.

CHAPTER 4

Zahid Anwar, Muhammad Gulfraz, and Muhammad Irshad, Agro-industrial lignocellulosic biomass a key to unlock the future bio-energy: A brief review, doi:10.1016/j.jrras.2014.02.003.

CHAPTER 5

Sergio C. Trindade (2011). Nanotech Biofuels and Fuel Additives, Biofuel's Engineering Process Technology, Dr. Marco Aurelio Dos Santos Bernardes (Ed.), ISBN: 978-953-307-480-1, InTech, DOI: 10.5772/16955.

CHAPTER 6

Lei Pei, Markus Schmidt, and Wei Wei (2011). Conversion of Biomass into Bioplastics and Their Potential Environmental Impacts, Biotechnology of Biopolymers, Prof. Magdy Elnashar (Ed.), ISBN: 978-953-307-179-4, InTech, DOI: 10.5772/18042.

CHAPTER 7

Samira Bagheri, Nurhidayatullaili Muhd Julkapli, and Sharifah Bee Abd Hamid, "Functionalized Activated Carbon Derived from Biomass for Photocatalysis Applications Perspective," International Journal of Photoenergy, Article ID 218743, in press.

CHAPTER 8

Anne De Poulpiquet, Alexandre Ciaccafava, Saïda Benomar, Marie-Thérèse Giudici-Orticoni, and Elisabeth Lojou (2013). Carbon Nanotube-Enzyme Biohybrids in a Green Hydrogen Economy, Syntheses and Applications of Carbon Nanotubes and Their Composites, Dr. Satoru Suzuki (Ed.), ISBN: 978-953-51-1125-2, InTech, DOI: 10.5772/51782.

Index